Microprocessor Technology

Stuart Anderson B. Ed. (Hons)

NEWNES

Newnes
An imprint of Butterworth-Heinemann Ltd
Linacre House, Jordan Hill, Oxford OX2 8DP

ℛ A member of the Reed Elsevier plc group

OXFORD   LONDON   BOSTON
MUNICH   NEW DELHI   SINGAPORE   SYDNEY
TOKYO   TORONTO   WELLINGTON

First published 1994

© J. S. Anderson 1994

British Library Cataloguing in Publication Data
Anderson, J.S.
    Microprocessor Technology
    I. Title
    621.3916

ISBN 0 7506 1839 6

Composition by Scribe Design, Gillingham, Kent
Printed and bound in Great Britain

Contents

Preface

This book covers the subject of microprocessor technology using the Z80 and 6502 processors. The text is suitable for BTEC NII and NIII Microelectronics, the C&G 726 Microprocessor Technology course, the micro portion of the C&G 224 Part II Electronics Servicing course and most other introductory microelectronics courses. Those studying at a higher level will also find much that is of use or interest, especially students on 'bridging' and micro appreciation courses where a measure of comparative studies is required.

In addition, the text describes in suitable detail how to design and construct a Z80-based microelectronic system. Since all the necessary background is covered in the main text, it is possible for the reader, whether at home, on a course or in industry, to produce a basic but powerful system which can be tailored to meet a variety of needs.

The inspiration for the book derives from my inability to find a single, up-to-date book that provides suitable material to accompany a Level II BTEC Microelectronics course. Several texts were available, many (but by no means all) dealing with the popular Z80 processor. Each made some contribution but, even collectively, none seemed to satisfy all requirements: there was no one book that could be recommended to students. The seed was sown; a suitable book needed to be written.

In planning the book it quickly became obvious that, with very few additions, it could be extended to cover the requirements of the microelectronics component of the C&G 224 Part II Electronics Servicing course and, with only minor adjustments and the inclusion of some further material, it could cater for students studying the C&G 726 Microprocessor Technology course and the Level III BTEC Microelectronics course as well.

To help the reader along, there are copious worked examples, a variety of questions including long answer and multiple choice and a large selection of assignments. Most of the questions come with answers. Where they are omitted, it is either because solutions may be considered correct if the programs called for actually work, or (in the case of Chapter 7) where verification may be obtained through the process of single-stepping. In some cases (particularly Chapter 2) some extra 'tutorial questions' are included which can be used by teachers or lecturers. Because the answers are not given, students must work these out and cannot simply look them up!

The game hasn't been given away entirely; there are other minor omissions in order to provide for assignment work. A comprehensive solution is not provided to every question, as this would preclude the essential opportunity for student investigations. However, all the main points in each area are covered in detail at the level required and, where additional material is called for, it will invariably be easy to access. Solutions to all the questions can be found in the text; assignment work may require some additional reading.

I hope that both students and lecturers will find this not only a comprehensive textbook but also one that is interesting, practical and easy to understand. Comments and suggestions will be received with interest.

Acknowledgements

The author wishes to thank Cecil Eccles who read the entire manuscript and offered many valuable and constructive criticisms. Thanks to Harry Kerridge for his comments and notes on structured programming and his story about the birth of 3.5" floppy discs! Thanks to Trevor Brown of the BATC for permission to reproduce parts of his article 'Beyond TTL' first published in *CQTV*. Trevor's design for 'Roger Bleep' formed the basis of the author's 'SAMSON' microelectronic system, and much of Chapters 16 and 17 was adapted from his excellent article.

Special thanks to Paul Oldnall who gave up so many of his weekends in order to work on the diagrams which form such a large portion of the book. There are more than 200 figures, all expertly and superbly produced from what were initially very rough sketches.

Thanks to Matthew Wilkes who supplied the EPROM programming equipment and offered some suggestions on advanced programming techniques. Thanks also to the technicians at Dudley College for providing so much excellent support on the use of the 'WordStar' word processing package and Bob Wild for freely supplying the original PC system upon which this book was written. Thanks to Mark Bircher for his technical expertise in fault finding which forms part of Chapter 18 and his assistance in programming the EPROMs used in the prototype SAMSON.

Thanks to Pete Boswell for his assistance in fault-finding on the prototype 'SAMSON' and the numerous hints and helpful advice he supplied.

Thanks to L J Technical Systems Ltd for permission to reproduce information on their MARC and EMMA microelectronic systems and ancillary equipment and Maplin Electronics for permission to reproduce data from their catalogues. All of the selected IC pin-outs which form Appendix II were taken from Maplin's catalogue from whom all the goods described are presently available. Thanks to Rapid Electronics for permission to reproduce details of their logic probe which is described and shown in Chapter 18.

Thanks to Graham and Russell Knight for their assistance with the maths in Chapter 2, together with their constant encouragement and support. And finally, thanks to my dear wife Anne, for putting up with my terrible temper and strange moods over the past 18 months, during which time this book was written.

Stuart Anderson
Dudley

Chapter 1
The microelectronic system

Introduction

This book provides an introduction to microelectronic systems, and one of the best ways to learn about them is to study not just microelectronics, but the architecture, use and programming of a microcomputer system. It will soon become obvious what the differences between the two are, and how a micro-electronic system can be set up to do a specific job. A microelectronic system can be represented as shown in Figure 1.1. Many people now have a personal computer (PC), and those who don't will probably know what one is. A PC is an example of a microelectronic system, specially configured as a micro-computer, with a 'typewriter style' keyboard, video display unit (VDU) and disc drive as a minimum.

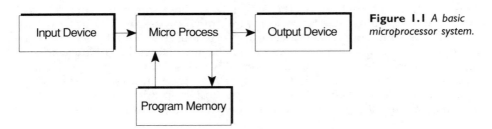

Figure 1.1 A basic microprocessor system.

In the system shown, there is no keyboard, VDU or interfaces for tape or floppy disc storage. It does not (necessarily) represent a microcomputer system (although it could do) because a microelectronic system is dedicated to performing a particular task programmed into its memory. It doesn't (usually) need a keyboard or VDU since it might be designed to control a motor car or a washing machine, for example.

To get the system to perform a different routine requires that the memory device be reprogrammed or substituted by a device with a different program in it. The microprocessor used in the car could be exactly the same as the one used in the washing machine; it is simply controlled by a different program and once programmed, the system performs one specific task.

A more detailed microelectronic system is shown in Figure 1.2. This is a simple central heating and air-conditioning system. There are three input devices: a humidity detector, a temperature setting control and a temperature sensor. The output consists of a heater and a fan. Both input and output devices require interfaces, electronic circuits which adapt the outputs of the input devices to make them suitable for inputting to the microprocessor, and adapt the outputs of the microprocessor to enable them to drive the fan and the heater. The memory chip holds the program which tells the microprocessor what to do.

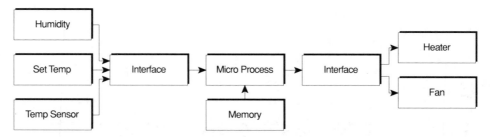

Figure 1.2 *A microprocessor controlled heating system.*

The program may turn the fan on only when the humidity reaches a specific level; the heater may be turned on when the temperature drops below a specific level and so on. The great flexibility of this system is demonstrated by the fact that the addition of more input sensors, or output devices, and the conditions under which they operate, can all be accommodated by changing the program in the memory.

Some program memories are on a silicon 'chip' which can be reprogrammed (e.g. an EPROM – erasable programmable read only memory); others are permanent. The choice of which to use depends upon the application and this will be discussed later.

The silicon chip

Most people will have heard of the 'silicon chip'. There are many such devices, but the words are usually taken to mean a microprocessor. The heart of any computer is something called a central processing unit (CPU), the micropro- cessor (MPU) is simply a CPU which has been considerably reduced in size using modern integrated circuit technology. This 'computer on a piece of silicon' is so small, it's probably no bigger than the nail on your little finger. How the microprocessor functions and how it is 'programmed' is what we are about to investigate. Machine code programming will be discussed – this is the code that the microprocessor actually 'understands'.

Some home computer users program in 'BASIC' but this is only for the users' convenience. An interpreter inside the computer converts the BASIC program into machine code. This slows down program execution hundreds of times and increases the cost of the system. Hence, using machine code programming speeds up operation dramatically and uses less 'firmware'.

The hardware

The book will be most useful to readers who have access to a microprocessor system such as the L J Technical Systems 'MARC' (Z80) or 'EMMA' (6502) Microcomputer (an 'embedded system'), but this is not essential. Much of the text refers to the Z80 CPU, and so can be read in conjunction with *any* Z80- based system, such as the simple 'Micro Professor', or a comparable 6502 system. Most IBM PCs can run Z80 emulators which are software packages enabling machine code programs to be created and run on a PC.

Alternatively, manufacturers such as L J Technical Systems supply micro- processor-based controllers as 'target systems' for a host PC. This means that the Z80/6502-based system is connected to the PC which is then used to create and run machine code programs. We concentrate on Z80 and 6502 CPUs simply because they are now the most widely used 8-bit processors in education and industry. They are commonly available at minimal cost and are adequate for thousands of applications where the increased speed and addressing capability of more modern processors is simply unnecessary.

In Chapter 17 we show how the reader can build a complete Z80- or 6502-based microelectronic system having hundreds of powerful applications. With only minor adjustments to mnemonics and codes, just about any other system may be used with this book since much of the material is applicable to all 8-bit microelectronic systems.

Starting with the basics

Anyone who has used a personal computer (PC) will probably be familiar with the BASIC programming language – beginners all-purpose symbolic instruction code. Not so long ago, nearly all home computer users programmed in BASIC (if they did any programming at all). The code allows simple instructions to be used; for example, in order to put 'ABCD' on the computer screen, you would type in:

```
10  PRINT  "ABCD"
```

When the program is 'RUN', 'ABCD' appears on the screen. All very simple and apparently almost instantaneous; in fact it is comparatively slow. The microprocessor (the so-called 'silicon chip') at the heart of the computer cannot understand 'PRINT', the instruction has to be changed into something the microprocessor can actually understand; it has to be changed into 'machine code' and this is made up entirely of only two numbers, one and zero.

An electronic 'block' inside the computer makes this conversion and it is called a BASIC Interpreter (see Figure 1.3).

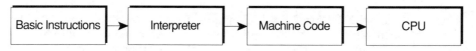

Figure 1.3 *Progress of instructions through a typical personal computer (PC), from BASIC instructions typed in at the keyboard to machine code.*

Taking the previous example, a BASIC instruction for printing 'ABCD' on the screen is simply as follows:

```
10  PRINT  "ABCD"  (Return)
    RUN
```

In 'Assembly language' – a simple method of producing machine code – the program may look like this:

```
        ORG    5C00H
        ENT    5C00H
VIDEO:  EQU    0E02AH
CLS:    EQU    01B9H
        CALL   CLS
        LD     HL,VIDEO
        LD     A,41H
        LD     B,4
LOOP:   LD     (HL),A
        INC    HL
        INC    A
        DEC    B
        JP     NZ,LOOP
        HALT
```

This is more than ten times longer than the equivalent BASIC program, but once converted into machine code, is hundreds of times faster. This is because the program is written in assembly language which can be directly translated into the machine code; a BASIC program is interpreted line by line every time the program is run; but once the machine code has been created (the 'object' code) the computer can run it directly, with a consequent increase in speed and reduction in cost.

Machine code programming

The microprocessor itself cannot understand words or even ordinary mathematics based on the familiar decimal system. It is a device which contains many thousands of tiny electronic circuits each of which can only ever be in one of two states – either ON or OFF. These two states can represent the numbers one and zero; this is the basis of logic circuits and of digital electronics generally. A simple switch and lamp circuit can be used to illustrate the ideas of ON and OFF, of logic '1' and logic '0', see Figure 1.4.

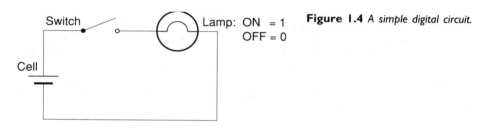

Figure 1.4 A simple digital circuit.

Compare this with Figure 1.5 which shows a simple analogue circuit. The switch has been replaced by a potentiometer connected as a simple variable resistor; as it is adjusted, the brightness of the lamp varies continuously. Figure 1.4 embodies the idea of digital electronics, the circuit is either ON or OFF and all the other states in between are 'not allowed'.

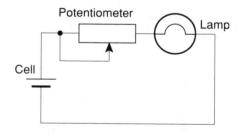

Figure 1.5 A simple analogue circuit. Here, the lamp varies in brightness continuously from OFF to fully ON.

In practice, ordinary switches are not used in computer systems as they would be too slow, too large and too unreliable. Transistors are used instead. Before they were developed, electronic valves were used in computers but they too were large and unreliable. By comparison transistors are much smaller, use less electricity and have greater reliability.

Because so many transistor switches are needed in a computer it was eventually decided to 'integrate' thousands of them together on a single chip of silicon – the element that most transistors are made from – hence the name 'silicon chip'. A microprocessor, therefore, contains thousands and thousands of tiny transistor switches, each capable of representing either a logic 1 or a logic 0.

Transistor operation

It is not necessary at this stage to deal with transistor theory or logic circuits in any detail; the purpose of the following is to give an overview, so don't be concerned if it is not totally comprehensible at the moment. The advice is: read and understand what you can now, even if only in a limited way, and come back to the details later. For more information see Chapter 14.

A transistor is an electronic component which has three electrodes, designated collector, base and emitter. If a small current is applied to its base, a much larger current is caused to flow between its collector and emitter (see Figure 1.6). It is rather like the huge flow of water available in a domestic supply; there is enormous pressure forcing the water out but it can nevertheless be controlled by a small tap.

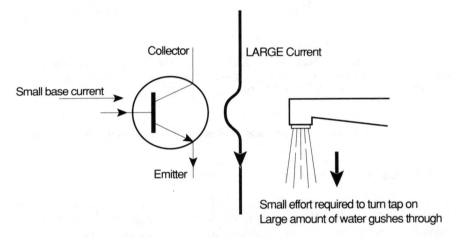

Figure 1.6 *On the left, an npn transistor; a small current flowing into the base terminal causes a large electric current to flow from collector to emitter. On the right, a water equivalent; a small effort is required to turn the tap on, but this could cause the flow of hundreds of gallons of water per minute.*

In the case of the water analogy, the flow of water is continuously variable from nothing at all (its minimum value), through a small trickle, to a gushing torrent, and then to the final deluge when the tap is fully open (its maximum value). This is an example of an analogue system – all the values between maximum and minimum are allowed. The transistor *can* also work in a similar fashion.

In an audio amplifier, collector/emitter current can be made continuously variable; the greater the current flowing into the base (the input) the greater will be the current flow between collector and emitter (producing the output). However in digital electronics only two states are allowed – fully on (a logic 1) or fully off (a logic 0). The transistor then operates as a rapid and reliable electronic switch. Figure 1.7 illustrates two simple transistor switches.

Digital computers

Modern computers are digital computers. Digital systems operate in a limited number of finite states, usually only two, as is the case with digital computers. Let us consider again precisely what is meant by digital. Take the example

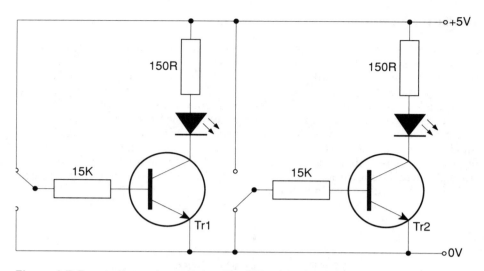

Figure 1.7 *Two simple transistor switches. The LEDs show the logic states; when the transistor is 'ON', the LED glows as current flows through the collector/emitter circuit; when the transistor is 'OFF' the LED is off. In this circuit, Tr1 is 'ON' and Tr2 is 'OFF'.*

of a clock used to tell the time. (A new meaning of 'clock' will be introduced later.)

Analogue and digital systems

In Figure 1.8 each clock shows 5.00, but what will each show next? The digital clock will show 17.01, but the analogue clock, in the meantime, will show all the seconds in between, and we could define hundreds or even millions of

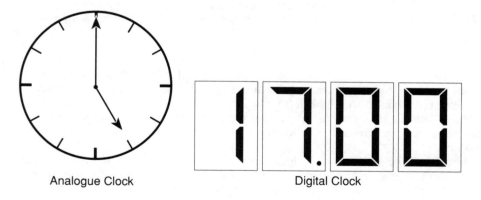

Analogue Clock   Digital Clock

Figure 1.8 *Both clocks show 5 o'clock, but what will each show next?*

different positions in between if we used a microscope to see them. Whether we see them or not, we know that those intermediate stages exist and that the 'minute hand' must pass through them all.

The digital display *jumps* directly from *zero* to *one* and *there is nothing in between.*

A digital computer

Digital computers operate using tiny transistor switches which are either ON or OFF. Hence, computers are digital devices utilising only two possible states. In digital electronics we refer to these states as logic 1 (ON) and logic 0 (OFF).

A system which has only two states and therefore two 'numbers' (0 and 1) would seem to have a very limited capability, but it will soon be shown that this is not true.

Since the digital computer has only two numbers to work with, we have to communicate with the computer only with those numbers. The numbers are simply 1 and 0, and we communicate using the *binary system*.

Binary mathematics

Chapter 2 of this book deals with binary mathematics in some detail but as an introduction, consider Table 1.1 which shows some decimal numbers and their binary equivalents. Note that binary numbers can only contain the digits 1 or 0; these two values can be 'stored' by having transistor switches which are either ON or OFF.

Table 1.1. Some decimal numbers with their binary equivalents. The list can of course be continued indefinitely

Decimal number	Binary number	Decimal number	Binary number
0	0	16	1 0000
1	1	17	1 0001
2	10	18	1 0010
3	11	19	1 0011
4	100	20	1 0100
5	101	21	1 0101
6	110	22	1 0110
7	111	23	1 0111
8	1000	24	1 1000
9	1001	25	1 1001
10	1010	26	1 1010
11	1011	27	1 1011
12	1100	28	1 1100
13	1101	29	1 1101
14	1110	30	1 1110
15	1111	31	1 1111
		32	10 0000

Storing binary numbers

In Figure 1.7 we illustrated two simple transistor switches. The left-hand transistor is turned ON because its base is connected to +5V via the 15K limiting resistor. The transistor conducts fully and the LED is ON. This can be taken as 'storing' a logic 1.

The reverse is true in the right-hand switch. The transistor base is at 0V, it does not conduct, so the LED does not light. This transistor switch may be regarded as 'storing' a logic 0.

A bistable or 'flip-flop' circuit can also take up either of the two logic states. Furthermore, it reacts to a logic pulse and 'remembers' its logic state even

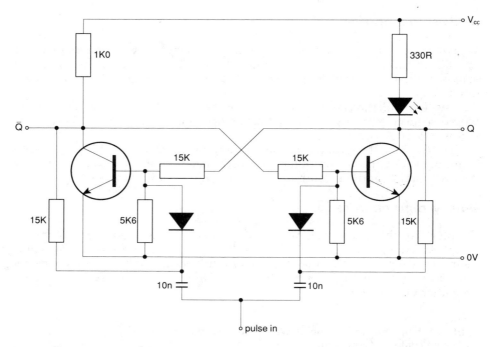

Figure 1.9 *A bistable or 'flip-flop' circuit with typical values. The transistors are BC108 or similar (almost any npn transistor works in this simple circuit); the diodes are 1N4148. When a pulse is applied, the bistable changes state, hence the LED indicator is ON in one state and OFF in the other. As long as power is applied, the circuit remembers which state it is in, it hence forms the basis of a computer memory unit.*

when the pulse that set it has been removed, as long as power is supplied to the circuit. The flip-flop is made up from two transistors only one of which can be conducting at any one time (see Figure 1.9).

When a pulse comes along, the flip-flop changes state (flips over). In one state the LED lights (it's storing a '1'), in the other state, the LED is off, and it's storing a logic '0'. Connecting certain types of flip-flop together produces a register or memory cell which can typically hold an 8-bit binary number. An 8-bit binary number can therefore either be stored in the computer's memory or in a register inside the microprocessor itself.

The state of the flip-flop can be changed by applying a square wave pulse (see Figure 1.10). This simply represents the two voltage levels available, equivalent to logic 1 (high voltage) and logic 0 (low voltage). After a pulse has been applied the state of each transistor reverses, i.e. if the left-hand transistor were originally high it would go low, and vice versa. The interesting thing is, that these transitions from low to high or from high to low are also pulses and they can be applied in turn to another bistable circuit.

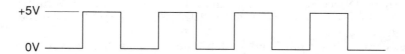

Figure 1.10 *A 'square' wave. Assuming transistor-transistor Logic (TTL) levels, 5V = 'ON' (logic 1) and 0V (zero volts) = 'OFF' (logic 0). Note: This waveform represents ON and OFF states only – there are no allowable states in between; it is a digital signal.*

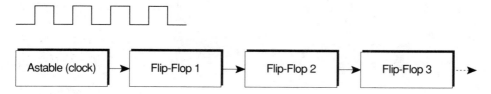

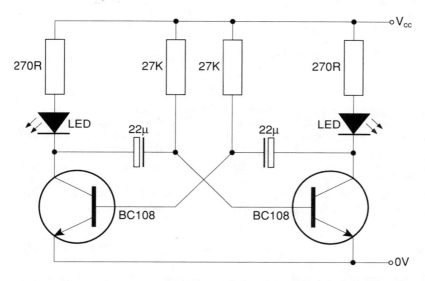

Figure 1.11 *Linked bistable operation. On the first pulse from the clock, flip-flop 1 changes state. On the next pulse, flip-flop 1 returns to its original condition and emits a pulse to flip-flop 2 allowing it to change state. The effect is carried through to any following flip-flops. Placed in reverse order (FF1 on the far right), such a system 'counts' up in binary.*

Figure 1.12 *An astable multivibrator. When power is applied, the two LEDs shown will flash alternately. The rate is determined by the values of C and R in the circuit. In a microprocessor system, the 'clock' is crystal controlled to ensure constant frequency and the 'flash' rate will be millions of times per second.*

The effect is described in Figure 1.11. It is not possible or appropriate to go into bistable operation in any detail here, however the upshot is, that a series of bistables connected as shown will count up in binary when 'clocked'. The word 'clock' is used here to describe the oscillator which provides the square wave pulses. A simple, two-transistor, astable 'clock' oscillator is shown in Figure 1.12.

A computer clock

The clock in this example is causing the system to count up in binary, but all computer systems have clocks to enable them to step through instructions and execute a program. For speed, the clock oscillator can operate at a rate of several million times a second, for example the clock frequency may be 4MHz, and in most modern systems is considerably higher. Figure 1.13 shows one possible design for a microcomputer, crystal-controlled clock oscillator; a Z80 CPU could be driven by such a circuit. The 6502 CPU has a clock oscillator already built into it.

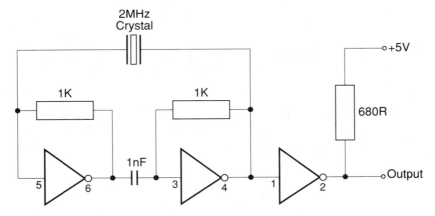

Figure 1.13 *A TTL microprocessor, crystal-controlled clock oscillator, which uses part of a 7404 hex inverter.*

In order to allow the bistable system to count up to ten, four bistables will be required (decimal 10 = 1010 in binary, one bistable is required for each digit in the binary number, i.e. four). If left to continue, all four bistables will be 'ON' giving a count of 1111 (binary) which is 15 in decimal. If there were two 'groups' of four (8 bistables in all) the system could count up to 255 (which is 1111 1111 in binary). Of course the LEDs are not necessary in a computer, but they are often included in similar circuits built for demonstration purposes which clock very slowly in order to make it easier to see the effect.

Registers

As we have shown, it is also possible to 'set' the state of bistables artificially so that the system can hold a particular number in binary. There are several different types of bistable which can be connected together in order to provide this facility. In a register, 8 of them are used, allowing any number from zero to 255 to be 'stored' in binary code. A typical register is illustrated in Figure 1.14.

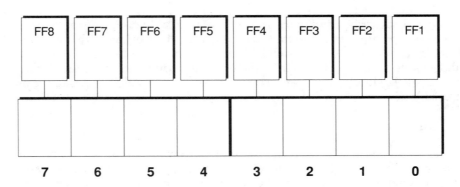

Figure 1.14 *A CPU register. Each element of the register may be considered as being one end of a bistable (flip-flop), FF1, FF2, etc. The bistables can be 'set' so the register can hold any number from 0 (0000 0000) to 255 (1111 1111), 256 numbers in all. Note that by convention, binary numbers are often written down in groups of four to make them easier to read, and the elements of the register are denoted 0 to 7 (rather than 1 to 8).*

As we shall see, there are several registers inside a microprocessor which can store numbers representing data or instructions. By a very complicated set of logical operations, the microprocessor can then actually *compute*.

Inside the microprocessor, however, all operations are carried out using small transistor circuits which are either ON or OFF. It is a digital system using the binary code, and before progressing it is important to understand the various codes that are in use, including not only binary itself, but also octal, hexadecimal and BCD systems.

Questions

1 What is the difference between a microelectronic system and a microcomputer?
2 Name one of the great advantages of using a microelectronic system to control an industrial process rather than using traditional, 'hard-wired' electronics.
3 What do the letters 'CPU' and 'MPU' stand for? What is the difference between them?
4 (a) What does BASIC stand for in terms of computing?
   (b) What is the main disadvantage of BASIC?
   (c) Why is BASIC still frequently used by home computer programmers in spite of this disadvantage?
5 Draw a block diagram to illustrate the progress of instructions through a typical personal computer from BASIC instructions typed in at the keyboard to machine code.
6 What is assembly language?
7 What are the only two numbers that a CPU can 'understand'?
8 Draw the symbol of an npn transistor and label the electrodes.
9 Draw a diagram of a simple transistor switch which may be used to represent a logic 1 signal.
10 Why do digital computers use the binary code?
11 What is the binary number 1010 in decimal?
12 Discuss the difference between analogue and digital systems.
13 Describe a bistable circuit in simple terms.
14 (a) What is the purpose of the crystal-controlled 'clock' oscillator in a microelectronic system?
   (b) Draw a diagram of a simple clock oscillator suitable for use in a microprocessor system.
   (c) Draw and label the kind of waveform to be expected at the output of the clock oscillator.
15 State a typical frequency for the clock oscillator in a microelectronic system.
16 What is a CPU register?
17 Sketch a CPU register showing how each element (bit) is numbered.
18 Why does a CPU require 'memory'?

Multiple choice questions

1 A motor car may be controlled by:
   (a) a microprocessor   ☐
   (b) a CPU   ☐
   (c) a microcomputer   ☐
   (d) a microelectronic system   ☐

2 A computer program may be stored in:
- (a) an input device ☐
- (b) an output device ☐
- (c) program memory ☐
- (d) a microprocessor ☐

3 The words 'silicon chip' are usually taken to mean a:
- (a) microprocessor ☐
- (b) memory device ☐
- (c) input device ☐
- (d) random access memory ☐

4 Many home computers users program in 'BASIC' because:
- (a) it's easy to understand ☐
- (b) it is very fast ☐
- (c) it runs on all computers ☐
- (d) there are only a few instructions ☐

5 A CPU actually 'understands':
- (a) BASIC ☐
- (b) machine code ☐
- (c) assembly language ☐
- (d) an interpreter ☐

6 A transistor has three electrodes designated:
- (a) collector, base and current ☐
- (b) voltage, current and base ☐
- (c) base, collector and emitter ☐
- (d) base, emitter and voltage ☐

7 The number of different values which can be represented by an analogue system is:
- (a) only two ☐
- (b) thousands ☐
- (c) millions ☐
- (d) infinite ☐

8 Compared to 'BASIC', machine code programs are:
- (a) twice as fast ☐
- (b) ten times faster ☐
- (c) hundreds of times faster ☐
- (d) millions of times faster ☐

9 The main purpose of a CPU clock is to:
- (a) regulate the speed at which programs are stored ☐
- (b) allow the CPU to step through a program ☐
- (c) keep a record of the time ☐
- (d) display the time on a VDU ☐

10 In decimal, the binary number 1100 is:
- (a) 3 ☐
- (b) 5 ☐
- (c) 11 ☐
- (d) 12 ☐

11 A simple, basic computer memory unit can be made up from a series of:
- (a) bistables ☐
- (b) monostables ☐
- (c) astables ☐
- (d) oscillators ☐

12 The term 'flip-flop' is another name for:
  (a)   an oscillator   ☐
  (b)   an astable   ☐
  (c)   a monostable   ☐
  (d)   a bistable   ☐
13 The Z80 CPU:
  (a)   doesn't need a clock   ☐
  (b)   has a clock already built into it   ☐
  (c)   requires an external clock oscillator   ☐
  (d)   generates a clock pulse from a bistable   ☐
14 The 6502 CPU:
  (a)   as a clock already built into it   ☐
  (b)   requires an external clock oscillator   ☐
  (c)   doesn't need a clock   ☐
  (d)   generates a clock pulse from a bistable   ☐
15 A crystal-controlled clock oscillator could be built from a:
  (a)   7404 chip   ☐
  (b)   7804 chip   ☐
  (c)   7805 chip   ☐
  (d)   7905 chip   ☐
16 An 8-bit register can store, in binary, any of the decimal
    numbers:
  (a)   1 to 256   ☐
  (b)   1 to 1000   ☐
  (c)   0 to 255   ☐
  (d)   0 to 999   ☐
17 The decimal number 16 is represented in binary as:
  (a)   00 0100   ☐
  (b)   00 1000   ☐
  (c)   01 0000   ☐
  (d)   10 0000   ☐
18 'Assembly language' allows the production of:
  (a)   machine code programs   ☐
  (b)   'BASIC' programs   ☐
  (c)   a computer system   ☐
  (d)   a computer memory system   ☐
19 Transistor switches are used in computers because they:
  (a)   never wear out   ☐
  (b)   consume no electricity   ☐
  (c)   are very quiet   ☐
  (d)   are very reliable   ☐
20 The abbreviation LED stands for:
  (a)   logic element digitiser   ☐
  (b)   light encased diode   ☐
  (c)   light emitting diode   ☐
  (d)   limiting electronic device   ☐
21 In standard TTL logic:
  (a)   0V = logic 0   ☐
  (b)   1V = logic 0   ☐
  (c)   10V = logic 0   ☐
  (d)   an open circuit = 0V   ☐

22 A typical CPU clock frequency might be:
   (a)   100Hz ☐
   (b)   1kHz ☐
   (c)   10kHz ☐
   (d)   10MHz ☐
23 The elements of a CPU register are denoted:
   (a)   0 to 7 ☐
   (b)   0 to 8 ☐
   (c)   1 to 7 ☐
   (d)   1 to 8 ☐
24 An EPROM is a:
   (a)   CPU ☐
   (b)   memory device ☐
   (c)   clock ☐
   (d)   bistable ☐
25 For a binary count up to ten, you need:
   (a)   only one bistable ☐
   (b)   two bistables ☐
   (c)   four bistables ☐
   (d)   ten bistables ☐

Assignment 1.1

(a) Make a components list for the astable circuit shown in Figure 1.12; obtain the components, then build and test the circuit.
(b) Make a components list for the bistable circuit shown in Figure 1.9; obtain suitable components, then build and test the circuit. It can be tested by 'clocking' it with the astable built previously. If you're working in a group, all the available bistables can be connected together to produce a simple binary counter.
(c) Obtain type numbers and details of integrated circuits containing bistables such as the dual J–K flip-flop ic.
(d) Obtain type numbers and basic details of static RAM memory chips.

Assignment 1.2 (MARC Z80)

(a) Obtain and make notes on the use of the MARC micro system (or another available to you). You should be able to set up the machine so that you can input a machine code program directly. We list below a program which is suitable for inputting to the MARC micro. If you use a different system, you will have to decide whether or not the code is compatible. Load and run the program to see what it does. You should obtain a message on the screen.

(b) Sample MARC MICRO program:

5C00	CD	B9	01	CD	C9	02	21	5B
5C08	E0	3E	4D	CD	83	5C	3E	49
5C10	CD	83	5C	3E	43	CD	83	5C
5C18	3E	52	CD	83	5C	3E	4F	CD
5C20	83	5C	3E	50	CD	83	5C	3E
5C28	52	CD	83	5C	3E	4F	CD	83
5C30	5C	3E	43	CD	83	5C	3E	45
5C38	CD	83	5C	3E	53	CD	83	5C
5C40	CD	83	5C	3E	4F	CD	83	5C
5C48	3E	52	CD	83	5C	21	DD	E0
5C50	3E	54	CD	83	5C	3E	45	CD
5C58	83	5C	3E	43	CD	83	5C	3E
5C60	48	CD	83	5C	3E	4E	CD	83
5C68	5C	3E	4F	CD	83	5C	3E	4C
5C70	CD	83	5C	3E	4F	CD	83	5C
5C78	3E	47	CD	83	5C	3E	59	CD
5C80	83	5C	76	77	23	23	C9	

Assignment 1.3 (EMMA 6502)

(a) Obtain and make notes on the use of the EMMA micro system (or another available to you). You should be able to set up the machine so that you can input a machine code program directly. We list below a program which is suitable for inputting to the EMMA micro. If you use a different system, you will have to decide whether or not the code is compatible. Load and run the program to see what it does; you should obtain a moving display on the machine's seven segment displays.

(b) Sample EMMA MICRO program:

0200	A9	1F	85	0E	A9	00	A2	07
0208	86	20	95	10	CA	10	FB	A9
0210	00	A6	20	95	10	A9	08	CA
0218	10	02	A2	07	95	10	86	20
0220	A2	03	20	0C	FE	CA	D0	FA
0228	F0	E5	00					

Chapter 2
Binary maths and number systems

The decimal system

The decimal system is so called because there are ten digits available, 0 to 9. All other numbers in the system are made up of combinations of these. (Dec means 10 as in decade – 10 years, December – 10th month (July and August have been added) and decathlon – 10 events, etc.) A number like 255 really means:

100s 10s   Units

  2   5   5

We can recognise 255 instantly because we are used to using decimal numbers, but in the binary system, 255 would be written as: 1111 1111. This is not so recognisable (unless we are used to it) although we could easily work it out. Similarly, we could work out the value of 255:

$$2 \times 100 = 200$$
$$5 \times 10 = 50$$
$$5 \times 1 = 5$$

then we add: 255

In 'index' notation (powers of 10) the column headings are:

10^3	10^2	10^1	10^0
(1000s)	(100s)	(10s)	(Units)
4	5	2	7

i.e.

$$4 \times 10^3 = 4000$$
$$5 \times 10^2 = 500$$
$$2 \times 10^1 = 20$$
$$7 \times 10^0 = 7$$
$$\overline{4527}$$

(Remember, incidentally, that $x^0 = 1$, whatever the value of x.)

Table 2.1 Decimal numbers 0 to 64 with their binary equivalents. Note how the convention of writing the numbers in groups of four is adhered to. (They're easier to read like this and each group of four can be represented by a single hexadecimal digit as we shall see later.) Frequently, the leading zeros are also written down, e.g. 62D would be written as 0011 1110. This is how it would appear in a CPU register or memory location

Decimal	Binary	Decimal	Binary
0	0	33	10 0001
1	1	34	10 0010
2	10	35	10 0011
3	11	36	10 0100
4	100	37	10 0101
5	101	38	10 0110
6	110	39	10 0111
7	111	40	10 1000
8	1000	41	10 1001
9	1001	42	10 1010
10	1010	43	10 1011
11	1011	44	10 1100
12	1100	45	10 1101
13	1101	46	10 1110
14	1110	47	10 1111
15	1111	48	11 0000
16	1 0000	49	11 0001
17	1 0001	50	11 0010
18	1 0010	51	11 0011
19	1 0011	52	11 0100
20	1 0100	53	11 0101
21	1 0101	54	11 0110
22	1 0110	55	11 0111
23	1 0111	56	11 1000
24	1 1000	57	11 1001
25	1 1001	58	11 1010
26	1 1010	59	11 1011
27	1 1011	60	11 1100
28	1 1100	61	11 1101
29	1 1101	62	11 1110
30	1 1110	63	11 1111
31	1 1111	64	100 0000
32	10 0000		

By using decimal numbers we have illustrated the precise format of writing numbers down, something which we are probably no longer conscious of. As we have seen, however, computers only work with binary numbers – and so must we. In order to attempt this, consider the following rules:

1   In all number systems, the column headings are in powers of the number of digits in the system. In decimal there are ten digits (0–9) so the column headings are in powers of 10. In binary, there are two digits (0 and 1) so the column headings are in powers of 2.

2   When all the digits in the system have been used, we write 10 (etc.), i.e. in decimal: 0, 1, 2, 3, 4, 5, 6, 7, 8, 9 and then 10. In binary, there are two digits (0 and 1), so we would write: 0, 1 (run out of numbers) and 10, i.e. 0, 1, 10.

3   To obtain the binary number sequence, use the decimal numbering system as a guide and select only those decimal numbers which contain either ones or zeros; you would then generate the following binary sequence: 0, 1, 10, 11, 100, 101, 110, 111, 1000 and so on. Table 2.1 lists the binary numbers from 0 to 64.

Converting from decimal to binary

Say we want to convert the decimal number 14 into its binary equivalent. This can be done by inspection, which, like most things, becomes easier with experience. We need to list the binary 'column headings' first, note the largest number which will go into the number to be converted, note the remainder and then repeat the process.

We begin by writing down the column headings:

2^7 2^6 2^5 2^4 2^3 2^2 2^1 2^0

(128) (64) (32) (16) (8) (4) (2) (1)

16 is obviously too large (again, experience will quickly lead us to write down only the columns we will actually need); so we start with 8 and write a '1' below it. We have therefore represented 8 by putting a '1' in that column, this leaves 6 (14 – 8) which is one 4 and finally one 2. Continuing like this it is easy to see that:

14 = (1 × 8), (1 × 4) and (1 × 2)

Hence:

2^3 2^2 2^1 2^0

1 1 1 0

So, 14 (decimal) is equal to 1110 (binary).

There are several conventions, but we shall use simply the letter D to indicate a decimal number, and the letter B to indicate a binary, i.e. 14D = 1110B.

Now let's convert 63D into binary; the method is illustrated in Figure 2.1.

The answers to two other problems are given below; try these for yourself until you find the correct solution:

92D = 101 1100   245 = 1111 0101

There is another (probably better) method of making these conversions known as successive division by 2. As an example, suppose we wish to convert 57D to binary. Doing this by successive division by 2, we first divide 57 by 2,

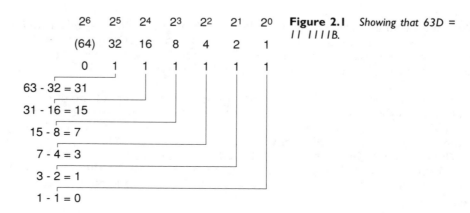

Figure 2.1 *Showing that 63D = 11 1111B.*

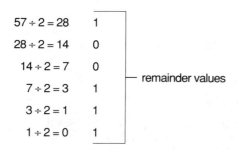

$57 \div 2 = 28$ 1
$28 \div 2 = 14$ 0
$14 \div 2 = 7$ 0
$7 \div 2 = 3$ 1   ── remainder values
$3 \div 2 = 1$ 1
$1 \div 2 = 0$ 1

Figure 2.2 *Showing that 57D = 11 1001B. This is because the remainder values have been taken in reverse order and written down to form the binary equivalent.*

obtaining 28 and a remainder of 1. These two numbers are noted down as shown in Figure 2.2. We then continue dividing by 2.

When the divisions have been completed, write down the remainder values in reverse order, producing 11 1001B.

Note the next example:

$92/2 = 46$ 0
$46/2 = 23$ 0
$23/2 = 11$ 1
$11/2 = 5$ 1
$5/2 = 2$ 1
$2/2 = 1$ 0
$1/2 = 0$ 1

Therefore: $92D = 101\ 1100B$

The answers to two other problems are given below; try these for yourself until you obtain the correct solution:

$179D = 1011\ 0011B$ $133D = 1000\ 0101$

Converting from binary to decimal

The easiest way of doing this is to write down the column headings as powers of 2 and/or the numbers they represent. Then write down the binary number below and add up the columns. The following examples should make this clear:

Convert 11001B to decimal:

2^4	2^3	2^2	2^1	2^0
(16)	(8)	(4)	(2)	(1)
1	1	0	0	1

$= 16 + 8 + 1 = 25$

i.e. $11001B = 25D$

Convert 11010B to decimal:

16	8	4	2	1
1	1	0	1	0

$= 16 + 8 + 2 = 26$

i.e. $11010B = 26D$

Convert 1110 1011B to decimal:

128	64	32	16	8	4	2	1
1	1	1	0	1	0	1	1

$$= 128 + 64 + 32 + 8 + 2 + 1 = 235$$

i.e. 1110 1011B = 235D

These numbers are fairly simple, but difficult to recognise and handle; for example, what is 1110 1011B in decimal? And what about a really large number like 64,424? This turns out to be 1111 1011 1010 1000 in binary! Of course, we now know how the conversions are made and we also know what a boring, time-consuming and tedious process it is. Furthermore, we have also learned that this is the only language the microprocessor understands and we're stuck with it. Imagine having to program a computer using only binary codes. In fact, this is how it used to be done and apart from the disadvantages just mentioned, the problem of avoiding errors was enormous. Clearly a better system of representing binary numbers is required. Such a system exists in the *hexadecimal code*.

Hexadecimal numbers

We have seen that it is usual to put binary numbers together in groups of four, so it would be useful to be able to represent each group of four with a single digit. However, the maximum number which can be represented by four binary digits is 15, i.e. 1111B = 15D, and 15 does, of course, contain two digits. For this reason, the hexadecimal code has been invented. The numbers 0 to 9 are used as in the decimal system and the remaining six numbers (10 to 15) are represented by the first six letters of the alphabet. These hexadecimal numbers are shown in Table 2.2.

We have seen that decimal 64,424 is equivalent to 1111 1011 1010 1000 in binary. In order to represent this number in hexadecimal we could simply make use of Table 2.2 as shown below:

1111	1011	1010	1000
F	B	A	8

hence: 1111 1011 1010 1000B becomes FBA8H which is a lot neater and can be input directly into most microelectronic systems. Generally, where this is possible, the computer will require you to include a leading zero whenever the hexadecimal number begins with a letter (to distinguish the number from a label) and to follow the number with the letter 'H' (to distinguish it from a decimal number). Hence, the correct format is 0FBA8H.

Some systems allow hexadecimal numbers to be input from a numeric keypad (for example, systems like the 'EMMA' which do not support a VDU). 6502 systems often represent a hexadecimal number by using the '&' prefix (though sometimes a '$' is used), hence 0FBA8H would be represented as #&FBA8 (or $FBA8). The '#' (hash) symbol indicates that the value is a number rather than a memory address. Obviously, the reader will need to become familiar with the conventions associated with the hardware being used; Z80 systems receive the most attention in this book, but substantial information is also presented for 6502 users.

Bits, bytes and words

We break off from number systems for a short while to discuss bits and bytes. A single binary digit is called a 'bit'. All information in a digital system is

Table 2.2 Introducing hexadecimal numbers

Decimal	Binary	Hex
0	0	0
1	1	1
2	10	2
3	11	3
4	100	4
5	101	5
6	110	6
7	111	7
8	1000	8
9	1001	9
10	1010	A
11	1011	B
12	1100	C
13	1101	D
14	1110	E
15	1111	F
16	1 0000	10
17	1 0001	11
18	1 0010	12
19	1 0011	13
20	1 0100	14
21	1 0101	15
22	1 0110	16
23	1 0111	17
24	1 1000	18
25	1 1001	19
26	1 1010	1A
27	1 1011	1B
28	1 1100	1C
29	1 1101	1D
30	1 1110	1E
31	1 1111	1F
32	10 0000	20
33	10 0001	21
34	10 0010	22
35	10 0011	23
36	10 0100	24
37	10 0101	25
38	10 0110	26
39	10 0111	27
40	10 1000	28
41	10 1001	29
42	10 1010	2A

represented by a sequence of bits. An 8-bit sequence is called a 'byte'; a 4-bit sequence is a 'nibble'; an 8-bit sequence in the Z80 system is also called a word because this is the number of bits which can be simultaneously processed. In other systems, the 'word' length may be 16, 32 or even 64 bits. Two or three bit sequences are called fields.

Hence: 1 and 0 are bits; 10, 101 and 110 are fields; 1111 and 1010 are nibbles; 1011 0100 is a byte; 1111 1001 in a Z80 and 6502 system is a word. In a 16-bit system, the word length is 16 so that 1111 1010 1100 0011 would also be called a word.

A single hexadecimal digit represents a nibble, e.g. 0AH, 0FH (or &0A, &0F, $0A, $0F), etc. Two hex digits comprise a byte, e.g. 0FFH, 12H, 20H, 0BAH, etc. Four hex digits represent a word in a 16-bit system, e.g. 0FFFFH, 1234H, 0BBC1H, &FFFF, $1234, etc.

In a series of bits like 1101 1001, the bit on the far right-hand side is called the 'least significant bit' (lsb). Similarly, the bit on the far left is called the 'most significant bit' (msb).

Now let's return to hexadecimal numbers.

Converting from binary to hex

Divide the binary number into groups of 4 bits (starting with the lsb on the right-hand side). For example, 11011101101B becomes:

0110	1110	1101
6	E	D

As you can see, once the binary numbers are arranged in groups of four, it is a simple matter to write down the hex equivalent of each nibble. (Binary numbers are usually written down in groups of four precisely for this reason.) The value in this case then becomes 6EDH (&6ED or $6ED).

Converting from hex to binary

This is of course the reverse process of converting from binary to hex. For example, take the number BBC2 (written as 0BBC2H, &BBC2 or $BBC2). Write down the numbers and then put the equivalent binary group of four below each one, thus:

i.e.	B	B	C	2
	1011	1011	1100	0010

Any hex number may be converted in a similar way.

From hex to decimal

As before, simply write down the relevant column headings. To see how easy it is, let's convert 1234H (&1234 or $1234) into decimal. The column headings will now be in powers of 16.

i.e.	16^3	16^2	16^1	16^0
	4096	256	16	1
	1	2	3	4

hence:

$$1234H = (4096 \times 1) + (2 \times 256) + (3 \times 16) + (4 \times 1)$$
$$= 4096 + 512 + 48 + 4$$
$$= \underline{4660D}$$

If the hexadecimal number contains letters, you will need to convert these too as shown in the following example, where we convert 0FBA8H into decimal:

4096	256	16	1
F	B	A	8
(15)	(11)	(10)	(8)

hence:

$$0FBAH = (4096 \times 15) + (256 \times 11) + (16 \times 10) + 8$$
$$= \quad 61440 \quad + \quad 2816 \quad + \quad 160 \quad + 8$$
$$= \quad 64424D$$

Octal numbers

Octal numbers are not in widespread use, but many micro and computing courses still require students to be familiar with them. One important application of these numbers, however, is in the use of programmable logic controllers (PLCs).

Converting from binary to octal

Using the same convention as previously would lead us to add the suffix O (for octal); clearly this would be unwise because of the possible confusion with zero, so the letter 'Q' is used to denote an octal number. To make the conversion, divide the binary number into groups of 3 bits (starting from the lsb on the right-hand side). As an example, let's convert binary 11011101101 to octal:

011	011	101	101	for each group the weightings are:
[3]	[3]	[5]	[5]	2^2, 2^1 and 2^0 or 4, 2 and 1

Hence: 11011101101B = 3355Q

Converting from octal to decimal

Let's convert 325Q to decimal. As before, simply write down the relevant column headings to give the weighting for each digit:

8^3	8^2	8^1	8^0
512	64	8	1
0	3	2	5

Hence:

$$325Q = (3 \times 64) + (2 \times 8) + 5$$
$$= \quad 192 \quad + \quad 16 \quad + 5$$
$$= \quad 213D$$

Converting from decimal to octal

This method has been used previously to convert from decimal to binary. The principle is the same except that now the number is divided by 8. The remainder values taken in reverse order give the octal number. For example, let's convert 57D to octal:

$57/8 = 7 \quad 1 \quad$ (remainder 1)

$7/8 = 0 \quad 7 \quad$ (remainder 7)

Hence: $\quad$ 57D = 71Q

Numbers with fractions

Not all readers will need to convert decimal numbers with fractions into binary, and few electronics courses require its use. However, for the sake of completion we now show how this is done by converting 34.625D into binary.

For the integer part (34) proceed as previously using successive division by two and noting the remainders. The binary number is then obtained by writing down these remainder values in the reverse order:

$34/2 = 17 \quad 0 \quad$ (no remainder)

$17/2 = 8 \quad 1 \quad$ (remainder 1)

$8/2 = 4 \quad 0$

$4/2 = 2 \quad 0$

$2/2 = 1 \quad 0$

$1/2 = 0 \quad 1$

The binary number is then: 100010

To convert the fractional part into binary, we progressively multiply the fraction by 2, removing and noting the carries; the carries taken in the forward direction form the binary equivalent.

$0.625 \times 2 = 1.25 \qquad$ with a carry of 1

$0.25 \times 2 = 0.50 \qquad$ with a carry of 0

$0.50 \times 2 = 1.00 \qquad$ with a carry of 1

The binary equivalent is therefore 0.101 and 34.625D is equal to 100010.101

Binary coded decimal (BCD)

It has been noted that binary numbers are separated into groups of four when written down to make it easier for humans to understand them; we don't have to do this for the benefit of computers, of course. It will be obvious that the maximum number which can be represented by a group of four binaries is 15 (1111). This can still be a problem, however, since we are so used to using the decimal numbering system we should much prefer to have the 'human' side of a computer give us numbers in decimal form.

Suppose we have an electronic display that gives decimal numbers (a digital clock would be a good example) the binary groups of four have to be converted into the corresponding decimal numbers. The digital clock may display the time 11:27 so that each unit of the display is fed with the appropriate binary sequence. This is shown in Figure 2.3.

0001	0001	0010	0111
1	1 :	2	7

Figure 2.3 *The binary sequences which provide the decimal readout.*

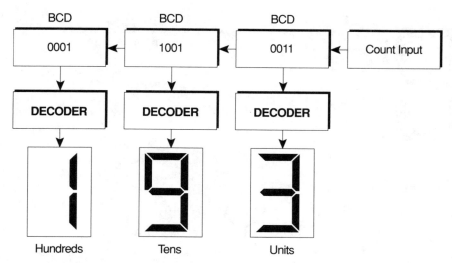

Figure 2.4 *A simple decimal counting system. Note that the central element is displaying decimal 9. You should now be able to see that this is the maximum it can display. If the counter proceeded to 10, the seven-segment display could not cope with it. Instead, the group of four binaries resets to zero on the tenth pulse and the hundreds element increments (increases by 1).*

Provided we can 'decode' each nibble electronically into its decimal equivalent (which of course we can) then this is OK. Decimal 8 is 1000 in binary and 9D is equal to 1001. However, if the computer output then goes onto binary 1010 (decimal 10) we are in trouble because it can't be displayed in a single digit by the clock display. Hence, the numbers 1010 to 1111 are 'illegal'. The binary coded decimal (BCD) system allows only the binary combinations from decimal 0 to 9.

There are several BCD codes, but the most common, which is also the one we shall use, is the 8421 code, so called because of the weighting of each digit in the nibble. This we have seen before.

Hence: 7D = 0111BCD, 3D = 0011BCD and 15D = 0001 0101BCD. The latter (15D) shows the difference between BCD and HEX. In the former, 15D = 0001 0101 and in the latter, 15 = 0FH which is 1111 in pure binary.

Seven-segment displays

Most decimal displays are implemented using seven-segment LED displays. The conversion from the BCD code to the decimal version displayed on the device is performed electronically. The devices are called 'decoders'. Figure 2.4 shows a simple system.

Seven-segment displays are standardised so that each section is given a small letter of the alphabet, a to g, in order to identify it. The eighth 'segment' (where it exists) is the optional decimal point. Hence, eight digits are required to represent a seven-segment display number. The accepted convention is shown in Figure 2.5. For outputting from a microprocessor system, the 8 bits would be held in the accumulator (or other register or memory location) in the order shown in Figure 2.6.

Simple binary arithmetic

Adding and subtracting binary numbers is similar to decimal arithmetic. If anything it is rather easier; all you need to know is the following:

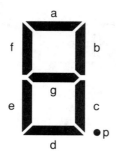

Figure 2.5
Convention for labelling the segments of a seven-segment display.

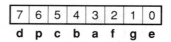

Figure 2.6 *A convention for holding the bits in a CPU register.*

$0 + 1 = 1;$ $1 + 1 = 10$ and $1 + 1 + 1 = 11$

$1 - 0 = 1;$ $10 - 1 = 1$ and $11 - 1 = 10$

The following examples should make the process clear:

```
1.    101 +      2.    1101       3.    1011 +
      110              1111             11011
     ----             -----           ------
     1011             11100           100110
```

When subtracting, if you need to take 1 from 0, 'borrow' a '10' and note that $10 - 1 = 1$. For example, subtract 1101 from 10011:

```
10011 -      1 - 1 = 0, and 1 - 0 = 1.
             0 - 1 can't be done, so we need
 1101        to borrow a 10. But there isn't
-----        one, so we have to progress along
 0110        the line until we find one.
```

The left-most digit therefore disappears, making the next 0 into 10. This in turn is reduced to 1 (since $10 - 1 = 1$) when 1 is borrowed from it. The difference then becomes:

```
0  1  10  1  1
   1   1  0  1
---------------
0  1   1  1  0
```

Two's complement

As far as the computer is concerned, there is no natural provision for identifying a negative number. For this reason, 'two's complement' notation has been developed. Using this system the msb of a binary number is used to indicate whether the number is positive or negative; if the msb is 0, the number is positive, if it's 1, the number is negative.

If an 8-bit register is being used, it means that only seven of the bits can be used to represent a number, the msb is used to give the sign as shown in Figure 2.7.

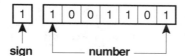

Figure 2.7 *Diagram of a typical register containing a number. The msb gives the sign, the rest of the digits represent the number.*

This arrangement obviously reduces the range of numbers that can be held in the register. They will be: 00 to 7FH (0 to 127D), plus a similar range of negative numbers (except zero).

Hence: 0 101 1011   is a positive number

and   1 011 1010   is a negative number

The negative number is obtained by what is called two's complementing:

e.g. 42 = 00101010   (ALL 8 digits)

invert: 11010101

add 1:   1
           _____
           11010110   (= −42)

Where it says 'invert' we simply change all the ones to zeros and all the zeros to ones. Whenever we wish to subtract a number from another, the number to be subtracted is two's complemented and then *added* to the number being subtracted from.

It is important to realise that confusion can arise if ordinary positive addition is attempted. For example, 0 111 1111 (7FH) is the maximum number allowed as a 'signed' number. Adding 01 (or more) to this number causes the 'overflow flag' to be set in the flag register (see Chapter 10).

In reality, the programmer will rarely be concerned with two's complement notation. All 8 bits can be used to make additions and subtractions in the CPU registers. The computer only requires two's complement negative numbers to be entered by the user in a limited number of instructions (relative jumps, in fact) otherwise all 8 bits can be used for a number. The calculation of relative jumps (or 'branches') is described in Chapter 5 but here's an example:

A program requires a backward jump of 17 locations:

          17 = 0001 0001

Complement = 1110 1110

        add 1 =   1
               _____
               = 1110 1111 =   0EFH (or &EF)

hence: −17 = 0EFH (&EF or $EF), so that's the number to be entered.

Alphanumeric codes

Any microprocessor must be capable of recognising and utilising a code which allows for all 26 letters of the alphabet, both upper and lower case, the numbers 0 to 9, special symbols such as & @ $ % #, etc. punctuation marks such as ().,;: etc. and 20 to 30 special characters and codes as well.

If only 30 special characters are used, the code described must have some 92 combinations. A 6-bit code has 64 combinations which would not be enough, however a 7-bit code has 128 combinations and this would be adequate for the purpose.

Many microprocessors are 8-bit machines, so that seven of the bits may be used for the data code whilst the eighth bit is assigned for error checking – called *parity*.

There are two popular alphanumeric codes, the 7-bit ASCII code – American Standard Code for Information Interchange – and the 8-bit EBCDIC code – Extended Binary Coded Decimal Interchange Code. The ASCII code is used by almost all microprocessors and is the one most commonly used.

The ASCII codes are listed in Table 2.3.

Table 2.3   The ASCII codes (values in hexadecimal)

Hex	Char	Hex	Char	Hex	Char	Hex	Char	Hex	Char
00	NUL	1A	SUB	34	4	4E	N	68	h
01	SOH	1B	ESC	35	5	4F	O	69	i
02	STX	1C	FS	36	6	50	P	6A	j
03	ETX	1D	GS	37	7	51	Q	6B	k
04	EOT	1E	RS	38	8	52	R	6C	l
05	ENQ	1F	US	39	9	53	S	6D	m
06	ACK	20	SP	3A	:	54	T	6E	n
07	BEL	21	!	3B	;	55	U	6F	o
08	BS	22	"	3C	<	56	V	70	p
09	HT	23	#	3D	=	57	W	71	q
0A	LF	25	$	3E	>	58	X	72	r
0B	VT	25	%	3F	?	59	Y	73	s
0C	FF	26	&	40	@	5A	Z	74	t
0D	CR	27	'	41	A	5B	[	75	u
0E	SO	28	(	42	B	5C	\	76	v
0F	SI	29	)	43	C	5D	]	77	w
10	DLE	2A	*	44	D	5E	^	78	x
11	DC1	2B	+	45	E	5F	–	79	y
12	DC2	2C	,	46	F	60		7A	z
13	DC3	2D	–	47	G	61	a	7B	{
14	DC4	2E	.	48	H	62	b	7C		
15	NAK	2F	/	49	I	63	c	7D	}
16	SYN	30	0	4A	J	64	d	7E	~
17	ETB	31	1	4B	K	65	e	7F	DEL
18	CAN	32	2	4C	L	66	f		
19	EM	33	3	4D	M	67	g		

The ASCII code

Each ASCII code consists of two hex numbers, usually from 00H to 7FH (128 different characters or codes), a decimal range of 0 to 127. The first 32 codes (00H to 20H) and the final group are non-printing characters, the rest are printing characters, including upper and lower case alphabetical characters, decimal numbers and some special characters.

Table 2.4 gives the meaning of some of the control codes. Many of them derive from the days when old mechanical 'teletype' machines were used to produce not only computer 'hard copy', but computer output itself (before we had 'VDUs'!). So the term 'CR' (carriage return) refers to the teletype carriage returning to the left-hand side of the machine, after completing a line of text, usually followed by an 'LF' (line feed). Although mechanical machines are rarely used nowadays, the code names have remained.

The format of the ASCII word itself is shown in Figure 2.8.

The ASCII word is contained in the lower 7 bits (b_0 to b_6), the msb (b_7) is used for parity check purposes.

We shall see later how the ASCII codes may be used to put letters on the screen of a microcomputer-controlled VDU.

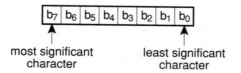

most significant   least significant
character   character

Figure 2.8   *The ASCII word is contained in the lower 7 bits (b_0 to b_6), the msb (b_7) is used for parity check purposes.*

Table 2.4   Meaning of some of the ASCII codes

NUL	All zero	SOH	Start of heading
STX	Start of text	ETX	End of text
EOT	End of transmission	ENQ	Enquiry
ACK	Acknowledge	BEL	Audible signal
BS	Backspace	HT	Horizontal tabulation
LF	Line feed	VT	Vertical tabulation
FF	Form feed	CR	Carriage return
SO	Shift out	SI	Shift in
DLE	Data link escape		
DC1, DC2, DC3, DC4,	Device controls		
NAK	Negative acknowledge		
SUB	Substitute		
ETB	End of transmission block		
CAN	Cancel	DEL	Delete
FS	File separator	ESC	Escape
GS	Group separator	RS	Record separator
US	Unit separator	SP	Space
SYN	Synchronous idle	EM	End of medium

Questions

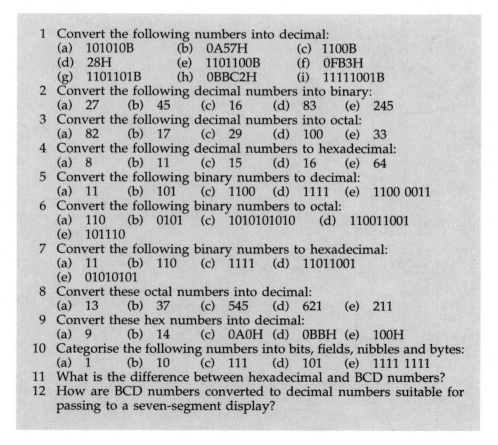

1 Convert the following numbers into decimal:
   (a) 101010B   (b) 0A57H   (c) 1100B
   (d) 28H   (e) 1101100B   (f) 0FB3H
   (g) 1101101B   (h) 0BBC2H   (i) 11111001B
2 Convert the following decimal numbers into binary:
   (a) 27   (b) 45   (c) 16   (d) 83   (e) 245
3 Convert the following decimal numbers into octal:
   (a) 82   (b) 17   (c) 29   (d) 100   (e) 33
4 Convert the following decimal numbers to hexadecimal:
   (a) 8   (b) 11   (c) 15   (d) 16   (e) 64
5 Convert the following binary numbers to decimal:
   (a) 11   (b) 101   (c) 1100   (d) 1111   (e) 1100 0011
6 Convert the following binary numbers to octal:
   (a) 110   (b) 0101   (c) 1010101010   (d) 110011001
   (e) 101110
7 Convert the following binary numbers to hexadecimal:
   (a) 11   (b) 110   (c) 1111   (d) 11011001
   (e) 01010101
8 Convert these octal numbers into decimal:
   (a) 13   (b) 37   (c) 545   (d) 621   (e) 211
9 Convert these hex numbers into decimal:
   (a) 9   (b) 14   (c) 0A0H (d) 0BBH (e) 100H
10 Categorise the following numbers into bits, fields, nibbles and bytes:
   (a) 1   (b) 10   (c) 111   (d) 101   (e) 1111 1111
11 What is the difference between hexadecimal and BCD numbers?
12 How are BCD numbers converted to decimal numbers suitable for
   passing to a seven-segment display?

13 Add the following binary numbers together:
    (a) 11011 + 11011   (b) 11110 + 101101   (c) 110111 + 101
    (d) 1011 + 1011100   (e) 100011 + 1010101
14 Subtract the following binary numbers from each other:
    (a) 10011 − 1001   (b) 10110 − 1111   (c) 101011 − 111
    (d) 100001 − 11101   (e) 1001101 − 11110
15 Draw a diagram showing a register containing a two's complemented number, indicating the sign of the number.

Multiple choice questions

1 The decimal value 83 is represented in BCD as:
    (a) 0011 0011 ☐
    (b) 0100 0011 ☐
    (c) 1000 0011 ☐
    (d) 1000 0101 ☐
2 The decimal value 27 is represented in BCD as:
    (a) 0010 0110 ☐
    (b) 0010 0111 ☐
    (c) 0100 0110 ☐
    (d) 0100 0111 ☐
3 The decimal value 69 is represented in BCD as:
    (a) 0110 1001 ☐
    (b) 0110 0111 ☐
    (c) 0110 1010 ☐
    (d) 01010 0110 ☐
4 The BCD value 1001 0101 is represented in decimal:
    (a) 85 ☐
    (b) 96 ☐
    (c) 86 ☐
    (d) 95 ☐
5 The decimal value 48066 is represented in hex by:
    (a) 0AF92H ☐
    (b) 0BF92H ☐
    (c) 0BBC1H ☐
    (d) 0BBC2H ☐
6 The hex value 10E9H is represented in decimal as:
    (a) 4329 ☐
    (b) 5815 ☐
    (c) 1101 ☐
    (d) 9821 ☐
7 The standard length of an ASCII character is:
    (a) 6 bits ☐
    (b) 7 bits ☐
    (c) 8 bits ☐
    (d) 9 bits ☐
8 The addition of 0FFH and 01H produces in hex:
    (a) 10H ☐
    (b) 100H ☐
    (c) AAH ☐
    (d) 0AAH ☐

9 Standard ASCII characters have decimal values in the range:
   (a)   0 to 127   ☐
   (b)   0 to 255   ☐
   (c)   32 to 127   ☐
   (d)   32 to 255   ☐
10 The binary number 101 is known as a:
   (a)   bit   ☐
   (b)   field   ☐
   (c)   nibble   ☐
   (d)   byte   ☐
11 The decimal number 100 in binary is:
   (a)   100 0000   ☐
   (b)   100 1000   ☐
   (c)   110 0100   ☐
   (d)   110 1000   ☐
12 The binary number 1000 0001 in decimal is:
   (a)   127   ☐
   (b)   128   ☐
   (c)   129   ☐
   (d)   130   ☐

Tutorial questions

1 Convert the following numbers into decimal:
   (a) 0B0BH   (b) 11101101B   (c) 64H   (d) 333Q
   (e) 2B7H   (f) 10010111B   (g) 111Q (h) 0F66AH
2 Convert the following decimal numbers into binary:
   (a) 77   (b) 392   (c) 415   (d) 511   (e) 1000
3 Convert the following decimal numbers into octal:
   (a) 123   (b) 39   (c) 415   (d) 511   (e) 1000
4 Convert the following decimal numbers to hexadecimal:
   (a) 99   (b) 255   (c) 517   (d) 1527   (e) 8822
5 Convert the following binary numbers to decimal:
   (a) 0110 0111   (b) 1110111   (c) 110 1111 0110
   (d) 1001 1100 0101 1010   (e) 1000 1111 1011 0101
6 Convert the following binary numbers to octal:
   (a) 1010001   (b) 1111111101   (c) 011000110
   (d) 10011001100110   (e) 101101110110101
7 Convert the following binary numbers to hexadecimal:
   (a) 1000000000000   (b) 11111111111111
   (c) 11011011001101   (c) 1001001001   (e) 1111000011110000
8 Convert these octal numbers into decimal:
   (a) 633   (b) 1253   (c) 3677   (d) 37722   (e) 1219
9 Convert these hexadecimal numbers into decimal:
   (a) 10AH (b) 0FAH (c) 80H   (d) 0FB1H (e) 0CC50H
10 Categorise the following numbers into bits, fields, nibbles and bytes:
   (a) 1010   (b) 1001 1100   (c) 0   (d) 1101   (e) 0110 1000
11 Convert the following decimal numbers into BCD code:
   (a) 3   (b) 7   (c) 9   (d) 10   (e) 16
   (f) 22   (g) 39   (h) 72   (i) 80   (j) 1000

12 Convert the following BCD numbers into decimal numbers:
   (a)   0101   (b)   1000   (c)   1000 0111   (d)   0011 0110
   (e)   1001 1001   (f)   0010 0111   (g)   1001 0011 0100
13 Why are BCD numbers required?
14 Add the following binary numbers together:
   (a)   101 + 1100   (b)   10110 + 11001   (c)   11011 + 1101
   (d)   1100 + 11011   (e)   1101110 + 101111
15 Subtract the following binary numbers from each other:
   (a)   11101 – 101   (b)   11001 – 111   (c)   100000 – 111
   (d)   1000 – 1100   (e)   100100 – 1101

Assignment 2.1

(a) Draw up a table containing the numbers I to 32 in: decimal, binary, octal, hexadecimal and BCD arranged as shown below:

Decimal	Binary	Octal	Hex	BCD
0	0000 0000	0	00	0000 0000
I	0000 0001	I	01	0000 0001
2	0000 0010	2	02	0000 0010
3	0000 0011	3	03	0000 0011
↓	↓	↓	↓	↓
31	0001 1111	37	IF	0011 0001
32	0010 0000	40	20	0011 0010

(b) Obtain details of integrated circuits which will interface between BCD signals and seven-segment displays.

Assignment 2.2 (MARC Z80)

(a) We reproduce the program given in Assignment 1.2. The message is produced by writing the appropriate ASCII codes to the Video Ram. The Video Ram start address was 0E05BH and this appears (in reverse order) at addresses 5C07H and 5C08H. Changing this value displays the message at a different part of the screen (provided you stay within the 1K memory allocated to the video). The ASCII characters are printed in bold. By consulting the table of ASCII characters given earlier, you can construct your own message. You should notice a pattern in the code; this may enable you to work out what's going on except that it's disturbed when the ASCII code for 'S' is given since in the original message this was printed out twice.

5C00	CD	B9	01	CD	C9	02	21	5B
5C08	E0	3E	**4D**	CD	83	5C	3E	**49**
5C10	CD	83	5C	3E	**43**	CD	83	5C
5C18	3E	52	CD	83	5C	3E	**4F**	CD
5C20	83	5C	3E	**50**	CD	83	5C	3E
5C28	**52**	CD	83	5C	3E	**4F**	CD	83
5C30	5C	3E	**43**	CD	83	5C	3E	**45**
5C38	CD	83	5C	3E	**53**	CD	83	5C
5C40	CD	83	5C	3E	**4F**	CD	83	5C
5C48	3E	**52**	CD	83	5C	21	DD	E0
5C50	3E	**54**	CD	83	5C	3E	**45**	CD
5C58	83	5C	3E	**43**	CD	83	5C	3E
5C60	**48**	CD	83	5C	3E	**4E**	CD	83
5C68	5C	3E	**4F**	CD	83	5C	3E	**4C**
5C70	CD	83	5C	3E	**4F**	CD	83	5C
5C78	3E	**47**	CD	83	5C	3E	**59**	CD
5C80	83	5C	76	77	23	23	C9	

Assignment 2.3

Assume that the convention for holding the data for a seven-segment display in a CPU register or memory location is as shown in Figure 2.9:

Figure 2.9 Convention for holding the bits of a seven-segment display in a CPU register or memory location.

| dp | g | f | e | d | c | b | a |

Work out the codes for the numbers 0 to 9. For example, zero needs all segments on except g and dp, i.e. 0011 1111 producing the hex digit, 3FH. The number 1 is 06H, 2 is 5BH. Confirm these and work out the rest.

Assignment 2.4 (EMMA 6502)

(a) refer to Assignment 1.3; the program is reproduced below. The hex code for the moving character is at address &0216 (&08). The convention for the EMMA displays is as shown in Assignment 2.3 (above). Having worked these hex values out, try changing the number at &0216 in the program given so that the moving display shows the number you have entered. The speed of the display can be varied by changing the value at &0221 (&03).

(b) EMMA MICRO program:

0200	A9	1F	85	0E	A9	00	A2	07
0208	86	20	95	10	CA	10	FB	A9
0210	00	A6	20	95	10	A9	**08**	CA
0218	10	02	A2	07	95	10	86	20
0220	A2	**03**	20	0C	FE	CA	D0	FA
0228	F0	E5	00					

Chapter 3
The CPU and a microprocessor system

Digital computers

In Chapter 1 we discussed digital electronics in a fairly general way, and noted that a digital electronic system (of which the microcomputer is presently the most important example) is a logic system which uses tiny transistor switches which can only ever be either ON or OFF and which can represent the logic states 1 and 0.

When digital computers were first constructed, the electronic switching elements were valves. These were very large and expensive, took up a lot of space and consumed a lot of electricity, becoming very hot in the process. The classical view of early computers is that of giant machines occupying whole rooms being tended by white-coated scientists and technicians. Only scientists and engineers could understand the complexities of programming and using these early machines and armies of technicians were required to keep them going.

The use of transistors produced a great step forward. These tiny semiconducting devices were physically very much smaller, consumed much less power and were much more reliable than valves. For the first time, relatively small, efficient computers could be made. It was not until the production of the first integrated circuit, however, that advances were made towards producing home computers.

The personal computer had to wait, not only for a reduction in size and cost of the equipment, but also for a new programming language that most people could understand, learn quickly and actually be able to use.

The machinery was known as the 'hardware' and eventually the computer programs used to operate them came to be known as 'software', by analogy. In Chapter 1 we also discussed the two-transistor circuits known as bistables. It was shown how such circuits could be made to count up in binary when 'clocked' by an astable multivibrator, and how, more importantly, they could be used as memory devices. With eight such circuits, it was shown how a byte (8 bits) of information could be stored.

The size of a computer memory is obviously very important and is always quoted in the sales literature, or even made to form part of the model number of the machine. 64K used to be a standard memory size, but now 128K, 256K and even larger internal memories are becoming commonplace. Chapter 1 introduced and Chapter 2 analysed in some detail the nature of binary mathematics; you should now see that 128 is 2^7 and 2^8 is 256. It therefore follows that 2^{10} is equal to 1024 and it is that number which is represented by the capital letter K. Hence, a 64K memory would be able to hold (64 × 1024) bytes of data.

The small letter 'k' is an abbreviation for kilo, which means 1000, so that a computer K is slightly larger (1024). Using the bistable memory system previously described, each bistable needs two transistors, each byte of memory requires 8 bistables and 8K of memory requires (8 × 1024) bytes. An 8K memory device – a memory chip – based on this system would therefore need (2 × 8 × 8 × 1024) = 131,072 transistors! It could hold 65,536 bits of information, and yet, by using integrated circuit technology, all of this can be made to fit in a tiny chip about half a centimetre by three. It is this kind of technology – the high 'packing density' – which has enabled the microprocessor to be produced. The heart of a computer – on a chip.

RAM and ROM memories

Computer memory devices come in two basic types, RAM and ROM. These words are not as helpful as they could be. RAM stands for 'random access memory', a term which can in fact be applied to both types. ROM is 'read only memory' which means that the computer system can obtain information stored in the memory ('read' it) but can't store anything new in it ('write' to it). RAM stores can both be read from and written to, but both ROM and RAM have the random access facility. Put simply, this means that any data byte stored in the memory chip can be obtained as easily as the next; it doesn't matter whereabouts in the chip it is.

Floppy disc systems offer the same facility. A program stored anywhere on the disc can (to all intents and purposes) be just as quickly accessed as any other. Compare this to information stored on magnetic tape. The first program on the tape is much easier to access than the last – if you want to do it automatically – so that a tape memory is not regarded as being random access.

ROM memory chips are programmed by the manufacturer; they contain the computer operating system and standard routines (mini programs) that may be used quite frequently; there may be a routine to clear the screen, for example. This information is stored permanently in the ROM chip and is not lost when power to the machine is switched off. It is termed a 'non-volatile' memory. There is no possibility at all of making future changes to the contents of a ROM chip. The ROM, as with all semiconductor memory devices, is in matrix form and the connections within are permanent.

RAM chips are used for storing the user's programs, you can write information to them and read it back again entirely at random according to your programming needs, but all the information is lost when the computer system is switched off; this is therefore termed a 'volatile' memory system.

A user would normally 'save' the contents of RAM (usually the program) on floppy or hard disc before turning the power off; such systems are called *memory backing stores*. It is possible to write a program and have it stored permanently in a ROM chip, but the process is quite complex and requires sophisticated equipment. For systems sold in high volume, a mask-programmable ROM is used which is programmed during manufacture and would be too expensive to produce in small quantities.

Programmable memories

A more accessible method is to use a PROM – a programmable ROM – which can be programmed once by the user. A special programming device (PROM-programmer) blows fusible links within the device. One of the most common is the EPROM (or UVEPROM), 'erasable, programmable, read only memory'.

It can be programmed by the user, but the whole program can be erased by illuminating a small quartz window above the silicon chip with ultra violet light. A special EPROM eraser is used. The binary coded program is stored as tiny electrical charges within the structure and the intense UV light disperses the charges; UV light is an ionising radiation; it is therefore very harmful to the eyes, so that this operation takes place in a special, light-tight box.

The EAROM is an electrically alterable ROM. It is a device which can be written to like an ordinary RAM chip but the data is not lost when the power is removed. It is of particular importance because individual locations can be erased and modified independently, whereas with an EPROM all the locations would be erased simultaneously. It is said that the EAROM was developed for missile guidance to enable re-targeting to be done.

The INTEL EPROM range, 2716, 2732, 2764 and 27128, is widely used and readily available. Various RAMS, EPROMS and ROMS are compatible, i.e. one can be unplugged and replaced directly by the other. Another term is 'pin compatible'. This allows the programmer to perfect his program in RAM and then, after 'burning' an EPROM or ROM, replace the RAM chip directly by the permanent ROM or EPROM.

For example, the old TMS 4008 static RAM is compatible with the 2708 EPROM or the TMS 4700 ROM and the 6264 RAM is compatible with the 2764 EPROM. Table 3.1 summarises this information. By analogy with software and hardware, the various types of ROM chip are called 'firmware'.

Table 3.1 ROM and RAM chips compared

Chip name	Volatile/non-volatile	Random access	Read/write
RAM	Volatile	Yes	Both
ROM	Non-volatile	Yes	Read only
PROM	Non-volatile	Yes	Read only
EAROM	Non-volatile	Yes	Both
EPROM	Non-volatile	Yes	Read/erasable
EEPROM	Non-volatile	Yes	Read/erasable

A typical 8K RAM chip is illustrated in Figure 3.1. This is the 6264 static RAM. There are 13 address lines and 8 data lines. There are two chip select pins, CS (pin 26) and $\overline{CS}$ (pin 20). CS is the only active high pin and this would normally be connected to the address bus. Which of CS or $\overline{CS}$ is used depends upon the nature of the system and the address decoding logic employed. This is discussed in Chapter 16. OE (Output Enable) and $\overline{WE}$ (Write Enable) are both active low, i.e. if CS is high when $\overline{WE}$ is low, it is possible to write to the RAM chip and store the byte on the data bus, in the memory location given by the address bus A_0 to A_{12}.

When $\overline{OE}$ is low and $\overline{WE}$ is high, a byte will be copied onto the data bus from the memory location given by the address bus A_0 to A_{12}. The $\overline{WE}$ command is generated by the CPU, so if the $\overline{WE}$ control signal is common to other chips in the system it must be carefully gated. This is because $\overline{OE}$ does not inhibit $\overline{WE}$ in the 6264 so it's possible to over-write some RAM memory when both $\overline{OE}$ and $\overline{WE}$ are low.

The microprocessor

The memory devices described above include just a few examples of integrated circuit technology. There are many such devices, including logic

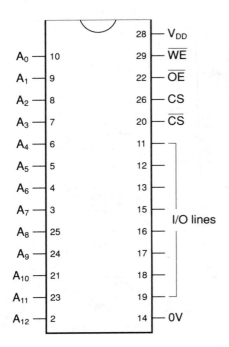

Figure 3.1   *A 6264 static RAM. This is a 28-pin device which holds 8K of RAM organised as 8192 × 8-bit locations. A CMOS device running from a single 5V supply; pin compatible with 64K EPROMS, i.e. 64K bits.*

gates, amplifiers and decoders, but possibly the most complex is the microprocessor itself, and it is to this 'wonder of the age' that we now turn our attention.

We have described memory chips (RAM and ROM, etc.) as being composed of basic bistable units, generally arranged in groups of eight each capable of holding a single byte of data. A similar arrangement is used in the composition of a register. This can be regarded as essentially the same thing. The CPU has several of these registers within its structure which enable it to hold numbers (as a binary sequence) for processing. Figure 3.2 shows a simple representation of a Z80 CPU, the most important parts of which are the instruction register, flag register, the other 'general purpose' registers, the ALU and the program counter.

In Figure 3.3 we show a simple representation of the 6502 CPU (as used in the PET, APPLE, 'EMMA' and BBC computers). The essential features are the same; there are fewer registers (only two apart from the accumulator) and, although it does exactly the same thing, the flag register is called the status register. Other 8-bit CPUs include the Intel 8080 and 8085, and the Motorola 6800.

Generally, a microprocessor contains:

1   an instruction register (IR);
2   an arithmetic and logic unit (ALU);
3   a set of 'general purpose' registers;
4   a set of 'special purpose' registers;
5   control and timing devices.

Control, address and data connections, internally and externally, are made via a system of wires called 'buses'. The data and control buses in 6502 and Z80 systems have 8 lines, whilst the address bus has 16 lines. The 6502 CPU has three general purpose registers: an accumulator and two index registers, X and Y.

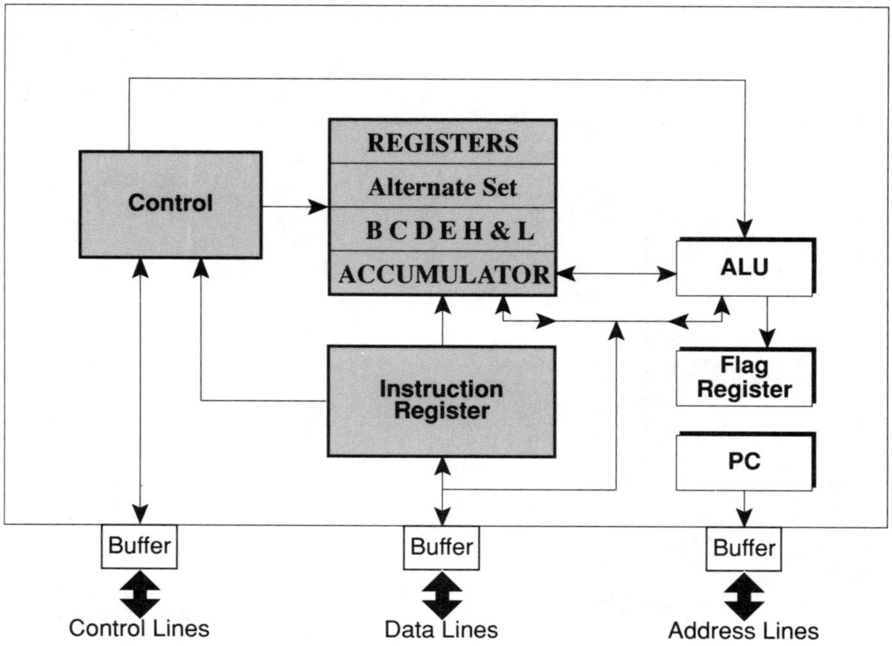

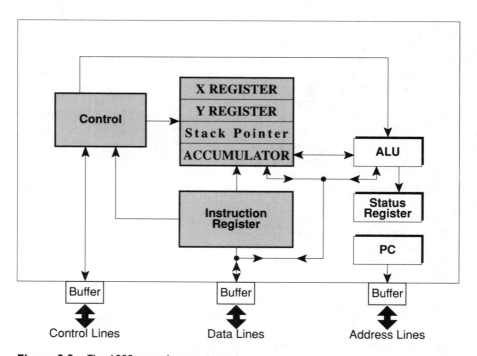

Figure 3.2   *The Z80 central processing unit.*

Figure 3.3   *The 6502 central processing unit.*

The flag register is called the processor status register and the stack register is 8 bits wide. Otherwise, as far as registers are concerned, the 6502 is very similar to the Z80. Since it has more registers, we use the Z80 as an example to describe the internal architecture of the CPU.

1 The instruction register and CPU control

The program stored in the computer RAM memory store consists of instructions and data carefully interleaved. We shall look at this in much greater detail later when we analyse the 'fetch–execute' sequence. For the present, suffice it to say that the CPU must look into the memory and fetch in the instructions and data and then act upon them according to carefully defined procedures. As each instruction is fetched from memory, it is placed in the instruction register and decoded. The control section performs this function and then generates and supplies all of the control signals necessary to read or write data from or to the registers, control the ALU and provide all required external control signals.

2 The arithmetic and logic unit (ALU)

The 8-bit arithmetic and logical instructions of the CPU are executed in the ALU. Although it is extremely complex, the ALU operates entirely on ordinary logic gates switching and changing their outputs according to the signals applied to them. Internally the ALU communicates with the registers and the external data bus on an internal data bus. The type of functions performed by the ALU include: add, subtract, logical AND, logical OR, logical EX–OR, compare, shift, increment and decrement. When the ALU performs an operation, the result is sent to the accumulator. The ALU may (and frequently does) also affect the contents of the flag register. Various bits inside this register are 'set' or 'reset' according to whether the previous ALU operation produced, for example, a negative or positive result, a carry or no carry, a zero or non-zero number, etc.

A simple, 1-bit ALU is shown in Figure 3.4 and a guide to how it operates is given in Table 3.2.

Table 3.2 The ALU operations

F_1	F_2	F_3	Operation	Name
0	0	0	$S \leftarrow 0$	Clear
0	0	1	$S \leftarrow \bar{A}$	Complement
0	1	0	$S \leftarrow A + C_{in}$	Increment
0	1	1	$S \leftarrow \bar{A} + C_{in}$	Negate
1	0	0	$S \leftarrow B$	Transfer B
1	0	1	$S \leftarrow A \ EOR \ B$	EOR
1	1	0	$S \leftarrow A + B + C_{in}$	Add
1	1	1	$S \leftarrow A + \bar{B} + C_{in}$	Subtract

3 The general purpose registers

All of these registers are 8 bits wide. The A register is called the accumulator because traditionally the result of a computation 'accumulates' there and the name has been retained. To take a simple example, if we wished to add two numbers together, one would be held in the accumulator and the other in a

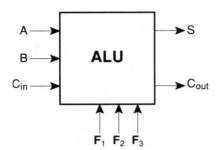

Figure 3.4 *A simple 1-bit ALU.*

Main Register Set		Alternate Register Set	
Accumulator A	Flags F	Accumulator A'	Flags F'
B	C	B'	C'
D	E	D'	E'
H	L	H'	L'

Figure 3.5 *The general purpose Z80 registers.*

different register, it doesn't much matter which, as long as it's not being used for anything else at the time. A simple instruction tells the ALU to make the addition and the result ends up in the accumulator, replacing the original number.

The Z80 has eight main registers, shown in Figure 3.5. These are all accessible by the programmer who can 'load' them with 8-bit binary numbers as required. The accumulator is rather special as recently noted, and so too is the F register. This is the 'flag' register which holds information about ALU operations. Each bit of the flag register has a particular significance. For example, one of the bits indicates whether a previous ALU operation resulted in a positive or negative number; another indicates whether the result was zero or not. The design and use of the flag register is fully covered in Chapter 10.

Another four registers are simply labelled with subsequent letters of the alphabet, B, C, D and E. These 8-bit registers can also be used in pairs and hence hold a 16-bit number. This obviously consists of two bytes; bits 0 to 7 are called the low byte and bits 8 to 15 the high byte. This led to their being referred to as the HL register pair and the label is still used. The A and F registers could also be used as a register pair, but they rarely are because of their special significance and use.

In addition to these general purpose registers, there is an additional matched set called the 'alternate register set'. Each set contains six 8-bit registers that may be used individually or as 16-bit register pairs. At any one time, the programmer can select either set of registers to work with through a single exchange command for the entire set.

4 The special purpose registers

These registers are shown in Figure 3.6. There are two independent index registers, IX and IY, which can hold a 16-bit base address that is used in indexed addressing modes. The instruction used contains an offset value which may be added to the address contained in the index register to produce

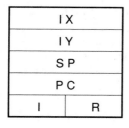

Figure 3.6   *The special-purpose registers.*

the target address. It is sufficient to say here that index addressing offers a powerful method of locating data in complex tables.

SP stands for the stack pointer which is a 16-bit register. It is used to hold the 16-bit address of the current top of a stack, a reserved area of memory which can be located anywhere convenient in the computer RAM. The stack is used to store data on a temporary basis. For example, it can hold the contents of the accumulator and the flag register, whilst those registers are being used for something else; when their temporary use has ended, the AF register contents can easily be retrieved from the stack. The stack is a LIFO file – last in, first out – but stack use and action is explained in greater detail in Chapter 9.

I is the interrupt page address register which is used for a mode of interrupts. This is an operation which halts normal program flow, often temporarily, in order to service an external device.

R is the memory refresh register. We have previously described memory elements as being composed of a great number of bistables (or flip-flops) which can be set or reset. This is called static RAM. Another type of memory stores the digital codes by charging tiny capacitors. These are called dynamic memories (DRAMs); however, the tiny capacitive elements lose their charge patterns which form the memory and have to be continuously 'refreshed'. This involves reading each memory location every 20ms or so, and then immediately writing it back to the same location in a continuous cycle.

Dynamic RAM was used for nearly all microprocessor applications when the Z80 first came onto the market (the Sinclair 'Spectrum' uses DRAM memory), so it was designed to enable dynamic memories to be used just as easily as static ones. Presently, static RAM chips tend to be used mostly; large capacity chips are readily available and their current drain is lower than that of DRAMs making them suitable for systems with battery backup.

The program counter (PC) holds the 16-bit address of the current instruction being fetched from memory. It is therefore closely associated with the instruction register. The PC is automatically incremented (increased by one) after its contents have been transferred to the address lines. Although CPU operation is generally a serial process, working through a program line by line, processing instructions and handling data, there will occasionally be program 'jumps'. This might occur when a subroutine is 'CALLed' for example. When the CPU encounters a program jump, the start address in memory is automatically placed in the program counter, overriding the incrementer.

5 Control and timing devices

This section was introduced earlier when the instruction register was discussed. The overall design of the Z80 CPU ensures that all timing requirements are organised automatically.

The control was shown in Figure 3.2. It is to this section that the crystal-controlled 'clock' is connected which enables other sections of the complete system to be synchronised. Its activities require some special consideration and thought since the control bus has some lines going into the CPU, some coming out, and others which are bi-directional. All this is explained in more detail in Chapter 8.

16-bit processors

One of the first 16-bit microprocessors was the Texas Instruments TMS9900, introduced in 1976. At 32768 words, it had a very limited address range, furthermore, all the working registers were external to the CPU so that execution times were very slow.

More recently, two 16-bit microprocessors have dominated the industry; the Intel 8086 series and the Motorola 68000 series. The Intel 8086 is used in all IBM PCs and IBM compatible clones. The CPU gives true 16-bit processing, with all the benefits of full 16-bit arithmetic operations, increased speed and a larger address range. However an 8-bit data bus is used because of the great cost of using a 16-bit system. This produces a satisfactory compromise between cost and speed.

It is possible to obtain suitable hardware to enable a user to program an IBM PC in machine code using 16-bit processing. Furthermore, the same machine can be used to program in Z80 code using software to emulate its use. It is even possible to run a real Z80 CPU by connecting it to a modern IBM compatible PC, using special hardware. The Z80 is still very widely used in industry and educational establishments so, although significant reference is made to the 6502 CPU throughout the text, especially where its operation and use is rather different, we shall concentrate mainly on the Z80.

A typical microprocessor system

Microprocessor systems are often categorised according to their word length. This indicates the number of bits which can be simultaneously manipulated within the system. For the Z80 system (and the 6502 amongst others), the word length is 8 bits (a byte) but there are more modern CPUs which have 16-, 32- and even 64-bit word lengths. The main advantage of having larger word lengths is the increased speed of processing. Word lengths of 8 and 16 bits are illustrated in Figure 3.7.

Referring again to Figure 3.2, it can be seen that there are three main connecting lines, the data, address and control buses. Every memory location (RAM or ROM) has a unique address which allows the CPU to access it for read or write operations. If the address bus which carries this information

Figure 3.7   *8- and 16-bit words.*

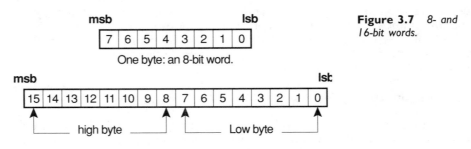

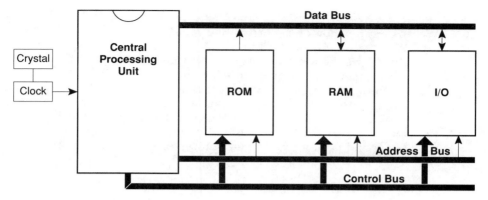

Figure 3.8   *A typical microprocessor system.*

consisted of only eight lines, the memory size would be only 256 bytes; you should recall that an 8-bit register can only hold the numbers 0 to 255, i.e. 0000 0000 to 1111 1111 in binary. With 16 address lines, 65536 locations can be directly addressed (64K), hence many systems will have a 16-line address bus.

The number of lines constituting the data bus depends upon the CPU word size, and for a 6502- or Z80-based system this is obviously eight. Because it would be difficult to read and interpret a circuit diagram with 16 address lines, eight data lines and the control lines all drawn in, the buses are usually simply shown as two parallel lines, inside which you are asked to imagine the individual wires exist.

Figure 3.8 shows a typical microprocessor system. The microprocessor or CPU itself is obviously the most important part. It controls all the automatic actions which take place, accesses the memory units, fetches and executes instructions, performs arithmetic and logic calculations, monitors external events and keeps track of where it has got to.

The memory of the system holds the program and various types of data. The I/O or input/output devices of the system allow the user to 'talk' to the computer in a familiar manner, for example, uses a typewriter-style keyboard, or in many modern systems a mouse in association with pull-down menus. A typical output system is of course the VDU – video display unit. The buses provide interconnections for the various parts of the system.

Questions

1 List the great advantages of using integrated circuits in micro-electronics and microcomputer systems.
2 State: (a) how many bits there are in a byte;
         (b) the name for 1024 bytes.
3 Assuming a memory device made up from basic flip-flop elements, how many transistors would be required to construct an 8K memory?
4 (a) Describe the differences between RAM and ROM.
   (b) What is meant by 'volatile' and 'non-volatile'?
5 Draw a simple diagram of a Z80 CPU (or a 6502 CPU) showing: registers, control, ALU, IR, flags and PC.

6 What is the function of the instruction register?
7 (a) What is the function of the ALU?
  (b) Where are the results of ALU operations sent to?
  (c) What other register may the ALU affect?
8 What is the 'stack' used for, and what is special about it?
9 Why might a 16-bit processor have an 8-bit data bus?
10 (a) How is a microprocessor system word length determined?
   (b) Draw and label 8-bit and 16-bit words
   (c) Draw and label a typical microprocessor system.

Assignment 3.1

(a) Looking at the memory chip pin-outs given in Appendix II, and using manufacturers' literature where necessary, select two RAM and two EPROM devices. Sketch the pin-out and state the function of each of the pins.
(b) Describe how an EPROM may be programmed and erased.

Assignment 3.2

Using Figure 3.8 as a guide, draw a *block* diagram of a typical microprocessor system, indicating the actual chips used. The clock may be implemented by using the circuit given in Chapter 1.

Chapter 4
Memory maps and memory organisation

Memory addressing

The Z80 CPU can support 64K of RAM since it has 16 address lines. With this number of lines, memory locations from zero to $(2^{16} - 1)$ can be addressed. Figure 4.1 shows how this can be achieved. The address lines are denoted A_0 to A_{15} so the lowest address value occurs when all address lines are set to zero and the highest address value occurs when all address lines are set to a one. This is an address range of zero to 1111 1111 1111 1111 binary, or 0000H to FFFFH in hexadecimal.

One line could address two locations (0 and 1), two lines can address four locations (0, 1, 10, 11) and three lines could address eight locations (0, 1, 10, 11, 100, 101, 110, 111), so the rule is:

no. locations addressable = $2^{\text{number of lines}}$

1 line: 2^1 locations = 2

2 lines: 2^2 locations = 4

3 lines: 2^3 locations = 8

and so:

16 lines: 2^{16} locations = 65536

Since the first address is zero, the address range is zero to 65535 $(2^{16} - 1)$. The address locations are most easily shown on a *memory map*.

A_{15}	A_{14}	A_{13}	A_{12}	A_{11}	A_{10}	A_9	A_8	A_7	A_6	A_5	A_4	A_3	A_2	A_1	A_0
0	0	0	0	0	0	0	0	0	0	0	0	0	0	0	0
1	1	1	1	1	1	1	1	1	1	1	1	1	1	1	1

Figure 4.1 *The range of possible values on a 16-bit address line. The address line can take up any values from all zeros to all ones, i.e. 0000 0000 0000 0000 to 1111 1111 1111 1111 which is zero to 65535, 0H to FFFFH.*

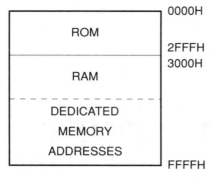

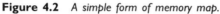

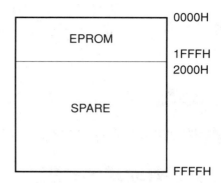

Figure 4.2 *A simple form of memory map.*

Figure 4.3 *Memory map for a small microelectronic system.*

Memory maps

A memory map shows, in pseudo-pictorial form, the memory layout of a microelectronic system. The map shows which parts of memory are made from ROM, EPROM or RAM, etc. as appropriate, with the start and finish addresses usually shown to the right. In some systems there may be a small amount of RAM, in others there may be none at all. Some systems may have ROMs, others EPROMs and some a combination of the two. The possibilities are endless so it is very useful to have such a device as a memory map which can show, very clearly, how a memory system is implemented. A simple basic memory map is shown in Figure 4.2. The map may be further divided in order to show how memory locations are allocated to the video screen, stack, text editor, floppy disc system, line printer, etc.

Obviously, the size of ROM (and RAM where it exists) will vary according to the application. The MARC Z80 microcomputer has the ability to address

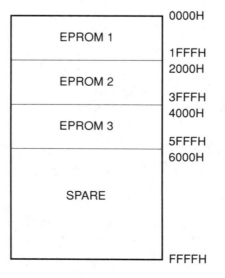

Figure 4.4 *Memory map of a slightly larger system.*

Table 4.1 Some common hexadecimal address values

Memory size (ROM, RAM, EPROM)	Size in hex	Last address
(1024b) 1K	400H	3FFH
(2048b) 2K	800H	7FFH
(3072b) 3K	0C00H	BFFH
(4096b) 4K	1000H	FFFH
(8192b) 8K	2000H	1FFFH
(12288b) 12K	3000H	2FFFH
(16384b) 16K	4000H	3FFFH
(20480b) 20K	5000H	4FFFH
(24576b) 24K	6000H	5FFFH
(28672b) 28K	7000H	6FFFH
(32768b) 32K	8000H	7FFFH
(49152b) 48K	C000H	BFFFH
(57344b) 56K	E000H	DFFFH
(65536b) 64K	10000H	FFFFH

64K of memory of which roughly 12K is ROM. The system can support EPROMs which can be addressed from software and deciding which are used is generally a matter for the user.

To take a simple example, suppose a microelectronic system has been devised to control a small industrial process. All of the program code is stored on a single 8K EPROM; the memory map would be very simple and look like that in Figure 4.3.

It is conventional where the system can support 64K of memory, for example, to show all the map and mark the unused areas as 'spare' or simply as 'unused', etc. Figure 4.4 shows a system with two more 8K EPROM memories and the resulting addresses.

It is usual to show the start and finish addresses in hexadecimal, by the side of the memory map. Following careful study of Chapter 2, you should have no problem in generating these addresses, but as an aid to constructing hexadecimal addresses on a memory map, Table 4.1 may prove useful.

Adding RAM to the system

A dedicated microelectronic system will have little need of a memory map. For more complicated systems, however, the map becomes essential. For a simple development system, RAM may be required. This has many advantages including allowing the use of a *stack*. The organisation and use of a stack is discussed in great detail in Chapter 9. For the moment it is enough to know that it is a dedicated area of RAM used for the temporary storage of addresses and CPU register contents. One of the most important features of the stack is that it is a LIFO file – last in, first out. This is rather like filling a lift with people; the first person in is usually the last one out. The stack takes this literally in that there is an orderly flow of information into the stack and a similarly orderly flow out, the last one in being the first one out as shown in Figure 4.5. A special purpose register called the 'stack pointer' indicates the address of the last piece of data pushed onto the stack.

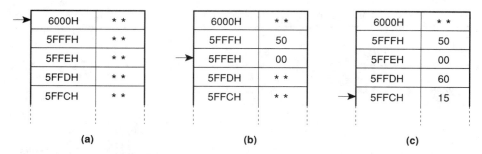

6000H	* *		6000H	* *		6000H	* *
5FFFH	* *		5FFFH	50		5FFFH	50
5FFEH	* *		5FFEH	00		5FFEH	00
5FFDH	* *		5FFDH	* *		5FFDH	60
5FFCH	* *		5FFCH	* *		5FFCH	15

(a)   (b)   (c)

Figure 4.5   *The top of the stack is at 6000H as shown in (a). The address 5000H has been pushed in at (b). When the next two bytes are pushed in (60 and 15), the stack pointer has moved further down the stack. When the data is retrieved, 15 and 60 come out first, followed by 00 and 50. Last in, first out.*

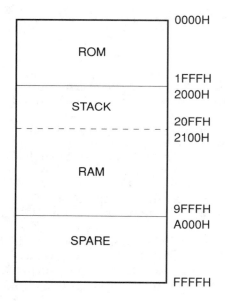

Figure 4.6   *Memory map indicating area used for stack in RAM.*

This is mentioned now because the stack may be shown on a memory map as a distinct part of the RAM, starting at a 'high' address and building down. A typical system could have 8K of ROM and 32K of RAM, 256 bytes of which may form the 'stack'. Such a system may have a memory map like the one shown in Figure 4.6.

As we have seen, we can introduce further complexities into the memory map by indicating not only the significant areas of RAM and ROM (including EPROM and similar devices) but also showing how RAM is divided up into different areas such as the stack. Other areas of RAM may be used in order to address the video screen. These are called dedicated addresses meaning that they are never used for any other purpose.

In a 6502 system, input and output are also memory-mapped, i.e. in order to read from or write to an external device, an appropriate area of RAM is addressed. This is not the case with the Z80, where input/output (I/O)

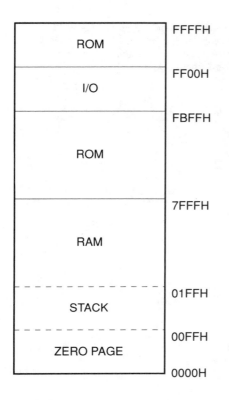

ROM	FFFFH
I/O	FF00H
ROM	FBFFH
RAM	7FFFH
STACK	01FFH
ZERO PAGE	00FFH
	0000H

Figure 4.7 *A possible memory map for a 6502 system showing the memory-mapped I/O. The position of the stack in RAM is also shown. Zero page describes the first 256 memory locations; each one has a one-byte address and so can be accessed with great rapidity, functioning with similar ease to CPU registers. It doesn't much matter whether the highest address is shown at the top of the map or at the bottom.*

devices, are accessed by using the special commands 'IN' and 'OUT' which transfers control from the CPU itself to a special interfacing device. A Z80 parallel input-output (PIO), is most frequently used for the purpose. This is covered in greater detail in Chapter 12; here we are concerned with the effect on the memory map.

In a 6502 system (like the BBC Micro), I/O addresses appear in the main memory map (a possible configuration is shown in Figure 4.7); in a Z80 system, the I/O addresses are quite separate. Where several exist, it is often useful to generate a 'mini memory map' just for the I/O circuitry. The 6502 I/O ports form just another section of RAM like the video RAM.

It is a useful exercise to look at the memory map for the system you are using. Figure 4.8 shows a complete memory map for the MARC micropro-cessor system. Chapter 16 covers the use of decoders to implement memory systems and shows how the resulting memory maps are formed.

Backing stores

Because of the volatile nature of RAM devices (all information being lost when power is removed), it is necessary to be able to store programs and data in a more permanent fashion. This is done by the use of 'backing stores'. This also means that a theoretically unlimited number of programs may be stored permanently and 'down-loaded' into RAM as and when required. There are various methods in common use.

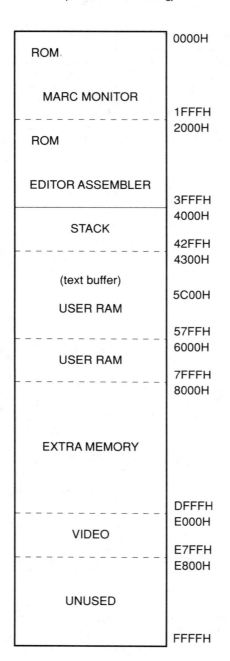

Figure 4.8 *Memory map for the MARC microcomputer system. (This is the EPROM operating system version; the disc-based system has RAM from 0000H to 3FFFH.)*

Magnetic tape and disc storage

Although there was a time when punched cards and paper tape were used to store computer programs and data, these are rarely (if ever) used nowadays. Magnetic tape and disc storage is by far the most common. Some large computer users still depend upon magnetic tape, and some use magnetic 'drums'. The latter find applications in aircraft control systems where speed

Table 4.2   Common tape and disc formats

Type	Access time	Capacity
Tape	0.1s to 10 mins	125Mb approx
Cassette	Up to 30 minutes and	Varies – but around
tapes	30 mins for all data	100Kb
	(but obviously depends	
	upon length of tape)	
3.5" disc (DD)	around 100ms	100Kb–800Kb
3.5" disc (HD)	around 100ms	1.44Mb
5.25" disc	80ms to 300ms	100Kb–500Kb
Drum stores	below 10ms	around 100Kb
Hard discs	20ms to 500ms	50Mb or more
CD ROM	Hitachi 300ms	around 600Mb
	Sony 340ms	
	Philips 375ms	

is of paramount importance when a crash seems imminent! Drums have particularly fast access times, but relatively small capacities. The fast access derives from there being several pick-up heads, the lower capacity being generally unimportant in the applications for which they are used. However, in both private and industrial environments, the use of hard discs or 'Winchesters' is rapidly taking over. Ease of storage, access time and memory capacity are the major factors influencing the choice, and the technology is constantly changing and improving.

Table 4.2 gives typical access times and storage capacities for the most common magnetic tape and disc systems.

Magnetic tape

The traditional long-term storage medium is magnetic tape, 0.5" (12.5mm) wide and up to 1km long. It is usual to find running speeds of 20" to 250" per second (0.5 to 5m per second). For example, a large, 9-track tape can store 10^9 data bits with retrieval at up to 10^7 bits per second if all tracks are used simultaneously.

The main disadvantage of using tape is the long access time. If the tape is at one end of its travel and the data required is at the other it could take about 3 minutes to get to the correct point (3.3 minutes if a 1km tape runs at 5m/s). Magnetic tape headers are used to provide information about the files stored on the tape. When home computers first came on the market, the long-term storage medium was cassette tape. This is highly unsatisfactory for many reasons – the length of time taken for saving and loading programs and the general unreliability of the system, for example – but it was used extensively because of its cheapness. Fortunately, most home computers and PCs now have disc stores (both floppies and hard discs), which are much faster, more convenient and more reliable.

Magnetic discs

Magnetic disc storage systems are managed within a microprocessor system by a device called a magnetic disc controller. A magnetic disc controller is used to:

(a) convert between serial and parallel data;
(b) read/write data to/from a disc;
(c) indicate the status of the disc device;
(d) format a disc.

Basically, the controller is used to provide an interface between the disc drive and the CPU. The magnetic disc directory holds the name and physical location of each file stored. The precise format varies from one system to another.

The two most common floppy disc formats are 5.25" and 3.5". The number of tracks they contain is now fairly standard at either 40 or 80. Formatting is a process which segments each of the tracks into a number of sectors. A sector is likened to a box which is capable of holding 256 bytes of information. The amount of sectors placed on a single track can vary; on the BBC Micro, the limit is 10 sectors.

The reason why 3.5" discs took over from their larger predecessors is said to be because a leading IBM software designer asked his engineers to produce a floppy disc that would fit more easily into his shirt pocket!

Many of the systems described, which are used to store computer data, have been around for some time. It is likely that most will remain in use for some time to come. The capacity of ROM and RAM chips seems to be continually expanding and even greater density hard discs and floppies are likely to be introduced in the near future. In an age where it is necessary to store ever-increasing amounts of data, and to access that data even more rapidly, computer storage systems of all types will continue to be improved upon. Compact disc (CD) ROM stores are the latest in a long list of hi-tec computer memory storage media. It is certainly one area of industry which seems bound to expand and continue its expansion for the foreseeable future.

Questions

1 Draw up a memory map for a small Z80-based system which contains two 8K PROMs and 8K of RAM. A stack, consisting of 128 memory locations, has been implemented in RAM which starts at address 4100H and builds down.

2 A small microcomputer system contains the following: 16K ROM, 32K RAM and 18K EPROM containing software for a light pen. The RAM must have an edit buffer, a 1K stack, 1K to address the video and the rest for a user to run programs. Construct a memory map for this system.

3 A microprocessor-controlled music system consists of seven, 8K EPROMs, each of which contains a different tune. A further 8K of RAM provides an environment for producing new tunes and subsequently blowing another EPROM. Construct a memory map for this system.

4 How many memory locations could be addressed by a system using an 8-bit address bus and what would be the range of addresses?

5 What is the memory size of 'zero page' in a 6502 system?

6 What type of memory backing store is a 'Winchester'?

7 State the memory storage capacity of two common 3.5" floppy disc types.

8 State four uses of a magnetic disc controller.

9 State two disadvantages of using magnetic tape for storing computer programs.

10 A microprocessor system has been designed which will take in 6502 machine code programs in ASCII code and convert them to Z80 machine code. The maximum code to be translated at one time is 12K. Decide what firmware will be required for such a system and construct a memory map for it.

Multiple choice questions

1   A microelectronic system has eight address lines. The number of
    memory locations it could address is:
    (a)   8   ☐
    (b)   64   ☐
    (c)   256   ☐
    (d)   65536   ☐

2   In a Z80- or 6502-based system, the number of memory locations
    which could be addressed is:
    (a)   8K   ☐
    (b)   16K   ☐
    (c)   32K   ☐
    (d)   64K   ☐

3   A system supporting 16K of memory would have hexadecimal
    addresses in the range:
    (a)   0000H to 0FFFH   ☐
    (b)   0000H to 1FFFH   ☐
    (c)   0000H to 2FFFH   ☐
    (d)   0000H to 3FFFH   ☐

4   A typical number of bytes of RAM allocated to a system
    'stack' may be:
    (a)   8 bytes   ☐
    (b)   256 bytes   ☐
    (c)   16384 bytes   ☐
    (d)   65536 bytes   ☐

5   6502 I/O ports are addressed in:
    (a)   RAM   ☐
    (b)   ROM   ☐
    (c)   PROM   ☐
    (d)   a PIO   ☐

6   A 'Winchester' drive is a form of:
    (a)   magnetic tape   ☐
    (b)   floppy disc   ☐
    (c)   hard disc   ☐
    (d)   drum store   ☐

7   A 3.5", DD, floppy disk has a storage capacity of:
    (a)   up to 64Kb   ☐
    (b)   64Kb to 100Kb   ☐
    (c)   100Kb to 800Kb   ☐
    (d)   up to 1.44Mb   ☐

8   A magnetic disc controller is *not* used to:
    (a)   convert between serial and parallel data   ☐
    (b)   format a disc   ☐
    (c)   convert ASCII characters to decimal form   ☐
    (d)   indicate status of a disc device   ☐

9 Magnetic tape headers provide information about:
   (a)   the files stored on the tape   ☐
   (b)   the length of the tape   ☐
   (c)   the access time of the tape   ☐
   (d)   the capacity of the tape   ☐

10 A memory backing store may be described as:
   (a)   a ROM chip   ☐
   (b)   a duplicate floppy disc which may be used if the
         original is lost or damaged   ☐
   (c)   an internal hard disc which stores all your programs ☐
   (d)   any system which stores a program permanently,
         even when power is removed from it.   ☐

11 In a memory map, the memory location immediately
   preceding address 2000H would be:
   (a)   1999H   ☐
   (b)   1AAAH   ☐
   (c)   01FFH   ☐
   (d)   1FFFH   ☐

12 A 'stack file' is generally organised as:
   (a)   LIFO   ☐
   (b)   FILO   ☐
   (c)   LILO   ☐
   (d)   LOLO   ☐

13 A 5.25" floppy disc has a storage capacity of:
   (a)   below 10Kb   ☐
   (b)   10Kb to 100Kb   ☐
   (c)   100Kb to 500Kb   ☐
   (d)   1Mb to 10Mb   ☐

14 A magnetic disc controller provides:
   (a)   an interface between a CPU and disc drive   ☐
   (b)   the disc operating system   ☐
   (c)   speed control of the disc drive motor   ☐
   (d)   a directory of files on the disc   ☐

15 A magnetic disc directory provides:
   (a)   information about access times   ☐
   (b)   the facility to copy files   ☐
   (c)   a list of files   ☐
   (d)   a text buffer between CPU and disc drive.   ☐

Assignment 4.1

Devise a small microcomputer system which uses memory-mapped I/O and construct a memory map for it.

Assignment 4.2

Devise a small microelectronic system which addresses I/O ports which are external to the main memory and construct a memory map for it.

Assignment 4.3

Using manufacturers' literature where necessary, select, name and describe suitable chips which could be used to implement the system asked for in Question 10.

Assignment 4.4

Using a small microprocessor system as a guide, produce for a typical system:
(a) a block diagram, indicating the function of each block
(b) a memory map, indicating the start and finish addresses of each section
(c) an I/O map, indicating the devices connected.

Assignment 4.5

Using computer magazines, and/or manufacturers' literature, etc., list and describe the various types of disc drives now available. Evaluate them in terms of capacity, access time, cost and availability.

Chapter 5
Machine code programming

Machine code programs

Computer programs consist of codes which represent instructions or data, interleaved throughout a program. Both instructions and data must be placed into the memory locations and CPU registers as a series of binary digits. Assuming the Z80 system, many of the standard codes and data will be 8 bits wide (a byte) and some may be 16 bits wide (they use two 8-bit registers or memory locations).

Originally, programmers had to input the binary codes directly into the computer. To make it easier, the machinery was adapted to take in hexadecimal codes. The system was further improved by using what are known as 'assembly mnemonics' which can be converted into machine code either by hand assembly or by the use of an assembler program.

Programming was improved even further by the invention of the so-called high level languages such as COBOL, FORTRAN, PASCAL and, finally, BASIC. BASIC is probably the simplest programming language to use and understand and it was on this foundation that the first home computers were sold to the general public. However, the CPU still runs on machine code, so that the BASIC program needs to be interpreted line by line into machine code, as it is being executed. It makes life easier for the programmer, but uses up much more memory space and slows down the program running time.

Further, in a microelectronic system, the program will normally be stored in ROM (or PROM, EPROM, etc.) and the codes in there will have to be machine code. One solution is to 'compile' the program. This involves writing the program in a high level language like BASIC and then running the

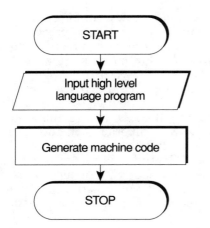

Figure 5.1   *A compiled program.*

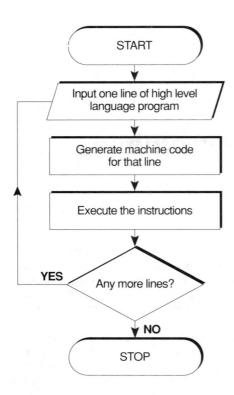

Figure 5.2 *A BASIC program during execution, being interpreted.*

program through a compiler; this converts the entire program into machine code which can then be RUN as many times as required at the faster speed of machine code. A BASIC program has to be run through the interpreter line by line every time it is RUN, whereas a compiled program is already in machine code and so will be very much faster. See Figures 5.1 and 5.2.

As far as we are concerned, machine code programs will be written in 'assembly mnemonics' which will be hand assembled using the relevant instruction set (mostly the Z80). Later on, it may be possible to use an automatic assembler. This will depend upon the nature of the course you are pursuing and the type of hardware available to you.

Programming design

It is wise to use a structured design for programming right from the start, so that even for the simplest of programs, we shall use a standard approach. It begins with writing down, in plain English, what the program is required to do; this is called the algorithm. From the algorithm, a flow chart, 'story-board', or a structured design, for example JSP (see later), is drawn up.

The use of flow charts as an aid to programming is regarded with some suspicion by many modern programmers. A system devised by Michael Jackson called Jackson Structured Programming (JSP) is now gaining in popularity. We acknowledge that JSP is probably a superior methodology, however we have already used flow charts to illustrate the sequence of compiling and interpreting (Figures 5.1 and 5.2) and will use a similar technique to illustrate the fetch–execute sequence (Chapter 7). In such cases, the flow chart method offers simplicity and clarity. We will describe only the basic principles of JSP here; the system is considered in much greater detail in Chapter 13.

Table 5.1 Steps in producing a machine code program

1 Write down the algorithm in plain English
2 Draw up a structured design, e.g. flowchart or JSP
3 Check operation of proposed design against software requirement
4 Write the program in assembly language mnemonics – this produces the 'source code'
5 Assemble the source code into object code – the object code is the machine code in hex
6 Debug, i.e. detect and correct any errors
7 Run completed program; store on disc or in PROM

In both systems, once the general structure has been produced, its operation should be checked against the software requirement. After this, the program can be written in assembly language. The mnemonics may be obtained by consulting the instruction set which also provides details of the actual machine code.

The assembly language program consists of source code; after hand or automatic assembly, the object code is obtained. Any errors are then detected and corrected (a process known as de-bugging); a monitor program may be used to step through the program line by line to assist in the process. The completed program may then be RUN, stored on disc or blown into an EPROM or similar device. Table 5.1 summarises the process.

Before proceeding, the reader is asked to review the relevant parts of Chapter 3 where the CPU architecture is described. In both the Z80 and the 6502, the main register is the accumulator. This is the most important register because it is generally the route into and out of the CPU and one of the registers involved in CPU operations. Once again, a simple example may help to make things clear.

Example I

Suppose we wish to add two numbers together, say 8 and 6, the procedure is as follows: one of the numbers is loaded into the accumulator using the 'LOAD' instruction (when writing the program down, the word 'LOAD' is abbreviated to LD). 'LD' is an example of an assembly language 'mnemonic' (an aid to the memory). There are similar abbreviations for most of the instructions and we shall be listing some of the most common ones soon; you learn them as you use them. The second number is added to the first using the ADD instruction. Unfortunately, there is no simple ADD instruction in the 6502 instruction set, only ADC which means add with carry. We are not yet ready to get involved with carries, so just accept that in this simple example we need to use the instruction 'CLC' in the 6502 program. This clears the carry flag which isn't needed in this example and ensures that we get the right answer.

In Z80 mnemonics the general format for loading a register is: LD r,n. In 6502 mnemonics it is: LDr n, where r is the name of the register, and n is the number to be loaded. In both cases the two portions of the instruction are called operator and operand; for the Z80, LD r is the operator and n is the operand. So if we want to load the accumulator with the number 8, the instruction is:

```
LD  A,8 (Z80)  or  LDA  #8 (6502)
```

The '#' (hash) sign in the 6502 instruction indicates that the value is a number and not an address. The next instruction simply adds the other data value to the contents of the accumulator, i.e. ADD A,n (Z80) or ADC n (6502), leaving the result in the accumulator.

Table 5.2   Some common Z80 instructions, mnemonics and codes

Function	Instruction	
	Mnemonic	Hex code
Data transfer		
Load accumulator with a number	LD A,n	3E xx
Load B register with a number	LD B,n	06 xx
Load C register with a number	LD C,n	0E xx
Load D register with a number	LD D,n	16 xx
Load E register with a number	LD E,n	1E xx
Load H register with a number	LD H,n	26 xx
Load L register with a number	LD L,n	2E xx
Load BC register pair with a number	LD BC,nn	01 ll hh
Load DE register pair with a number	LD DE,nn	11 ll hh
Load HL register pair with a number	LD HL,nn	21 ll hh
Contents of memory location to A	LD A, (nn)	3A ll hh
Contents of A to memory location	LD (nn) ,A	32 ll hh
Register B to accumulator	LD A,B	78
Accumulator to register B	LD B,A	47
Register E to accumulator	LD A,E	7B
Accumulator to register E	LD E,A	5F
Memory (pointed to by HL) to A	LD A, (HL)	7E
A to memory (pointed to by HL)	LD (HL) ,A	77
A to memory (pointed to by BC)	LD (BC) ,A	02
Arithmetic and logic		
Add a number to accumulator	ADD A,n	C6
Add B contents to accumulator	ADD A,B	80
Add E contents to accumulator	ADD A,E	83
Subtract B contents from accumulator	SUB B	90
Increment accumulator	INC A	3C
Increment B register	INC B	04
Increment C register	INC C	0C
Increment BC register pair	INC BC	03
Increment HL register pair	INC HL	23
Decrement accumulator	DEC A	3D
Decrement B register	DEC B	05
Decrement C register	DEC C	0D
Decrement D register	DEC D	15
Decrement HL register pair	DEC HL	2B
Decrement value held in memory location pointed to by HL reg. pair	DEC (HL)	35
Test branch and control		
Jump relative to specified location	JR n	18 n
Jump unconditionally to specified address	JP nn	C3 ll hh
Jump to specified address on zero	JP Z,nn	CA ll hh
Jump to specified address, non zero	JP NZ,nn	C2 ll hh

Table 5.3 Some common 6502 instructions, mnemonics and codes

Function	Mnemonic	Hex code
Data transfer		
Load accumulator with a number	LDA #n	A9 xx
Load X register with a number	LDX #n	A2 xx
Load Y register with a number	LDY #n	A0 xx
Load accumulator with memory	LDA nn	AD ll hh
Load A with memory plus X contents	LDA nn,X	BD ll hh
Store accumulator in memory	STA nn	8D ll hh
Store A with memory plus X contents	STA nn,X	9D ll hh
Store X register in memory	STX nn	8E ll hh
Store Y register in memory	STY nn	8C ll hh
Transfer register X to accumulator	TXA	8A
Transfer register Y to accumulator	TYA	96
Transfer accumulator to X register	TAX	AA
Transfer accumulator to Y register	TAY	A8
Arithmetic and logic		
Clear carry flag	CLC	18
Add to accumulator with carry	ADC #n	69 xx
Add memory location contents to accumulator (zero page)	ADC n	65 xx
Add memory location contents to accumulator (absolute)	ADC nn	6D ll hh
Subtract memory from accumulator with borrow	SBC	E9
Compare accumulator contents with n	CMP #n	C9 xx
Increment memory by one (zero page)	INC	E6
Increment memory by one (other)	INC	EE
Increment X register by one	INX	E8
Increment Y register by one	INY	C8
Decrement memory by one (zero page)	DEC	C6
Decrement memory by one (other)	DEC	CE
Decrement X register by one	DEX	CA
Decrement Y register by one	DEY	88
Test branch and control		
Jump unconditionally to specified address	JMP	4C ll hh
Branch to relative address, on zero	BEQ	F0 xx
Branch to relative address, non zero	BNE	D0 xx
Breakpoint	BRK	00

```
(Z80)          (6502)
LD  A,8        LDA  #8
ADD A,6        CLC
               ADC  #6
```

Once the instructions are written in assembly mnemonics we can look in the instruction set and find out what the machine code values are, or use an assembler program which will do it for us. We will start by hand assembling but instead of turning to the complete instruction set, let's use Table 5.2 (Z80) or Table 5.3 (6502) which lists some of the most common instructions and makes the process somewhat easier at this early stage.

When using these tables, note the following conventions:

x = a single hexadecimal digit (e.g. 5H or 0AH)
n = a hexadecimal byte of data, i.e. 8 bits (e.g. 05H or 0A5H)
nn = two hexadecimal bytes of data, i.e. 16 bits (e.g. 5A05H)
ll = low byte of data or address
hh = high byte of data or address
     e.g. in 5D00H: 00 = ll and 5D = hh
     so: LD BC,5D00H becomes: 01 00 5D after assembly.

The machine code is usually written down by writing the code in up to three columns to the left of the mnemonics, and after the address in memory where the program is placed. In the Marc Z80 system this is often at address 5C00H; in the Emma 6502 system, 0200H; Hence for the Z80:

```
5C00H       3E      08      LD  A,8

5C02H       C6      06      ADD A,6
```

and for the 6502:

```
0200H       A9      08      LDA #8

0202H       18              CLC

0203H       69      06      ADC #6
```

Careful examination of the above should reveal how the 'hand assembly' has been carried out. The code for an LD A,n instruction is looked up and written down (3E), the next piece of code is the number itself, 08 in this case. Finally, the code for an ADD A,n is obtained, (C6) followed by the number to be added, (06). The 6502 mnemonics are assembled in a similar manner, and all the hex codes are placed in consecutive memory locations.

In order to make hand assembly easier, the use of coding sheets is recommended; a suitable format, with space for comments, is shown in Table 5.4.

Flow charts

The symbols shown in Figure 5.3 should be used when designing a flow chart.

For very simple programs, a flow chart may be perfectly adequate, but for more complicated programs, Jackson Structured Programming (JSP) is generally regarded as being superior. JSP is a 'top down' approach. Such a design begins with a statement about the whole program and then goes on to define smaller and smaller parts until the simplest components of the program have been defined.

Jackson Structured Programming

The programmer starts from the premise that the complete program is too difficult to comprehend in its entirety. The problem is broken down into a sequence of more manageable components. Each such component is then broken down into smaller components, and so on, until a level is reached which cannot be further simplified.

Without this technique, a programming problem can be too large and complex in its complete form. The above process produces a program structure resembling a tree diagram, as shown in Figure 5.4.

Table 5.4   Layout of a typical coding sheet

Micro coding sheet				Page of
Program:		Version:	Author:	Date:
Address	Machine code	Label	Operator & operand	Comments

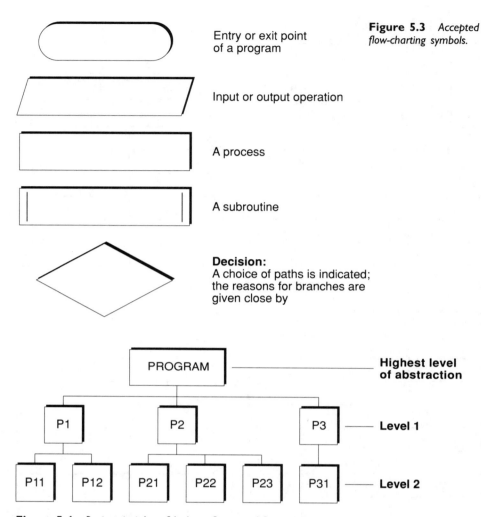

Figure 5.3 *Accepted flow-charting symbols.*

Entry or exit point of a program

Input or output operation

A process

A subroutine

Decision:
A choice of paths is indicated; the reasons for branches are given close by

Highest level of abstraction

Level 1

Level 2

Figure 5.4 *Basic principles of Jackson Structured Programming.*

The elementary components

Each level lists the procedures (processes) that describe the complete program. Level 1 describes the program, but this level is still too high a level of abstraction. Each procedure (P) in level 1 is therefore further decomposed into procedures, giving level 2. Notice that the program consists of procedures P1, P2 and P3 in that order. Procedure P1 consists of a sequence, procedures P11 (pronounced P one,one) followed by P12. Procedures P2 and P3 also consist of further procedures. Figure 5.5 shows a JSP design for a simple operation involving the input of data, the processing of it and the subsequent output.

Apart from one or two simple examples of JSP, we shall only use flow charts in this chapter. For further details of JSP, the reader is referred to Chapter 13.

Now it's possible to draw up complete, structured, design documentation for our simple addition program according to Table 5.1.

1   *Algorithm*
Program which adds two single integer numbers together

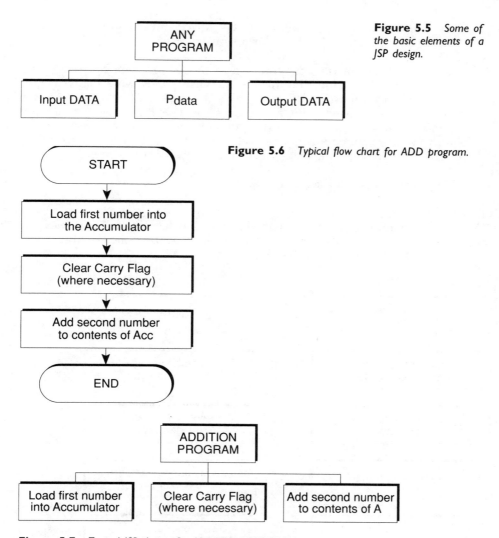

Figure 5.5 *Some of the basic elements of a JSP design.*

Figure 5.6 *Typical flow chart for ADD program.*

Figure 5.7 *Typical JSP design for ADDITION PROGRAM.*

2   *Flow chart/JSP*
    The flow chart is given in Figure 5.6 and a JSP design in Figure 5.7.
3   *Check operation*
    We can be fairly sure in this simple example that the proposed design will
    meet the software requirement.
4   *Assembly mnemonics*
    (Z80)   (6502)
    LD A,8   LDA #8
    ADD A,6   CLC
                    ADC #6
5   *Machine code*
    (Z80)   (6502)
    3E 08   A9 08
    C6 06   18
                    69 06

Note: Steps 4 and 5 would normally be carried out using a coding sheet such as that shown in Table 5.4 or a similar format.

6   *Debug*

This is not really necessary with such a simple program but supposing that in the Z80 version, the third and fourth codes were inadvertently interchanged. You will note from Table 5.2 that 06 is also the code for loading the B register with a number. The program would run quite happily, but would not, of course, give you the result you were expecting. Careful examination of the code should lead you to the problem; debugging techniques are described in greater detail in Chapter 11.

7   *Run*

The program operates according to the design specification.

Example 2

We'll add two numbers together again, but this time let's assume that we want to retain each of the numbers as well as the result of the addition. The documentation, in line with Table 5.1, may be as follows:

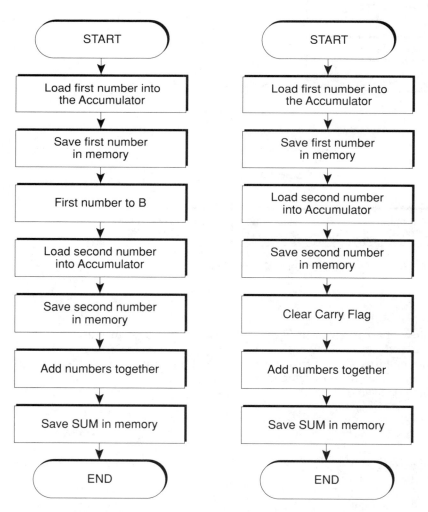

Figure 5.8   *Flow charts for Example 2. On the left, Z80; on the right, 6502.*

1 *Algorithm*
Program which accepts two numbers, adds them together, and saves the two numbers and their sum in three consecutive memory locations.

2 *Flow chart*
The flow charts are given in Figure 5.8.

3 *Check*

4 *Assembly mnemonics* and 5 *Machine code*
In this example, let's use the numbers 11 and 15. Since both are greater than 9, we need now to specify whether they're decimal or hexadecimal. Assume they are hex numbers so that we can continue to get used to using the notation: 11H and 15H (Z80), &11 and &15 (6502).

Z80

Machine code	Mnemonics	Comments
3E 11	LD A,11H	;first number into A
32 00 5D	LD (5D00H) ,A	;save in memory
47	LD B,A	;save in B reg
3E 15	LD A,15H	;second number into A
32 01 5D	LD (5D01H) ,A	;save in memory
80	ADD A,B	;add numbers together
32 02 5D	LD (5D02H) ,A	;save in next memory
76	HALT	;halt processor

6502

Machine code	Mnemonics	Comments
A9 11	LDA #&11	;first number into Acc
8D E0 02	STA &02E0	;store in memory
A9 15	LDA #&15	;second number into Acc
8D E1 02	STA &02E1	;store in memory
18	CLC	;clear carry flag
6D E0 02	ADC &02E0	;perform addition
8D E2 02	STA &02E2	;store result in memory
00	BRK	;breakpoint

6 *Debug* and 7 *Run*
The reader can proceed with 6 if required. After RUNning the program, examine the appropriate memory locations to see that the correct effect has been achieved.

Comments on Z80 version

We have used memory locations 5D00H, 5D01H and 5D02H. Referring to Table 5.2 the first instruction, LD A,n becomes: 3E 11. The second instruction (to save in memory) uses: LD (nn),A. The machine code is given as: 32 ll hh. 'll' represents the low byte, and 'hh' the high byte, and these are coded backwards. The address (nn) is 5D00H, so the machine code becomes: 32 00 5D and NOT: 32 5D 00.

We now transfer the first number from the accumulator to the B register to save it. The second number has to be loaded into the accumulator in order subsequently to place it in a memory location. You can't load B (C, or any other) register contents directly into memory locations; it can only be done from A. So we now do another LD (nn),A which becomes:

```
LD  (5D01),A  i.e. 32  01  5D
```

The ADD A,B instruction performs the add, and a final LD (5D02),A puts the answer in memory (the sum always ends up in the accumulator).

Comments on 6502 version

Here, we are storing the numbers and their sum in memory locations 02E0, 02E1 and 02E2. The first instruction, LDA (LoaD the accumulator), is A9 in machine code; the value is #&11 ('#' means a number, '&' means it's hexadecimal) and this follows the LoaD instruction. The address 02E0 is assembled backwards, since the code for LDA nn is 80, the instruction becomes: 80 E0 02 in machine code (NOT: 80 02 E0).

Example 3

The previous examples have shown how numbers can be moved in and out of CPU registers, manipulated and stored in selected areas of memory. For one or two numbers, these simple techniques are adequate, but with larger groups of numbers, more sophisticated methods are desirable.

We now wish to examine the numbers contained in ten consecutive memory locations, move them in the same order to a different section of memory (a block move) meanwhile adding them together and storing the sum in the first memory location following the new position. Produce the appropriate documentation; write and run the program. To make it perfectly clear, the two blocks of memory are shown in Table 5.5 (Table 5.6 shows the 6502 version).

Make an attempt to produce the documentation before studying the solution. As a guide, note how numbers were moved in and out of the CPU and added together in Example 2. An extended form of that program may be used for Example 3.

Example 3 – Solution

1   *Algorithm*
     Ten numbers in consecutive memory locations are to be examined, added together and moved to ten different consecutive memory locations, together with their sum, stored in the eleventh location.
2   *Flow chart*
     A full solution is not given here; as a guide, we present a simple 'story board' layout in Figure 5.9.

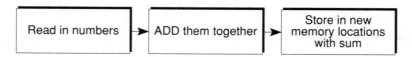

Figure 5.9 *A 'story board' technique to help in the production of a machine code program. This may precede, or, in some cases, replace a flow chart or JSP design.*

3   *Check*
4   *Assembly mnemonics* and 5 *Machine code*
     See Tables 5.7 (Z80) and 5.8 (6502).
6   *Debug* and 7 *Run*

Table 5.5 Data for Example 3. We want to move the data from block 1 to block 2, adding up the numbers on the way and putting the sum at the end of block 2, i.e. at 5E0AH

Block 1 Memory location	Data	Block 2 Memory location	Data
5D00H	00	5E00H	**
5D01H	05	5E01H	**
5D02H	0A	5E02H	**
5D03H	0F	5E03H	**
5D04H	14	5E04H	**
5D05H	19	5E05H	**
5D06H	1E	5E06H	**
5D07H	23	5E07H	**
5D08H	28	5E08H	**
5D09H	2D	5E09H	**
5D0AH	**	5E0AH	??

Table 5.6 Data for Example 3, 6502 version

Block 1 Memory location	Data	Block 2 Memory location	Data
&02D0	00	&02E0	**
&02D1	05	&02E1	**
&02D2	0A	&02E2	**
&02D3	0F	&02E3	**
&02D4	14	&02E4	**
&02D5	19	&02E5	**
&02D6	1E	&02E6	**
&02D7	23	&02E7	**
&02D8	28	&02E8	**
&02D9	2D	&02E9	**
&02DA	**	&02EA	??

Readers will note that we have not completed Tables 5.7 and 5.8. Each becomes rather repetitive after a while which, to the inexperienced, may seem rather tiresome. In fact, it is the key to its own solution. A process which is repeated in this way can be contained within a loop and the amount of code correspondingly reduced. This is an example of an iterative process.

Basic programmers may be aware of subroutines, procedures and 'FOR–NEXT' loops; we are talking about the same sort of thing here, although you don't need to be a BASIC programmer to understand the process. Figure 5.10 shows a flow chart of an iterative process – one which is repeated over and over again. We can use this idea to produce a better program – one that is much neater and uses less code.

In Table 5.7 (Z80) and Table 5.8 (6502) you will see that the following code is being continually repeated:

```
(Z80)                           (6502)
47          LD  B,A             AD  D3  02    LDA  &02D3
3A  03  5D  LD  A,(5D03H)       8D  E3  02    STA  &02E3
32  03  5E  LD  (5E03H),A       18            CLC
80          ADD  A,B            6D  EA  02    ADC  &02EA
                                8D  EA  02    STA  &02EA
```

Table 5.7 Part of the machine code of a suggested solution to Example 3 (Z80). The reader should note that the code becomes rather repetitive after a while . . .

Machine code	Mnemonics	Comments
3A 00 5D	LD A,(5D00H)	;first number into A
32 00 5E	LD (5E00H),A	;first number to new loc
47	LD B,A	;save in B reg
3A 01 5D	LD A,(5D01H)	;second number into A
32 01 5E	LD (5E01H),A	;second number to new loc
80	ADD A,B	;add numbers together
47	LD B,A	;save running total
3A 02 5D	LD A,(5D02H)	;third number into A
32 02 5E	LD (5E02H),A	;third number to new loc
80	ADD A,B	;add running total
47	LD B,A	;save running total
3A 03 5D	LD A,(5D03H)	;fourth number into A
32 03 5E	LD (5E03H),A	;fourth number to new loc
80	ADD A,B	;add to running total
47	LD B,A	;save running total
3A 04 5D	LD A,(5D04H)	;fifth number into A
.	.	.
.	.	.
.	.	.
76	HALT	;halt processor

Table 5.8 Part of the machine code of a suggested solution to Example 3 (6502). The reader should note that the code becomes rather repetitive after a while . . .

Machine code	Mnemonics	Comments
AD D0 02	LDA &02D0	;first number into A
8D E0 02	STA &02E0	;first number to new loc
8D EA 02	STA &02EA	;store total
AD D1 02	LDA &02D1	;second number into A
8D E1 02	STA &02E1	;second number to new loc
18	CLC	;clear carry flag
6D EA 02	ADC &02EA	;add to running total
8D EA 02	STA &02EA	;save running total
AD D2 02	LDA &02D2	;third number into A
8D E2 02	STA &02E2	;third number to new loc
18	CLC	;clear carry flag
6D EA 02	ADC &02EA	;add to running total
8D EA 02	STA &02EA	;save running total
AD D3 02	LDA &02D3	;fourth number into A
8D E3 02	STA &02E3	;fourth number to new loc
18	CLC	;clear carry flag
6D EA 02	ADC &02EA	;add to running total
8D EA 02	STA &02EA	;save running total
AD D4 02	LDA &02D4	;fifth number into A
.	.	.
.	.	.
.	.	.
00	BRK	;breakpoint

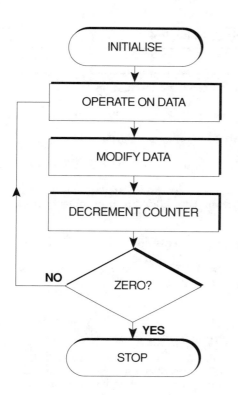

Figure 5.10 *Diagram showing an iterative process.*

only the addresses are being changed. It would be very useful to write out the code once, and repeatedly run through it, simply updating the address values. We shall do just this in Example 4.

Example 4 – Z80 version
Referring to Figure 5.11, to initialise means to set the starting values of addresses and numeric data, etc. The rest of the diagram is discussed in greater detail below.

We'll load the HL register pair with the Block 1 start address (see Table 5.5) and the BC register pair with the Block 2 start address. We'll need another register to count the number of addresses we're going to step through (ten in this case) and a final register to hold the running total. The program, with explanatory comments, is shown in Table 5.9.

After careful study, you should understand the meaning and function of every mnemonic given. The ORG and ENT instructions are only used in automatic assembly, but it's thought worthwhile to introduce them now in order to become familiar with them. The comments given should be sufficient for you to understand what's going on. Note the use of increments and decrements, essential to the design of this iterative process. The 'DEC D' and 'JP NZ,5C0AH' instructions are worthy of further comment.

The D register initially contains 0AH. On each pass of the loop, this value is decremented (reduced by 1). As long as it's NON-ZERO (any number but zero), the JUMP occurs. When, after ten passes, the D register contains zero, the JUMP fails and the CPU proceeds to the next instruction (LD A,E).

The CPU knows when the D register contains a zero because the JUMP instruction causes the processor to check the flag register on each pass, but where does 5C0AH (in the JP NZ,5C0AH instruction) come from? In preparing

Table 5.9 Program solution with comments for Example 4

Mnemonics	Comments
ORG 5C00H	;ORIGIN: the address in memory where the first line of the ;program resides
ENT 5C00H	;ENTRY POINT: address where program is to run from
LD HL,5D00H	;LoaD the HL register pair with the Block 1 start address
LD BC,5E00H	;LoaD the BC register pair with the Block 2 start address
LD D,0AH	;LoaD D register with 0AH (10 in decimal) being the number of ;data items in the block
LD E,00H	;LoaD E register with zero. This clears the register so that it can ;have the contents of the accumulator added to it in order to create ;the running total
LD A,(HL)	;LoaD the accumulator with the contents of the memory location ;whose address is held in the HL register pair
LD (BC),A	;LoaD the accumulator contents INTO the memory location whose ;address is held in the BC register pair
ADD A,E	;create running total
LD E,A	;save running total in E register
INC HL	;INCREMENT HL: add 1 to the contents of the HL register pair
INC BC	;INCREMENT BC
DEC D	;DECREMENT D: subtract 1 from the contents
JP NZ,5C0AH	;JUMP NON-ZERO; JumP to memory location given if result of ;previous ALU operation (DEC D) produced a non-zero result; in ;this program the jump takes program execution back to the LD ;A,(HL) instruction
LD A,E	;LoaD the running total into the accumulator
LD (5E0AH),A	;LoaD memory location 5E0AH with running total

this example, we started at 5C00H as we usually do (and as indicated by the ORG statement), so we know that the LD A,(HL) instruction is held in address location 5C0AH. We've left the addresses out in this example to leave room for the comments which are presently thought to be more important.

In fact, we could have used a label here; probably the word 'LOOP' which is often used on these occasions. We'd create another column for the label (see Table 5.4) so in this case the program would become:

```
LOOP:   LD  A, (HL)    ;comments
        LD  (BC),A     ;comments
        .
        .
        .
        JP  NZ,LOOP
```

When you come to assemble the program by hand into machine code, you must insert the actual address. An automatic assembler does it for you, and we consider both methods later.

The ALU performs the decrement (as well as all other arithmetic and logic operations), and flags are set (or reset) according to the result. If an ALU operation produces a negative result, then it sets the appropriate flag. In the present case, if the decrement produces a zero result, then the zero flag is set. The CPU looks at the result of the previous ALU operation, so (in this program) the decrement instruction *must* come just before the JUMP instruction. The following would not work for example:

```
INC BC

DEC D

INC HL

JP NZ,5C0AH
```

The code is still present but the change in order means that the loop goes on forever; the HL register pair would always contain a non-zero number (unless the program cycled through all the available values and ended up back at zero), so the program would have failed.

Incidentally, incrementing or decrementing any of the Z80 register pairs does not affect the flags. We'll encounter this in the next example. There's a lot more about flags in Chapter 10, but we have all we need to know for the present.

A flow chart for this program is shown in Figure 5.11 and the coded program is shown in Table 5.10. In an attempt to make the flow chart compatible with the 6502 version, placing the running total into the appropriate memory location at the end (LD (5E0AH),A) is not shown.

In the Z80 program, we needn't have used the E register to store the running total at all, and could in fact have put it into the memory location directly. However, this can only be done using the HL register pair (ADD A,(HL) then LD (HL),A) and it's already being used.

Alternatively, we could have added another line after LD E,A, i.e. LD (5E0AH),A, which would also obviate the necessity for the last two lines of code, but this would increase program running time unnecessarily. In this

Table 5.10 Complete coded Z80 program for Example 4

Address	Machine code	Mnemonics	Comments
		ORG 5C00H	;ORIGIN
		ENT 5C00H	;ENTRY POINT
5C00H	21 00 5D	LD HL,5D00H	;block 1 start address
5C03H	01 00 5E	LD BC,5E00H	;block 2 start address
5C06H	16 0A	LD D,0AH	;load counter
5C08H	1E 00	LD E,00H	;clear running total
5C0AH	7E	LD A,(HL)	;load accumulator with ;first number in block 1
5C0BH	02	LD (BC),A	;put number into new loc
5C0CH	83	ADD A,E	;create running total
5C0DH	5F	LD E,A	;save running total in E
5C0EH	23	INC HL	;point to next number
5C0FH	03	INC BC	;point to next location
5C10H	15	DEC D	;decrement counter
5C11H	C2 0A 5C	JP NZ,5C0AH	;jump non-zero
5C14H	7B	LD A,E	;load A with running total
5C15H	32 0A 5E	LD (5E0AH),A	;load memory location with ;running total
5C18H	76	HALT	;halt processor

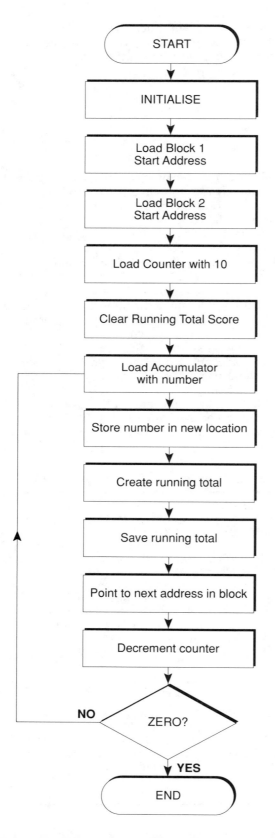

Figure 5.11 *Suggested flow chart for Example 4.*

example, the increased running time would be of little consequence, but in larger programs it would be regarded as a great disadvantage. After all, one of the main reasons for using machine code is to increase the program's speed of execution and this must always be a prime consideration.

Having said all that, either method of programming (and many other alternatives not considered here) would be entirely satisfactory in practice.

Example 4 – 6502 version

Referring to Figures 5.10 and 5.11, to initialise means to set the starting values of addresses and numeric data, etc. The rest of the diagram is discussed in detail below.

First of all, we load the X index register with zero. We're going to increment this value later to enable us to 'point' to each memory location in turn. The Y register is loaded with decimal 10, which is the number of locations we're going to step through. Hopefully, the rest of the program is adequately explained in Table 5.11, though two of the instructions may require further explanation.

'LDA &02E0,X' means 'load the accumulator with the contents of memory location &02E0 plus the contents of the X register'. If X contains 0 (as it does to start with) then the accumulator is loaded with the contents of address &02E0 + 0 which is &02E0, of course. After X has been incremented (INX) to one, then LDA &02E0,X becomes: LDA &02E0 + 1, which is &02E1. At some point X will contain 5, so that LDA &02E0,X becomes: LDA &02E5 and so on. In this way, the program steps through each of the memory locations in turn. This is an example of the indexed addressing mode. Much more is said about the various addressing modes in the next chapter.

Table 5.11 Program solution with comments for Example 4 (6502 version)

Mnemonics	Comments
ORG &0200	;ORIGIN: the address in memory where the first line of the ;program resides
ENT &0200	;ENTRY POINT: address where program is to run from
LDX #&00	;LoaD X index register with zero
LDY #&0A	;LoaD Y register with &0A (10 in decimal) being the number of ;data items in the loop, thus setting up the 'counter'
LDA #&00	;LoaD accumulator with zero –
STA &02EA	;then store in memory location where running total is to be kept
LOOP	
LDA &02D0,X	;LoaD the accumulator with the contents of memory location ;&02D0 + contents of X register, i.e. &02D0 + 0 = &02D0, which ;is the first location in the block
STA &02E0,X	;Store number in new location
CLC	;clear carry flag
ADC &02EA	;add to running total
STA &02EA	;store new running total
INX	;INCREMENT X register
DEY	;DECREMENT Y register (acting as counter)
BNE LOOP	;BRANCH NOT EQUAL: jump back a given number of memory ;locations if result of previous ALU operation (DEY) produced a ;non-zero result; in this program the jump takes program ;execution back to the LDA &02D0,X instruction
BRK	;breakpoint

Table 5.12 Complete coded 6502 program for Example 4

Address	Machine code	Mnemonics	Comments
		ORG &0200	;ORIGIN
		ENT &0200	;ENTRY POINT
&0200	A2 00	LDX #&00	;index X
&0202	A0 0A	LDY #&0A	;counter
&0204	A9 00	LDA #&00	;zero for running total
&0206	8D EA 02	STA &02EA	;store in memory
	LOOP		;loop marker
&0209	BD D0 02	LDA &02D0,X	;first number in block 1
&020C	9D E0 02	STA &02E0,X	;store in new location
&020F	18	CLC	;clear carry flag
&0210	6D EA 02	ADC &02EA	;add to running total
&0213	8D EA 02	STA &02EA	;store running total
&0216	E8	INX	;increment index
&0217	88	DEY	;decrement counter
&0218	D0 EF	BNE LOOP	;branch not equal
&021A	00	BRK	;halt processor

The use of increments and decrements is essential to the design of this iterative process; this is also worthy of comment. The Y register initially contains &0A. On each pass of the loop, this value is decremented (reduced by 1). As long as it's non-zero (any number but zero), the BRANCH occurs. When, after ten passes, the Y register contains zero, the BRANCH fails and the CPU proceeds to the next instruction which terminates the program. The CPU knows when the Y register contains zero because the BNE instruction causes the processor to check the processor status (flag) register.

The ALU performs the decrement (as well as all other arithmetic and logic operations), and flags are set (or reset) according to the result. If an ALU operation produces a negative result, then it sets the appropriate flag. In the present case, if the decrement produces a zero result, then the zero flag is set (becomes '1'). The CPU looks at the result of the *previous* ALU operation, so (in this program) the decrement instruction *must* come just before the JUMP instruction. The following would not work, for example:

```
STA &02EA

DEY

INX

BNE LOOP
```

The code is still present but the change in order means that the loop goes on forever; the X register would always contain a non-zero number (unless the program cycled through all the available values and ended back at zero), and the program would have failed.

The final coded program is shown in Table 5.12.

Relative jumps and branches

Before leaving Example 4, it's important to discuss the JUMPS and BRANCHES. The essentials are the same in both 6502 and Z80 programs.

In Table 5.12, line &0218 gives BNE LOOP and the corresponding code is: D0 EF. D0 is the code for a BNE, the value that follows is the displacement which is taken away from or added to the program counter. 'But', you might be saying, '&EF is hexadecimal for 239, and we're only jumping back a few addresses (17 in fact)'. Correct, but this is one of the few occasions when we need to use a negative number obtained by the use of two's complement arithmetic, mentioned in Chapter 2.

After executing:

```
&0218     D0
&0219     EF
&021A     **
```

the program counter (PC) points to location &021A. We want to jump back to location &0209, so we need to count back from location &0219 to &0209 (BD) which is 17 locations.

$$17 \; = \; 0001\ 0001$$
$$\text{Complement} \; = \; 1110\ 1110$$
$$\text{add } 1 \qquad \qquad 1$$
$$= \; 1110\ 1111 \; = \; \&EF$$

In the Z80 program (see Table 5.10), we used a conditional jump (JP). However, we could have used a conditional jump relative (JR). We couldn't use a conditional jump in the 6502 program because no instructions exist for it.

Using a JR NZ,LOOP instruction in the Z80 program would produce a revised program as shown in Table 5.13. The theory for calculating the relative jump displacement is the same as that described for the 6502.

Table 5.13 Revised Z80 program for Example 4, using a relative jump

Address	Machine code	Mnemonics	Comments
		ORG 5C00H	;ORIGIN
		ENT 5C00H	;ENTRY POINT
5C00H	21 00 5D	LD HL,5D00H	;block 1 start address
5C03H	01 00 5E	LD BC,5E00H	;block 2 start address
5C06H	16 0A	LD D,0AH	;load counter
5C08H	1E 00	LD E,00H	;clear running total
		LOOP	
5C0AH	7E	LD A,(HL)	;load accumulator with first ;number in block 1
5C0BH	02	LD (BC),A	;put number into new loc
5C0CH	83	ADD A,E	;create running total
5C0DH	5F	LD E,A	;save running total in E
5C0EH	23	INC HL	;point to next number
5C0FH	03	INC BC	;point to next location
5C10H	15	DEC D	;decrement counter
5C11H	20 F7	JR NZ,LOOP	;jump non-zero, LOOP
5C13H	7B	LD A,E	;load A with running total
5C14H	32 0A 5E	LD (5E0AH),A	;load memory location with ;running total
5C17H	76	HALT	;halt processor

Towards the end of the chapter, Table 5.14 gives a solution for Example 5. But remember, it is only an *example*, and in it we have shown *both* versions of the JUMP; the JP NZ, 5C06H and the relative jump. *You wouldn't normally see both like this in a program listing*. Further, we have shown the use of a 'LABEL' (referred to earlier) in this example called 'LOOP'. This is an acceptable (and very useful) technique, used in automatic assembly and more will be said about this in Chapter 13. The relative jump is from 5C0FH to 5C06H, i.e. nine locations. (It's 5C0FH because the PC increments after the CPU takes in the displacement value (F7) at address 5C0EH.) This is calculated as follows:

$$9 \ = \ 0000\ 1001$$
$$\text{Complement} \ = \ 1111\ 0110$$
$$\text{add 1} \qquad\qquad 1$$
$$\overline{1111\ 0111} \ = \ 0\text{F7H}$$

Normally, a separate column is used for labels as in Table 5.4. It would be a useful exercise to copy this program onto a coding sheet, using the correct format, and thinking carefully about each instruction before trying it out on a computer. Remember that the correct format is to put 'LOOP' (above LD A,(HL) here) in the label column.

After studying the previous examples carefully, the reader should be familiar with the basic techniques of moving data into and out of CPU registers and memory locations, manipulating the data in a variety of ways, and storing it in different memory locations.

Although this may appear to have no obvious practical use at the moment, we'll see later that much data processing, and similar tasks undertaken by computers, involves just this. Word processors are constantly taking in data and storing it in memory locations; transferring it, comparing it (as in 'spell-check' programs) and moving it around. What we're really doing at the moment though is getting used to the CPU hardware, the relevant instruction set and the correct procedures for documenting a program. A thorough understanding of the principles discussed and illustrated so far is essential before progressing, and a good grasp of these principles is an excellent start to achieving competence in machine code programming.

Summary

You should now be familiar with:

1   Program documentation
2   Using assembly language mnemonics and converting them to machine code
3   Using correct formats
4   Flow charting and introductory JSP symbols and techniques
5   Some machine code instructions in Z80 and/or 6502 instruction sets
6   Program loops and iterative processes.

Now, one more example before concluding this chapter.

Example 5 (for Z80 systems)
For systems with a VDU, it is frequently necessary to clear the screen before displaying new data. In BASIC programming, the instruction 'CLS' (clear screen) is used; similar routines are available in most micro systems, the MARC, for example, has a subroutine in the operating system which may be called in order to clear the screen (CLRSCR: at 01B9H in memory).

For our final example, let's see how we can write a program to clear the VDU screen. What this involves in fact is filling the video RAM with a series of blanks. To do this, each video memory location is loaded with the ASCII code for a blank space. All the ASCII codes are listed in Chapter 2 (Table 2.3) and we can see from this that the code for a blank space is 20H.

In the MARC system, the first video RAM location is 0E000H. The screen is made up from 24 lines each capable of holding 42 characters, so the number of locations is: 42 × 24 = 1008. (1K is allocated to the video ram, so it's not quite all needed.) In hex, 1008 is 3F0H, so we can put 0E00H in the HL register pair and 3F0H in the BC pair. If the accumulator is loaded with 20H (the ASCII code for a blank space) you should be able to see, by analogy with the last example, that you can increment HL to step through all the video locations, meanwhile decrementing BC to count them. Note, however, that when BC reaches zero, it does not affect the flag register (we mentioned this a short while ago) so we must use an alternative method to detect the zero state.

A method commonly used is as follows. We load the accumulator with the contents of the B register and then logical OR the accumulator with the C register. This effectively ORs B with C and only when both B and C contain zero will the operation produce zero. This will set the zero flag in the flag register and program execution may be diverted.

You've got to put B register contents into the accumulator because logic operations are performed on that important register. 'OR C' means 'logical OR the contents of the C register with the contents of the accumulator'.

Table 5.14 Complete coded Z80 program for Example 5. This time, we include the label column to show how the documentation should look.

Address	Machine code	Label	Ops	Comments
			ORG 5C00H	;ORIGIN
			ENT 5C00H	;ENTRY POINT
5C00H	21 00 E0		LD HL,0E000H	;start of video RAM
5C03H	01 F0 03		LD BC,03F0H	;number of video locations ;this is the counter
5C06H	3E 20	LOOP:	LD A,20H	;load A with ASCII blank
5C08H	77		LD (HL),A	;store in first video ;memory location
5C09H	23		INC HL	;point to next video ;memory location
5C0AH	0B		DEC BC	;decrement counter
5C0BH	78		LD A,B	;put contents of B into A
5C0CH	B1		OR C	;logical OR to see if both ;B and C registers contain ;zero
5C0DH	C2 06 5C		JP NZ,5C06H	;jump non-zero
5C0DH	20 F7		JR NZ, LOOP	;(or use a relative jump) ;jump relative, non-zero ;F7 = −9: jump backwards ;9 addresses
5C0FH	76		HALT	;halt processor

In the Mnemonics column group, "Label" and "Ops" are sub-columns.

Because the accumulator has necessarily been used for the logical operation, don't forget to put the ASCII blank back into the accumulator before proceeding with the next pass of the loop.

Produce the appropriate documentation and run the program before consulting the solution which is shown in Table 5.14.

The end of chapter questions which now follow suggest other programming ideas you might like to try. The next chapter introduces addressing modes which involves further programming examples. Other ideas are discussed in the chapters on the flag register, stack, and input/output (I/O) ports. Finally, Chapter 13 explores some more advanced programming techniques. We have not really touched on 'subroutines' using the CALL and RETURN commands. We'll see in Chapter 13 that JSP leads programmers to use subroutines. For readers who feel they can handle a bit more at this introductory stage, Questions 16 onwards suggest some possibilities.

Questions

NOTE: In answering some of the questions that follow, you will normally choose between Z80 and 6502 versions where appropriate (though you can try both if you like) and must decide, where memory addresses are quoted, if these are suitable for the system you are using. Generally, the questions are written assuming that the reader is using the MARC for Z80 programming, and an EMMA for the 6502. However most of the codes are identical for all other comparable systems, and the reader will have to adapt where necessary. Where a question is prefaced by 'MARC Z80' it means that the questions were originally written with this system in mind. You will have to decide whether or not a suitable program can be created and run on your system, if it is other than a MARC.

1 Name some high level computer languages. State some advantages and some disadvantages of high level languages.
2 Flow charting is called a 'bottom up' approach to software design. What is 'JSP' and what sort of approach does it use?
3 List the stages in producing a machine code program.
4 State the two processors described in this chapter. List and explain the meaning of some assembly mnemonics from the instruction set of each processor.
5 (a) (Z80) If you wanted to load the accumulator with the number held in memory location 5E05H, what instruction would you use?
   (b) (6502) If you wanted to load the accumulator with the number held in memory location &2E05, what instruction would you use?
6 Write down the following mnemonics; supply the appropriate machine code, and explain what the instructions mean and what they do:

	(Z80)	(6502)
(a)	LD B,42H	LDX #&42
(b)	LD (5E08H),A	STA &2E08
(c)	LD A,B	TXA
(d)	INC A	INX
(e)	DEC C	DEY

7 Consider the following algorithm:
ALGORITHM
Load three numbers into consecutive memory locations; move the three numbers to consecutive memory locations, over 250 locations away. Put the sum of the three numbers in the memory location following that containing the last number.

Draw up a suitable flow chart, convert to assembly mnemonics and write down the machine code program. Run, check and debug the program (where necessary), if possible.

8 Write a program, with full documentation, that will read in the upper-case characters of the alphabet to 26 consecutive memory locations. The program must then put in the next memory location the ASCII code for a blank space, followed by the codes for the word 'MICROPROCESSOR'. The tables below suggest some addresses:

Z80			6502		
Address	Code	Letter	Address	Code	Letter
5E00H	**	ASC 'A'	&2E00	**	ASC 'A'
5E01H	**	ASC 'B'	&2E01	**	ASC 'B'
..			..		
..		..	..		..
5E??H		ASC 'Z'	&2E??		ASC 'Z'
5E??H		ASC ' '	&2E??		ASC ' '
5E??H		ASC 'M'	&2E??		ASC 'M'
5E??H		ASC 'I'	&2E??		ASC 'I'
..		..	..		..
5E??H		ASC 'R'	&2E??		ASC 'R'

9 What is meant by an iterative process?
Fully document a program that will put the numbers zero to 64H (&64) into consecutive memory locations, using an iterative process.

10 Fully document a program which will put 00 and FFH (&FF) into twenty alternate memory locations as shown in the diagram:

Z80	6502	**
5F00H	&2F00	00
5F01H	&2F01	FF
5F02H	&2F02	00
5F03H	&2F03	FF
5F04H	&2F04	00
..	..	
5F13H	&2F13	FF

11 Write a program in (a) Z80 and (b) 6502 code, using an iterative process, which decrements a register in order to produce a time delay. (By using two registers, the delay can be made longer.)

12 (MARC Z80) Write a program which will put the numbers 1 to 20 in two lines, suitably spaced, on the video screen. You can use the time delay created in Question 11 to pace the display.

13 (MARC Z80) Write a program that will put a row of As, followed by a row of Bs, and so on through the alphabet, to a row of Xs, on the video screen. Example 5 contains some valuable hints on how this can be achieved. Y and Z cannot be put on the screen because there are only 24 lines.

14 (MARC Z80) Write a program which will 'move' an asterisk back and forth through a set of zeros, as shown in the diagram. Incorporate a time delay so that the 'movement' can be observed.

```
*   0   0   0   0   0   0   0
0   *   0   0   0   0   0   0
0   0   *   0   0   0   0   0
0   0   0   *   0   0   0   0
0   0   0   0   *   0   0   0
0   0   0   0   0   *   0   0
0   0   0   0   0   0   *   0
0   0   0   0   0   0   0   *
0   0   0   0   0   0   *   0
0   0   0   0   0   *   0   0
0   0   0   0   *   0   0   0
0   0   0   *   0   0   0   0
```
etc.

Some of the relative positions are shown in the diagram but the idea of the program is to have just *one* line and make the asterisk appear to travel along it. You can increase the number of zeros or asterisks or even use some other characters if you wish; if you're feeling really adventurous, you could fill the screen with a moving display of your own design.

15 (MARC Z80) Write a program to put a message on the screen like the one shown below:

```
┌─────────────────────────────┐
│                             │
│      MICROPROCESSOR         │
│      TECHNOLOGY             │
│                             │
│                             │
│                             │
│                             │
│                             │
│                             │
└─────────────────────────────┘
```

You can, of course, choose your own message if you like. When writing the program, try first of all to use as many of the instructions given in Table 5.2 as you can. When this has been successfully completed, rewrite the program to make it as short as possible.

Assignment 5.1

Subroutines and the use of CALL and RET (RETurn) for the Z80, JSR and RTS (Jump to and Return from Subroutine) for the 6502 have not really been considered so far; they're referred to in Chapter 9 (Organisation and use of the stack) and Chapter 13. For readers who can't wait, assign yourself the task of using subroutines (the use of delays is very cumbersome without them), and rewrite some of your programs incorporating them. The programs suggested in questions 12, 13, 14 and 15 are particularly suitable.

If you're using a system such as the 'EMMA' (or something similar) which has no VDU, many more interesting programs can be written using the I/O ports. It's probably more appropriate in such cases to wait until the chapter on the use of the I/O ports has been reached, when more programming opportunities will present themselves.

Assignment 5.2

Compare the instruction sets of the Z80 and 6502 and discuss the relative merits of the two processors.

Chapter 6
Machine code instructions and addressing modes

CPU instructions

We have seen from the previous chapter that CPU instructions generally consist of one, two or three bytes, usually composed of two parts: the operator (given as an 'OPCODE') and the operand. The opcode may be, for example, the instruction to load the accumulator with a number and the operand could be either the number to be loaded or the address in memory where it can be found:

Opcode	Operand
LD A,n	n
LD A,(nn)	nn

In explaining this, we have also shown that there is more than one way of loading a number into a register. The operator could be followed by the actual number to be loaded – this is called 'immediate' addressing – or it could be followed by the address of the memory location where the number to be loaded may be found, and this is called 'direct' (or 'absolute') addressing. In a Z80-based system, the number may also be contained in another register, giving rise to REGISTER addressing. These are all examples of addressing modes many of which apply to both Z80 and 6502 systems. However, there are some major differences (and some which are more subtle), so that for clarity we are presenting this information separately for each CPU.

Most readers will probably be interested in either Z80 or 6502 systems (rather than both of them). One of the major difficulties in writing a book of this sort has been in making the text equally suitable for these two processors. Fortunately, much of the information is common to both, including the addressing modes. However, 6502 addressing modes differ sufficiently from those for the Z80 and we think it's worthwhile to consider them separately. Some readers may find it interesting or useful to make comparisons.

To begin with, some Z80 addressing modes are shown in Table 6.1. (For 6502 addressing modes, see Table 6.4, but we recommend that you look at the following section on operator and operand first.)

The opcode

This part of the instruction, which always comes first, tells the CPU what operation is to be carried out. Most opcodes require only one byte, but some

Table 6.1 Some Z80 addressing modes

Instruction		Addressing mode
LD A,64H	(1)	Immediate
LD B,27H	(1)	Immediate
LD BC,5DF0H	(1a)	Immediate extended
LD A,B	(2)	Register
LD C,D	(2)	Register
JR NZ,LABEL	(3)	Relative
LD A,(5DFEH)	(4)	Direct (also called Absolute)
LD (5DFFH),A	(4)	Direct
LD A,(HL)	(5)	Register Indirect
NOP	(6)	Implied
INC A	(6)	Implied
LD A,(IX+30H)	(7)	Indexed
CP (IX+1)	(7)	Indexed

of the 16-bit loads, for example, require two. Some Z80 opcodes are shown below:

LD A,n	LoaD the accumulator with a number
LD B,A	LoaD the B register with the contents of A
LD BC,nn	LoaD the BC register pair with a two-byte number
LD HL,nn	LoaD the HL register pair with a two-byte number
INC A	Increment the accumulator
DEC C	Decrement C register

For the 6502:

LDA	LoaD the accumulator
ADC	Add with carry
CLC	Clear the carry flag
INX	Increment X register
DEY	Decrement Y register
TXA	Transfer contents of X to accumulator

The operand

This is the part of the instruction which supplies the rest of the information; in the examples below, the number to be loaded or the address of a memory location which contains the number to be loaded (see Table 6.4 for some 6502 instructions).

Opcode	Operand
LD A,n	2FH
LD A,(nn)	5D66H
LD B,A	–
LD BC,nn	1234H
LD HL,(nn)	5DFEH
INC A	–

The opcode: 'LD A,n' means load the accumulator with a number. The code is always the same, regardless of the number (assuming it's not too large to fit into the accumulator); so the complete instruction for the first example would be: LD A,2FH.

Table 6.2 Instructions broken up into operator and operand

Operator	Operand
LD A,	2FH
LD A,	(5D66H)
LD BC,	1234H
LD HL,	(5DFEH)

In describing the opcode part of an instruction, we have to state 'LD A,n' (rather than just 'LD A') to indicate that we wish to load the accumulator with a number. Consulting the instruction set reveals that the code is: 3E. So the code for LD A,2FH would be: 3E 2F. An 'LD A,B' (load the accumulator with the contents of the B register) instruction has the code: 78.

LD B,A is an example of register addressing and all the information is contained within the opcode. The instruction is therefore only 1 byte in length. The same applies to increments and decrements, etc. LD A,(nn) is an example of direct addressing. Where the number is enclosed in brackets, it is the address of a memory location. So this instruction means: 'Load the accumulator with the number stored in memory location 5D66H'. In the case of 'LD HL,(5DFEH)', register L would be loaded with the contents of memory location 5DFEH and register H would be loaded with the contents of memory location 5DFFH. Table 6.2 shows some of the instructions broken up into operator and operand.

There are seven distinct addressing modes used in Z80 programming as follows:

Z80 Addressing modes

1   Immediate addressing
1a Immediate extended addressing
2   Register addressing
3   Relative addressing
4   Direct addressing
5   Register indirect addressing
6   Implied addressing
7   Indexed addressing

We now take a look at each of these in some detail.

I Immediate addressing

In this mode of addressing, the byte which follows the opcode contains the operand. The mode is used to perform arithmetic or logical operations with constants, for example:

```
LD  A,45H  ;LoaD 45H into the accumulator

ADD A,60H  ;ADD 60H to the contents of the accumulator
```

In general, the first byte of an instruction in the Z80 is the opcode, although some instructions have two bytes as opcodes. The second byte of the 'LD A,45H' instruction is called the immediate data value, in this case 45H. This value is therefore in the instruction itself, rather than in a memory location.

1a Immediate extended addressing

This mode is simply an extension of immediate addressing in that the operand now consists of two bytes, enabling a register pair to be loaded with a 16-bit number, e.g. LD BC,0E065H. This instruction means that the BC register pair is to be loaded with the number 0E065H. Note that the number is in hexa-decimal (the trailing H indicates this) and that because the number begins with a letter, a leading zero is added. When such a number is hand assembled, the least significant byte goes in first:

```
01 65 E0    LD BC,0E065H
```

Whenever a program is assembled, all 16-bit data is reversed in this way. When using an assembler program, it is done automatically, but in hand assembly, or when checking or modifying a machine code program, you will need to be aware of this format.

2 Register addressing

In this mode, it is possible to load any register with the contents of another register. In the instruction set, the general format is LD r,r' and examples include: LD A,B and LD C,A. The first means 'LoaD the accumulator with the contents of the B register', the other means 'LoaD the C register with the contents of the accumulator'. The great advantage of this mode is that only one byte is needed to perform the operation. Other examples include ADD A,D and SUB C.

3 Relative addressing

This mode of addressing is used *only* for relative JUMP instructions; no other types of instruction use this mode of addressing, including CALLs. Relative addressing uses one byte of data following the opcode to specify a displace-ment from the existing program to which a program jump can occur. The value of relative addressing is that jumps to nearby locations can be specified in only two bytes. The displacement is limited to the range between +127 and −128 from (A+2), the address where the first byte of the instruction is located, plus 2. This allows for a total displacement of +129 to −126 from the jump relative opcode address. Such a limitation is not usually a disadvantage because program jumps tend to be to nearby address locations in most programs. The reason why it's A+2 is because the PC increments as soon as the present instruction has been fetched (see next chapter); so the jump takes place from that point, i.e. the instruction the CPU would do if it didn't do the instruction jumped to.

4 Direct addressing

This addressing mode is also often referred to as 'absolute addressing'. It is the mode by which the contents of a memory location may be loaded into a register or register pair:

```
e.g. LD A,(5DFEH)    ;LoaD A with contents of 5DFEH
or LD (5DFEH),A     ;LoaD the contents of the accumu-
                     lator into memory location 5DFEH
```

Note that *only* the accumulator may be addressed in this way. It is not possi-ble to use the other 8-bit registers directly. However, the register pairs HL,

BC, DE, SP, IX and IY can be loaded or stored in this way. Loading the HL register pair directly is the most useful because it requires only three bytes; all the others require four:

```
2A 00 50   LD  HL,(5000H)   ;LoaD  register  L  with
                             contents of memory location
                             5000H and register H with
                             the contents of memory
                             location 5001H

ED 4B 05 55 LD  BC,(5505H)  ;LoaD  B  from  5505H  and  C
                             from  5506H  (similar  to  the
                             above)
```

Direct addressing is also used with CALLs and JUMP instructions. All CALLs use direct addressing, and all JUMPS are direct, except of course for the RELATIVE JUMPS already described:

```
C3 00 50   JP  5000H        ;JumP  to  address  5000H
C2 00 51   JP  NZ,5100H     ;Jump  to  address  5100H  if  the
                             result  of  a  previous  ALU
                             operation produced a non-zero
                             result

CD 00 5D   CALL  5D00H      ;CALL  subroutine  starting  at
                             address  5D00H
```

The opcode tells the CPU what sort of JUMP is being used (conditional or unconditional, and if the former, what the condition is) or if it's a CALL. The operand formed by the next two bytes gives the address of the JumP or CALL location. When the instruction is executed, the relevant address is placed directly into the program counter, replacing the incremented value which would normally be there.

5 Register indirect addressing

It has already been pointed out that direct addressing can only be used with the accumulator or with register pairs. However, it is possible to load the B register, for example, with data in a memory location by using the HL register pair as a 'pointer'. This means that the relevant memory location address is loaded into the HL register pair, and the B register loaded from that information. The following examples should make this clear:

```
LD  HL,5000H   ;LoaD  HL  with  memory  location  address

LD  B,(HL)     ;LoaD  B  with  contents  of  5000H

LD  (HL),B     ;LoaD  memory  location  contained  in  HL
                register pair with contents of B reg

INC (HL)       ;INCrement  contents  of  memory  location
                whose  address  is  held  in  the  HL  regis-
                ter pair
```

In these examples it would be common to say: 'Load the B register with the contents of the memory location "pointed to" by the HL register pair' or 'load the memory location "pointed to" by the HL register pair with the contents of the B register'.

6 Implied addressing

In this addressing mode, the operand is inherent in the instruction itself, e.g. CCF, INC A, INC HL, HALT or NOP.

7 Indexed addressing

This is a very powerful addressing mode which allows a base address to be specified in one of the index registers, IX or IY, to which is added a displacement. This mode allows easy reference to data tables, which, if standardised, can be looked up with common offset values. Only the offset need then be changed in order to access different data. The following example assumes that the IX register contains 5D00H:

```
LD  A,(IX+20H)    ;LoaD the accumulator with the
                  contents of memory location
                  5D00H+20H, i.e. 5D20H
```

BIT addressing

This mode was not included in the original list because BIT addressing uses other modes, i.e. register, register indirect and indexed. In BIT addressing, the individual bits of a register or memory location may be inspected, set or reset. These final examples should clarify:

```
BIT  4,A          ;the zero flag is set if bit 4 of
                  register A is zero, and is reset
                  otherwise.
                  (Register addressing).

BIT  4,(HL)       ;if HL contains 5D00H and bit 4 of
                  memory location 5D00H is zero, then
                  the zero flag flag is set.
                  (Register indirect addressing)

BIT  4,(IX+2H)    ;if IX contains 5D04H and bit 4 of
                  memory location 5D06H is zero, then
                  the zero flag is set.
                  (Indexed addressing)
```

The instructions SET and RES, set or reset the appropriate bit in a manner analogous to the BIT instructions given above.

Examples

Table 6.3 shows a typical Z80 program utilising the different addressing modes. The modes are indicated in the comments column. Another column is marked 'LABEL' which did not appear in all the examples given in the previous chapter. The use of labels is a great aid to programming where automatic assembly is being employed. We are assuming that the reader is presently using only hand assembly, so that the format introduced in the last chapter would be sufficient. In this example, we now show how the labels would be entered into an assembler program. Note that each label in the LABEL column is followed by a colon. More about programming with an automatic assembler is given in Chapter 13.

Table 6.3   Typical Z80 program showing addressing modes

Machine code	Label	Mnemonics	Comments
		ORG 5C00H	
		ENT 5C00H	
	VIDEO:	EQU 0E033H	;screen location
06 0A		LD B,10	;immediate
3E 41		LD A,41H	;immediate
21 33 E0		LD HL,VIDEO	;immediate extended
77	START:	LD (HL),A	;register indirect
23		INC HL	;implied
05		DEC B	;implied
20 FB		JR NZ,START	;relative
01 65 E0		LD BC,0E065H	;immediate extended
3E 20		LD A,20H	;immediate
C6 42		ADD A,42H	;immediate
0A		LD A,(BC)	;register indirect
46		LD B,(HL)	;register indirect
32 62 E0		LD (0E062H),A	;direct
00		NOP	;implied
76		HALT	;implied

6502 addressing modes

The 6502, developed by MOS technology, has 56 basic instructions, seven internal registers, an 8-bit stack pointer and six basic addressing modes.

The index registers are 8 bits wide (rather than 16 in the Z80) so the 6502 requires a larger number of addressing modes in order to achieve a full 64K indexed address range.

The X and Y registers in the 6502 are actually index registers; register addressing is not allowed in the 6502. Further, there are no register pairs available to the programmer within the CPU. However, access to zero page memory in 6502 systems is rather like having 256 working registers on which to operate, so the low number of CPU working registers and their (apparently) limited use, is not such a problem in reality. The use of zero page addressing gives rise to some addressing modes not found in a Z80 system.

There are six main addressing modes for a 6502 CPU as follows:

1   Immediate addressing
2a   Absolute addressing
2b   Zero page absolute addressing
3a   Implied addressing
3b   Accumulator addressing
4   Relative addressing
5   Indexed addressing
6a   Indirect addressing
6b   Pure indirect addressing
6c   Indirect indexed (post-indexed) addressing
6d   Indexed indirect (pre-indexed) addressing

The comments on opcode and operand given for the Z80 are equally applicable here, i.e. the opcode dictates the operation to be performed, whilst the operand supplies the data required, usually in the form of data values or memory address locations. Some 6502 instructions and addressing modes are

Table 6.4 Some 6502 addressing modes

Instruction		
Opcode	Operand	Addressing mode
LDA	#&64	Immediate
ADC	#27	Immediate
CLC		Implied
ASL	A	Accumulator
LDA	&5200	Absolute
JMP	(&2000)	Pure indirect
LDA	(&40,X)	Indexed indirect
LDA	(&40),Y	Indirect indexed
BEQ	&9A	Relative

given in Table 6.4. In these instructions, the '#' (hash) sign indicates a number (rather than an address) and the suffix '&' indicates a hexadecimal number. If the '&' is missing, the CPU assumes a decimal number.

1 Immediate addressing

In this addressing mode, the byte which follows the opcode contains the operand. Immediate addressing is used to load numbers into registers or to perform arithmetic and logical operations on them. Some examples are given below:

```
LDA  #25    ;LoaD 25 (decimal) in the accumulator

LDA  #&25   ;Load 25 (hexadecimal) in the accumulator
ADC  #30    ;Add with carry, 30 (decimal) to the
             accumulator

OR   #11    ;OR 11 (decimal) with contents of accumula-
             tor
```

In all of these examples, the resulting values end up in the accumulator.

2a Absolute addressing

In this addressing mode, the contents of a memory location may be loaded into a register, or the contents of a register stored in a memory location. The operand value is therefore a memory location, not simply a number. There are two versions: one for zero page (which requires only one byte to specify the address, so requires less memory space to store and is therefore quicker to execute) and another for all the other addresses.

Examples of absolute addressing, not zero page, are given below:

```
LDA  &45FF  ;load the accumulator with the contents of
             memory location &45FF

STA  5000   ;store the accumulator contents in memory
             location 5000 (decimal)
```

2b Zero page absolute addressing

```
LDA  &22    ;load the accumulator with the contents of
             memory location 22 (hexadecimal)
```

```
LDA  34    ;load the accumulator with the contents of
              memory location 34 (decimal)

STA  &42   ;store accumulator contents in memory
              location 42 (hexadecimal)

STA  66    ;store accumulator contents in memory
              location 66 (decimal)
```

Since &22 = 34D and &42 = 66D, each pair of instructions would do exactly the same thing.

3a Implied addressing

In this addressing mode, the operand is inherent in the instruction itself, e.g. INX, INY, DEX, DEY, RTS and NOP. No further information is required, since the appropriate register or memory location is implied in the opcode.

3b Accumulator addressing

This mode is used where the data to be acted upon is in the accumulator, e.g. ASL, LSR, ROL and ROR. These shift and rotate instructions can be used in conjunction with memory locations, but execute most quickly when the accumulator is used. Examples of the format include: ROL A and ASL A.

4 Relative addressing

This mode of addressing is used *only* for relative JUMP instructions; no other types of instruction use this mode of addressing, including CALLs. Relative addressing uses one byte of data following the opcode to specify a displacement from the existing program to which a program jump can occur. The value of relative addressing is that jumps to nearby locations can be specified in only two bytes. The displacement is limited to the range between +127 and −128 from (A+2), the address where the first byte of the instruction is located, plus 2. This allows for a total displacement of +129 to −126 from the jump relative opcode address. Such a limitation is not usually a disadvantage because program jumps tend to be to nearby address locations in most programs.

5 Indexed addressing

In this mode, the target address may be specified by a value stored in an index register (X or Y) which is then added to a predetermined base address. Since the index registers are 8 bits wide, the offset value is limited to 255.

The X register may be used in this way to index address any memory location. The Y register cannot be used if the memory location is in zero page, unless the X register itself is being loaded from, or stored in, zero page, e.g. LDX &42,Y or STX &66,Y. To help clarify indexed addressing, other examples are given below:

```
LDA  &42,X ;Load the accumulator with the contents of
              memory location (&42 + contents of X reg)
              e.g. if X contains &05 then the accumula-
              tor would be loaded with the contents of
              memory location &47.
```

```
LDA &2000,Y    ;Load the accumulator with the contents
                of memory location (&2000 + contents
                of Y reg) e.g. if Y contains &06 then
                the accumulator would be loaded with
                the contents of memory location &2006.
```

6a Indirect addressing

In this mode, the address required is contained in a memory location and not the instruction itself. The code instructs the CPU to go to that memory location, first of all to retrieve the data and then to act upon it.

6b Pure indirect addressing

There is only one instruction in the 6502 instruction set which uses pure indirect addressing, and that's the jump instruction JMP. (In the Z80 set there are many others.) Instead of stating the address to be jumped to in the instruction, the memory location (or locations) where the address may be found is specified. To clarify this, study Figures 6.1 and 6.2.

2000	00
2001	1A
2002	AF
2003	80
2004	FF
2005	10
2006	45
2007	20
2008	45

Figure 6.1 *The diagram shows a section of memory with the values stored at those particular locations.*

0D07	FF
0D08	4C
0D09	1E
0D0A	FF

Figure 6.2 *Another section of memory showing the values being stored.*

Figure 6.1 shows nine memory locations starting at address &2000. At the side, the content of each memory location is given, i.e. memory location &2000 contains &00, and &2001 contains &1A. If the CPU now encounters the instruction: JMP (&2000), program execution is not diverted to address 2000, but instead to the address contained within memory locations &2000 and &2001, in this case to address &1A00. Note the usual convention of storing addresses low byte first, high byte second.

Given the contents of the appropriate memory locations, therefore:

```
JMP (&2000)    ;diverts program execution to address
                &1A00

JMP (&2002)    ;diverts program execution to address
                &80AF

JMP (&2005)    ;diverts program execution to address
                &4510
```

42	00
43	02
44	DF
45	02
46	00
47	E1
48	02
49	E0
4A	02

Figure 6.3 *A section of memory in zero page.*

```
JMP  (&2007)    ;diverts program execution to address
                 &4520

JMP  (&0D08)    ;diverts program execution to address
                 &1E4C  (see Figure 6.2)
```

It would be useful to be able to use other instructions in this mode, e.g. LDA (&44). This would mean: 'Load the accumulator with the contents of the memory location pointed to by memory locations &44 and &45'. So if &44 and &45 contained 02DF (see Figure 6.3) and memory location &02DF contained 99, then 'LDA (&44)' would load the accumulator with 99. Although this instruction does not exist as such, we can choose from two similar instructions that do:

```
LDA  (&44,X)  and   LDA  (&44),Y
```

Note that we are using memory locations in zero page to store the target address. It is essential that we do this for correct operation of these instructions, which are described in detail next.

6c Indirect indexed (post-indexed) addressing

In this mode, the Y register is used to provide an offset value which is added to the address in memory 'pointed to' by the instruction. The address required must be placed in zero page memory. Referring to Figure 6.3 will help to explain the following examples:

```
JMP  (&42),Y   ;If the Y register contains &05, memory
                location &42 contains &00, and &43
                contains &02, then the JUMP would be
                made to address &0205   i.e. &0200 +
                &05.

LDA  (&47),Y   ;If the Y register contains &06, memory
                location &47 contains &E1, and &48
                contains &02, then the accumulator
                would be loaded with the contents of
                memory location &02E7, i.e. &02E1 +
                &06.
```

6d Indexed indirect (pre-indexed) addressing

In this mode, the offset value is contained within the X register. This value is added to the address pointed to by the instruction. The computed address may be any suitable memory location, but the pointer must be in zero page. The format is: LDA (&40,X). Using the section of memory shown in Figure 6.3, and assuming the X register contains &03, we now see:

Address in instruction:	&40
Value in X register:	&03
Computed address:	&43

So: LDA (&40,X) becomes: LDA (&43)

and this points to: LDA &DF02.

The effect of using this instruction is that the accumulator is loaded with the contents of memory location &DF02. It would be a simple matter to change the value in the X register by loading or incrementing in order to access a different memory location. Table 6.5 shows a typical 6502 program with the addressing modes indicated.

We have still not covered subroutine CALLs in any detail. In answering the questions that follow, some students may wish to make use of CALLs and this is, of course, quite in order, although adequate solutions to all the questions can be obtained without them. At this stage we are still mainly getting used to the appropriate instruction set and addressing modes.

Table 6.5 Typical 6502 program showing addressing modes

Machine code	Label	Mnemonics	Comments
		ORG &0200	
		ENT &0200	
A9 33	.START	LDA #&33	;immediate
A0 A5		LDY #&A5	;immediate
8D E1 02		STA &02E1	;absolute
8C E0 02		STY &02E0	;absolute
A9 00		LDA #00	;immediate
8D E2 02		STA &02E2	;absolute
8D E3 02		STA &02E3	;absolute
A2 03		LDX #03	;immediate
8E E4 02		STX &02E4	;absolute
A1 44		LDA (&44,X)	;indexed indirect
8D E5 02		STA &02E5	;absolute
A1 46		LDA (&46,X)	;indexed indirect
8D E6 02		STA &02E6	;absolute
A0 04		LDY #&04	;immediate
B1 47		LDA (&47),Y	;indirect indexed
8D E7 02		STA &02E7	;absolute
A2 00		LDX #&00	;immediate
EA	.HOLD	NOP	;implied
A1 49		LDA (&49,X)	;indexed indirect
8D E8 02		STA &02E8	;absolute
00		BRK	;implied

Questions

1   What is meant by 'opcode' and 'operand'?
2   (Z80) List the seven addressing modes available to the Z80 programmer and give two examples of each.
3   (6502)   List the addressing modes available to the 6502 programmer and give two examples of each.
4   (Z80)   Give the addressing mode for each of the following instructions:

(a) LD A,0FH	(b) LD A,B	(c) LD A,(HL)
(d) ADD A,B	(e) ADD A,0B0H	(f) INC A
(g) LD (BC),A	(h) RLA	(i) INC (IX+5H)
(j) LD HL,5DD0H	(k) JR NZ,LOOP	(l) NOP

5   (6502)   Give the addressing mode for each of the following instructions:

(a) LDA #&0F	(b) TXA	(c) LDA &2E00
(d) ADC &33	(e) ADC #&B0	(f) INX
(g) STA &2E01	(h) ROL A	(i) INC (&40,X)
(j) LDA (&44),Y	(k) BNE LOOP	(l) NOP

6   (Z80)   Give the addressing modes in the 'Print ABCD' program (see Chapter 1, p. 3).
7   Write a program, using and identifying mixed addressing modes, to put the numbers 1 to 20 on a video screen, or in a block of memory of your choice.
8   Write a program using and identifying mixed addressing modes to sort the following numbers into numerical order:
15H, 27H, 00H, 80H, 45H, 18H, A0H, A5H, 01H, 0FFH, 33H, 72H.
9   (MARC Z80)   Write a program, indicating the addressing modes used, to produce a time delay suitable for slowing down the video display of Question 7.
10   Write a program which stores 00 in the first of 20 consecutive memory locations, incrementing the value and storing it in the following location, until the final location contains 19 (decimal). Having stored each value in memory, read it back immediately as you progress, and use it to create a running total which is then stored in the twenty-first memory location. The value will be the sum of the numbers 0 to 19 inclusive. If the value is correct (work it out first) the program has run successfully and you have created a simple program to test the integrity of a block of RAM.
  Of course, you could write another program which will add up the numbers 0 to 19 (0 + 1 + 2 + 3 + . . . + 19) for you.

Tutorial questions

11   (Z80)   Write a program, identifying the addressing modes used, which will satisfy the following algorithm:
  Load memory location 5C55H with 20H and 5C56H with 18H
  Multiply each number by 4
  Add the result of each multiplication together and place in the D register. Add 5H to this result.

Load the HL register with 5C5AH and put the result of the last sum into memory location 5C5BH.

If you're using a 'MARC' micro, or something similar, you might like to try using some OS (operating system) user routines such as: HLDSP to display the address, WRSPCE to write a space, and MBYTDS to display the result. You can also incorporate your own clear screen routine (described in the last chapter) or use the ROM routine, CLRSCR.

12 (6502) Write a program, identifying the addressing modes used, which will satisfy the following algorithm:

Load memory location &2E55 with &20 and &2E56 with &18
Multiply each number by 4

Add the result of each multiplication together and place in a zero page memory location; add &05 to this result, and store the result in memory location &2E57. It's always useful, in a program like this, to work out by hand what the result should be and compare it with what you get.

13 (6502) Write a program, indicating the addressing modes used, to carry out the following sequence of operations:
   (a)   load the data: J = &04, K = &16 and L = &A4 into suitable memory locations in zero page RAM;
   (b)   put J + K + L into the X register;
   (c)   put J + L into the Y register;
   (d)   put 2J + L into the accumulator.

14 (Z80) Write a program, indicating the addressing modes used, to carry out the following sequence of operations:
   (a)   load the data: X = 04H, Y = 16H and Z = A4H into the Z80 registers B, C and D respectively;
   (b)   put X + Y + Z into the E register;
   (c)   put X + Z into the H register;
   (d)   put 2Y + Z into the L register.

15 (6502) Write a program, indicating the addressing modes used, to carry out the following sequence of operations:
   (a)   load the data J = &0A, K = &11 into suitable zero page memory locations
   (b)   (i)   Starting at memory location &2E00 load the following values into consecutive memory locations:
         (ii)   J, K, (J + K), (K − J), (K * J), (2K + J)

16 (Z80) Write a program, indicating the addressing modes used, to carry out the following sequence of operations:
   (a)   load the data X = 0AH, Y = 11H into suitable memory locations
   (b)   (i)   Starting at memory location 5E00H load the following values into consecutive memory locations:
         (ii)   X, Y, (X + Y), (X − Y), (X * Y), (2X + Y)

17 Work out the two's complement relative jumps for the values −1 to −16. Make a list of these for future reference. A start has been made below:

Backward jump	Two's complement value
−1	FF
−2	FE
.	.
.	.
.	.
−16	F0

18 (a) Write a program, identifying the addressing modes used, to put the following letters into alphabetical order:
D, X, Y, A, C, F, B, Z, E.
(b) Where possible (i.e. where your equipment allows), have the program print out the letters as shown in the question and then print them below, in order, on the VDU screen.

19 (MARC Z80) Write a program, indicating the addressing modes used, which will print out the numbers given below, followed by the same number of asterisks, on the video screen.

2, 8, 14, 3, 9, 15, 16, 10, 5.

The display should look something like this:

```
2       **
8       ********
14      **************
3       ***
9       *********
15      ****************
16      *****************
10      **********
5       *****
```

20 Write a program of your own choice which uses all the addressing modes available (Z80 or 6502; both if you like). If you can't think of anything, consider loading all the registers and some memory locations with different numbers, adding them together and storing the total in a different location. Where your system allows, you could also display the numbers and their sum on the screen.

Multiple choice questions (Z80)

1 The addressing mode for which LD A,88H is an example is:
(a) immediate   ☐
(b) indexed   ☐
(c) indirect   ☐
(d) indeterminate   ☐
2 An example of register addressing is:
(a) LD A,27H   ☐
(b) LD A,(5DD0H)   ☐
(c) LD B,27H   ☐
(d) LD A,B   ☐
3 The instruction 'ADD A,64H' is an example of:
(a) register addressing   ☐
(b) immediate addressing   ☐
(c) implied addressing   ☐
(d) indirect addressing   ☐

4 The correct format for loading the accumulator with the contents of memory location 5000H is:
   (a) LDA 5000H ☐
   (b) LDA (5000H) ☐
   (c) LD A,5000H ☐
   (d) LD A,(5000H) ☐

5 The correct format for the instruction: 'Jump Non-Zero to address 5000H' is:
   (a) JPNZ 5000H ☐
   (b) JPNZ (5000H) ☐
   (c) JP NZ,5000H ☐
   (d) JP NZ,(5000H) ☐

6 An example of register indirect addressing is:
   (a) LD A,(HL) ☐
   (b) LD A,(B) ☐
   (c) LD A,(5000H) ☐
   (d) LD A,5000H ☐

7 The two's complement value for –7 is:
   (a) F7 ☐
   (b) F8 ☐
   (c) F9 ☐
   (d) FA ☐

8 Two's complement values are needed for:
   (a) direct jumps ☐
   (b) conditional jumps ☐
   (c) relative jumps ☐
   (d) subroutine CALLs ☐

9 The only invalid Z80 instruction in the list below is:
   (a) LD A,(HL) ☐
   (b) LD A,(5D00H) ☐
   (c) LD HL,A ☐
   (d) LD (HL),A ☐

10 One example of immediate extended addressing is:
   (a) LD HL,0E066H ☐
   (b) LD (HL),0E066H ☐
   (c) LD HL,(0E066H) ☐
   (d) LD (0E066H),HL ☐

11 Relative addressing is only used for:
   (a) direct jumps ☐
   (b) subroutine CALLs ☐
   (c) comparing registers ☐
   (d) relative jumps ☐

12 The direct jump JP NC,5DFFH is an example of:
   (a) direct addressing ☐
   (b) indirect addressing ☐
   (c) conditional addressing ☐
   (d) unconditional addressing ☐

13 The instruction INC A is an example of:
   (a) register addressing ☐
   (b) relative addressing ☐
   (c) indexed addressing ☐
   (d) implied addressing ☐

14 The only valid Z80 instruction in the list below is:
   (a) ADC B ☐
   (b) ADA B ☐
   (c) AD C,B ☐
   (d) ADD C,B ☐
15 An instruction in the list below which has the general format LD r,r' is:
   (a) LD A,8 ☐
   (b) LD A,0FH ☐
   (c) LD A,B ☐
   (d) LD A,0BH ☐

Multiple choice questions (6502)

1 The addressing mode for which LDA #&88H is an example is:
   (a) immediate ☐
   (b) indexed ☐
   (c) indirect ☐
   (d) indeterminate ☐
2 The correcting format for the absolute addressing mode is:
   (a) LDA &45FF ☐
   (b) LDA (&45FF) ☐
   (c) LD A &45FF ☐
   (d) LDA #&45FF ☐
3 The instruction 'ADC #&64' is an example of:
   (a) register addressing ☐
   (b) immediate addressing ☐
   (c) implied addressing ☐
   (d) indirect addressing ☐
4 The correct format for loading the accumulator with the contents of memory location &02E0 is:
   (a) LDA &02E0 ☐
   (b) LDA (&02E0) ☐
   (c) LD A,&02E0 ☐
   (d) LD A,(&02E0) ☐
5 The correct format for the instruction 'Branch Non-Zero to address &02E0' is:
   (a) B NZ,&02E0 ☐
   (b) BNZ &02E0 ☐
   (c) BNE &02E0 ☐
   (d) BNE (&02E0) ☐
6 An example of zero page addressing is:
   (a) LDA #&30 ☐
   (b) CMP #&30 ☐
   (c) LDA &0030 ☐
   (d) CMP &30 ☐
7 The two's complement value for –7 is:
   (a) F7 ☐
   (b) F8 ☐
   (c) F9 ☐
   (d) FA ☐

8 Two's complement values are needed for:
   (a)   direct jumps   ☐
   (b)   conditional jumps   ☐
   (c)   branches   ☐
   (d)   subroutine CALLs   ☐

9 The only invalid 6502 instruction in the list below is:
   (a)   LDA &40,X   ☐
   (b)   LDA &40,Y   ☐
   (c)   LDA &4000,X   ☐
   (d)   LDA &4000,Y   ☐

10 One example of pure indirect addressing:
   (a)   LDA &30,X   ☐
   (b)   LDA &30,Y   ☐
   (c)   JMP (&2E00)   ☐
   (d)   BNE (&2E00)   ☐

11 Relative addressing is only used for:
   (a)   direct jumps   ☐
   (b)   subroutine CALLs   ☐
   (c)   comparing registers   ☐
   (d)   branches   ☐

12 The direct jump JMP &02FF is an example of:
   (a)   direct addressing   ☐
   (b)   indirect addressing   ☐
   (c)   conditional addressing   ☐
   (d)   absolute addressing   ☐

13 The instruction INY is an example of:
   (a)   accumulator addressing   ☐
   (b)   implied addressing   ☐
   (c)   indexed addressing   ☐
   (d)   immediate addressing   ☐

14 The only valid 6502 instruction in the list below is:
   (a)   ADC (X)   ☐
   (b)   ADC X   ☐
   (c)   ADD X   ☐
   (d)   ADC &30,X   ☐

15 The instruction TAX means:
   (a)   transfer accumulator contents to X   ☐
   (b)   transfer X contents to accumulator   ☐
   (c)   put accumulator onto stack   ☐
   (d)   take accumulator off stack   ☐

Chapter 7
The fetch–execute sequence

Executing machine code programs

When a machine code program is 'run', the CPU is repeatedly cycling through the fetch–execute sequence for each instruction. This involves:

(a) fetching the instruction (opcode) from memory into the CPU instruction register;
(b) executing the instruction within the CPU – the control unit examines and implements the opcode and initiates the fetching of any data required.

In the last chapter we looked at the way instructions are coded and a program is built up; it is important to understand all this before looking at the fetch–execute cycle in any detail. How instructions and data are moved in and out of the CPU, and how this is related to the CPU clock, will be covered in the next chapter. All CPU timing is controlled by a crystal-controlled oscillator called a clock oscillator and designated: Φ. The significance of the crystal is that it ensures that the clock operates at a precise and constant frequency.

Fetching the first instruction

The CPU will know where to look for the first instruction in memory because it will be initially told the 'start address' or origin (ORG), and after that, the program counter (PC) will increment (increase by one) so that the CPU always knows where it has got to.

Remember that the program in memory is a series of 8-bit codes, usually entered into memory in pairs of hexadecimal digits. Every operation that the CPU can carry out (e.g. ADD, SUBTRACT, etc.) has an instruction code and they are all listed in the CPU's instruction set. The instruction set for the 6502 will be quite different from the Z80. Each CPU has its own set of instructions; there are about 60 for the 6502 and about 160 for the Z80. In practice, only a

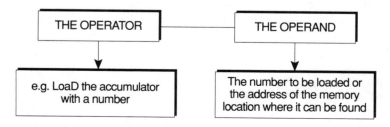

Figure 7.1   *A typical machine code instruction.*

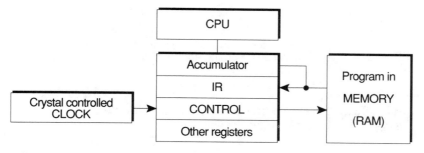

Figure 7.2   *Simplified diagram of a CPU system.*

few of them are normally needed, and at the moment we will only be dealing with a handful of them.

The instruction is generally in two parts, the operator and the operand. This is shown in Figure 7.1. A simplified form of the CPU system is shown in Figure 7.2 for easy reference in this section.

The first piece of program code in the memory goes into the instruction register (IR) and this is decoded by the control unit. This decides whether any more data is needed. If it is, another 'fetch' is executed and so on, until everything required for the operation has been fetched from memory. The instruction is then executed.

While this is happening, the PC is being automatically incremented at an appropriate point in the cycle, so that it always 'points' to the next instruction or next piece of data in memory that will be required for program execution. The precise nature of 'operator' and 'operand' will be discussed more fully later. Assume for the moment that the program to be executed is already stored in memory and that the CPU knows its origin.

Data movement

The ones and zeros stored in the machine code program represent either data or instructions and the CPU will automatically know which is which (if the program has been correctly written) since each binary pattern has only one meaning to the IR logic. The IR will *know* when to expect 'instructions' and when to expect 'data', because the first byte of a program is always assumed to be an instruction; after that, the system only works if codes and jumps are correct.

Most computing time is spent moving data from one place to another. In machine code programs, data is flowing into and out of the registers almost constantly. This data is often described as being transient; compare this to data in RAM which may be described as active and data stored on floppy disc which is the most permanent. Whatever the program, data is being continuously moved between CPU registers, from registers to memory and from memory to registers, with ALU operations, flag register updates and PC increments carefully executed in between.

In the Z80 instruction set, the general term for moving data from memory to a CPU register, or from a register to a memory location, is 'LOAD'. In the 6502 instruction set, a register is also loaded, but register contents are STOREd in memory. The data LOADed into a CPU register is operand data, actual numbers, values, addresses, ASCII codes for alphabetic data, etc. rather than operator or opcode data which represents the instruction itself. This code goes into the instruction register waiting to be decoded by the control unit.

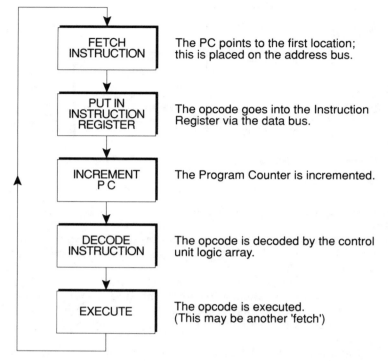

Figure 7.3   *The fetch–execute sequence as a flow chart.*

The fetch–execute sequence is shown as a flow chart in Figure 7.3.

Example I

Beginning very modestly, all we want to do is to place a number in the accumulator, say 15H (&15). In a real program this simple instruction would probably be followed by many others, but ignoring that for the moment, the program would look like this:

Program
1   Load the accumulator with a number
2   The number to be loaded (15H or &15)
3   Next instruction

Before looking up the codes in the instruction set, we write out the program in mnemonics, a short-hand representation, designed to be reasonably recognisable, of the program we are writing. Machine code mnemonics were first shown in Chapter 1 in the 'Print "ABCD" ' program. For this program, the mnemonic would look like this:

    (Z80) LD A,15H   (6502) LDA #&15

This would be read as 'load the accumulator with the hexadecimal number 15H'. The next task is to look in the instruction set and find the opcode for loading the accumulator with a number. This opcode becomes the first instruction in the program. The second item of code is the number itself.

Turning to the Z80 instruction set we find that the code for loading the accumulator with a number is 3AH. The program therefore consists of just two codes: 3AH followed by 15H. In 6502 code we obtain: A9   15.

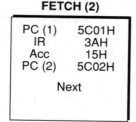

Figure 7.4 *The fetch–execute cycle for the example.*

The remainder of this chapter is devoted to the Z80. 6502 users need only substitute the appropriate addresses and codes where necessary.

In Chapter 4 we discussed memory maps. Using this information a suitable area of RAM may be chosen in the microprocessor system being used in which to place the program. For the MARC system, the usual start address or 'origin' (ORG) is 5C00H, so the program would be stored like this:

```
5C00H       3A
5C01H       15
5C02H       XX
```

The 'XX' is used to indicate that the next piece of code is presently unknown. In a real program it will depend upon what happens next and the code for that happening will be at address 5C02H. Assuming that these two lines of code are the extent of our simple program (which it is), we can now analyse the fetch–execute cycle as in Figure 7.4.

The box labelled 'Fetch (1)' shows the program counter pointing to address 5C00H. The code at that memory location is fetched into the CPU and placed in the instruction register. The PC is then incremented so that it now points to the next memory location, 5C01H. The instruction is then decoded by the control logic which 'realises' that the instruction means 'load the accumulator with a number', so it initiates another fetch in order to obtain the number. If you haven't programmed correctly, the CPU control will simply fetch the next piece of code, whatever it is, and so it is your duty to ensure that you have put in the correct number, in the correct place, in your program. If you haven't then we have what is called a program 'bug'. The computer only does what you tell it to do, not what you think you've told it to do – but don't worry! – generally, it is a relatively simple process to write programs that will run satisfactorily; it only takes practice.

The box labelled 'Fetch (2)' shows what happens next in the sequence. The PC is now at 5C01H, the IR still contains 3AH and the accumulator has been loaded with 15H. The PC then increments ready to point to the next memory location (5C02H) and look at the next instruction.

The fetch–execute sequence continues in this way until all the instructions in the program have been fetched and executed.

Illustrating the fetch–execute cycle

Now that the fetch–execute cycle has been explained, let's try a few examples. To illustrate the cycle, we now modify the method of representing it so that it looks like Figure 7.5. To make the layout simpler and more understandable we now write the PC value at the top of the box; inside the box we put the contents of the IR and any other registers that are involved in the operation. All the registers to be used should be listed from the beginning and be labelled 'XX' (meaning contents unknown) to show that the contents have not yet been

1) PC 5C00H

IR	3AH
A	XX
PC	5C01H

2) PC 5C01H

IR	3AH
A	15H
PC	5C02H

3) PC 5C02H

IR	??
A	XX
PC	5C03H

Figure 7.5 *Illustrating the changing register contents during the fetch–execute cycle – first example.*

modified by program execution. The incremented PC value is placed at the bottom of the register list even though it is usually incremented after an instruction or piece of data is fetched, and before it's executed. Remember, after execution, the CPU looks at the PC for the next memory location and fetches from it.

In Figure 7.5 we see that the PC initially points to location 5C00H. The CPU looks into that location and discovers the code 3AH there. This code is copied into the instruction register. Although the process is called an instruction fetch, the code (3AH in this case) is not removed from its memory location, it is copied. It remains there until the programmer purposely wishes to change it. At this time, we don't know what's in the accumulator; the next stage is to show the incremented PC value, which is then used to head the next box.

Box (2), therefore, is headed with PC 5C01H. The CPU looks into this memory location. It already knows that it wants a data value because it has decoded the instruction 3AH and found that it means 'load the accumulator with a number'. When the contents of 5C01H have been fetched, therefore, the CPU puts the value into the accumulator. The PC is incremented to 5C02H and the CPU then looks for another instruction, whatever it might be.

ADDRESS	Machine Code	Meaning	Hex
5C00H	0011 1110	Load the accumulator with n	3EH
5C01H	0000 0101	n = 5H	05H
5C02H	0000 0110	Load B register with n	06H
5C03H	0000 0111	n = 7H	07H
5C04H	1000 0000	ADD contents of B to A	80H

Figure 7.6 *Programming details of simple ADD.*

The instruction codes and data values are held in registers and memory locations as a series of binary digits as shown in Figure 7.6. To make it easier to program the computer in machine code, we use pairs of hex digits as we have already seen. Some instruction sets will give you the codes in hex, others will give binary values; in some cases you may have to work out the code for yourself.

A full fetch–execute sequence

Let's use a simple program to study a complete fetch–execute sequence. To make it easy, we'll use the program that simply adds two numbers together

| 5C00H | 3E | 05 | 06 | 07 | 80 | XX | XX | XX |
| 5C01H | XX | XX | XX | XX | XX | XX | XX | XX |

Figure 7.7 *The program in memory.*

1) PC 5C00H

A	XX
B	XX
PC	5C01H
IR	3E

2) PC 5C01H

A	05
B	XX
PC	5C02H
IR	3E

3) PC 5C02H

A	05
B	XX
PC	5C03H
IR	06

4) PC 5C03H

A	05
B	07
PC	5C04H
IR	06

5) PC 5C04H

A	0C
B	07
PC	5C05H
IR	80

6) PC 5C05H

A	0C
B	07
PC	5C06H
IR	XX

Figure 7.8 *Illustrating the changing register contents during the fetch–execute cycle – further example.*

and puts the result in the accumulator. The program details are shown in Figure 7.6. The first number (say 05H) is loaded into the accumulator, the second number (say 07H) is loaded into the B register. The contents of B are then added to the contents of A. Figure 7.7 shows how the program would be arranged in computer memory. Figure 7.8 illustrates the sequence.

Follow through the sequence in Figure 7.8. The initial conditions are shown in Box 1; the PC points to the first memory location, 5C00H, an instruction fetch is performed and 3E appears in the IR. In box 2, the data (05) has been placed in the accumulator. Similar actions occur for the LD B,07H instruction in boxes 3 and 4. In box 5, the ADD A,B instruction code, 80H, is fetched *and* the addition is carried out. The result, 0CH, appears in the same box, there being no need to fetch any more instructions or data for the operation.

ADDRESS	Machine Code	Meaning	Hex
5C00H	0 0 0 0 0 1 1 0	LOAD B, n	06H
5C01H	0 0 0 0 1 1 1 1	n = 15	0FH
5C02H	0 0 0 0 1 1 1 0	LOAD C, n	0EH
5C03H	0 0 0 1 1 0 0 1	n = 25	19H
5C04H	0 1 1 1 1 0 0 0	LOAD A, B	78H
5C05H	1 0 0 0 0 0 0 1	ADD A, C	81H

Figure 7.9 *Programming details for Example 2.*

1) PC 5C00H

A	XX
B	XX
C	XX
PC	5C01H
IR	06

2) PC 5C01H

A	XX
B	0F
C	XX
PC	5C02H
IR	06

3) PC 5C02H

A	XX
B	0F
C	XX
PC	5C03H
IR	0E

4) PC 5C03H

A	XX
B	0FH
C	19H
PC	5C04H
IR	0EH

5) PC 5C04H

A	0FH
B	0FH
C	19H
PC	5C05H
IR	78H

6) PC 5C05H

A	28H
B	0FH
C	19H
PC	5C06H
IR	81H

Figure 7.10   *Illustrating the changing register contents for Example 2. In box 6, instruction 81H (ADD A,C) is fetched and the execute part of the cycle performs the ADD A,C instruction, putting 28H in the A register.*

Execution of the third instruction involved the ALU accepting the contents of the A and B registers and adding the two bytes together. The result has ended up in the accumulator replacing the number put there originally. The number in the B register remains unchanged and the PC has incremented, but in box 6 we don't know what the next instruction is. Since there were no more instructions in our simple program, the next instruction may well be a HALT.

Example 2

Programming details appear in Figure 7.9. Try this for yourself.

The B register is loaded with 15D, the C register with 25D, then both numbers are added together. Note that there are no instructions for adding B to C, so you must use the accumulator. The solution appears in Figure 7.10, but don't look until you've had a go.

Example 3

Sometimes, it is required to fetch data from a RAM address which does not form part of the program. In the previous example, the data to be loaded into the B register followed the instruction at address 5C01H. Suppose we wish to load the accumulator with the data value held in memory location 5C50H; we could use a LD A,(5C50H) instruction, or use the HL register pair as a pointer with the instruction: LD HL,5C50H and then follow it with LD A,(HL) which would have a similar effect. Let's look at the fetch–execute sequence for such a program:

1   Load the accumulator with 25D (19H)
2   Load the HL register pair with 5C50H
3   Load the contents of the accumulator into memory address 5C50H using the HL register pair as a pointer
4   Load the contents of address 5C50H into the B register.

Work out the codes and show A, B, HL registers, PC and IR contents for each pass of the fetch–execute cycle.

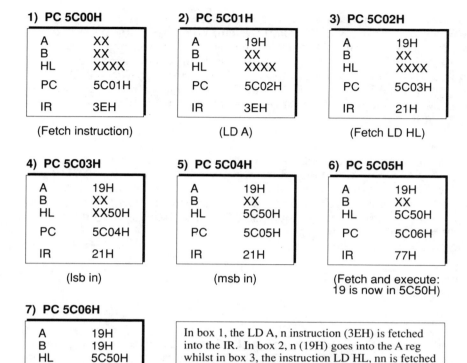

1) PC 5C00H

A	XX
B	XX
HL	XXXX
PC	5C01H
IR	3EH

(Fetch instruction)

2) PC 5C01H

A	19H
B	XX
HL	XXXX
PC	5C02H
IR	3EH

(LD A)

3) PC 5C02H

A	19H
B	XX
HL	XXXX
PC	5C03H
IR	21H

(Fetch LD HL)

4) PC 5C03H

A	19H
B	XX
HL	XX50H
PC	5C04H
IR	21H

(lsb in)

5) PC 5C04H

A	19H
B	XX
HL	5C50H
PC	5C05H
IR	21H

(msb in)

6) PC 5C05H

A	19H
B	XX
HL	5C50H
PC	5C06H
IR	77H

(Fetch and execute:
19 is now in 5C50H)

7) PC 5C06H

A	19H
B	19H
HL	5C50H
PC	5C07H
IR	46H

(Fetch and execute:
19 is now in B reg)

In box 1, the LD A, n instruction (3EH) is fetched into the IR. In box 2, n (19H) goes into the A reg whilst in box 3, the instruction LD HL, nn is fetched and put in the IR. In 4, the lsb goes in followed by the msb. In 6, LD (HL), A is fetched and executed. Finally, LD B, (HL) is fetched and executed.

Figure 7.11   *Illustrating the fetch—execute sequence for Example 3.*

Solution

1	LD A,19H	3E, 19
2	LD HL,5C50H	21, 50, 5C
3	LD (HL),A	77
4	LD B,(HL)	46

The full sequence is given in Figure 7.11.

The fetch—execute sequence for an I/O operation

The organisation and use of the I/O ports is fully covered in Chapter 12, but it's convenient to include a typical I/O operation now. A brief description is given below for clarity.

The Z80 CPU system can cope with a large number of I/O ports (over a hundred) each of which can address an external device; a bank of LEDs or other indicators, motors, robot arms, printers and disc drives when outputting, sensors such as microphones, thermistors, light-dependent resistors, etc. when inputting.

Each I/O port has to be 'set up' to indicate whether the port is configured for input or output or a combination of the two. This is done by loading the accumulator with the appropriate codes and sending them to the port's control register. In the MARC Micro System, Port 1 (the first port) is designated Port A, so its control register is denoted CRA. Once the CRA has been sent the code for output, the accumulator is then loaded with the data we wish to output. This is then sent to Port A's data register, denoted DRA.

As you will see in Chapter 12, the 6502 system has memory mapped I/O so that there's not much difference between storing data in memory and outputting it to a port.

Example 4

The basic command for outputting bytes to external devices via the I/O ports is: OUT (n),A, where (n) is the address of the DRA, in this case 00H. As an example, we'll load the accumulator with the contents of memory location 5000H and then output the contents of the accumulator to Port 1. We'll assume that memory location 5000H contains 0FFH and that the output port has already been set up.

The program is as follows:

```
5C00H       LD       A,(nn)     3AH
5C01H       (n)      00H        00H
5C02H       (n)      50H        50H
5C03H       OUT      (n),A      0D3H
5C04H       (n)      00H        00H
```

You will note that we have chosen to load the data byte from a memory location. Before the CPU can find that data byte the memory location address needs to be read into the CPU a byte at a time. Since the address is 5000H, it should be obvious that two fetches will be required. But where is this data kept? The answer is that there are various temporary registers in the CPU which can store this information. There are many other features of the CPU, such as this, which are not generally shown on CPU architecture diagrams, quite simply for clarity.

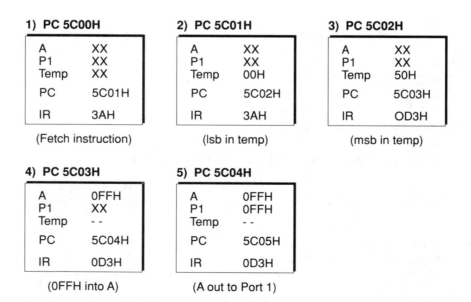

1) PC 5C00H

A	XX
P1	XX
Temp	XX
PC	5C01H
IR	3AH

(Fetch instruction)

2) PC 5C01H

A	XX
P1	XX
Temp	00H
PC	5C02H
IR	3AH

(lsb in temp)

3) PC 5C02H

A	XX
P1	XX
Temp	50H
PC	5C03H
IR	0D3H

(msb in temp)

4) PC 5C03H

A	0FFH
P1	XX
Temp	- -
PC	5C04H
IR	0D3H

(0FFH into A)

5) PC 5C04H

A	0FFH
P1	0FFH
Temp	- -
PC	5C05H
IR	0D3H

(A out to Port 1)

Figure 7.12 *The fetch–execute sequence for Example 4.*

It is said that the only person who knows exactly what's inside the CPU is the man who designed it, and he's never there when you ring him up! For the present we are keeping our CPU diagram fairly simple for the obvious reason that it's easier to understand. In any case, apart from the odd buffer register and some temporary stores, it is sufficient for our needs. We do introduce a more complete CPU diagram in the next chapter, but for the moment simply be aware of the *principal* components of the CPU and bear in mind that in reality there are many, many more. We shall describe the use of temporary registers as we go along, but meanwhile just accept that they exist and we'll discuss them in greater detail later.

The fetch–execute sequence for the program in Example 4 is shown in Figure 7.12.

The format we have used to illustrate the fetch–execute sequence is not sacrosanct; it can be changed or modified to suit an individual's requirements. We do feel, however, that whatever method of illustrating the sequence is chosen, a structure similar to the one proposed and illustrated here will be superior to one that merely attempts to explain what is happening in words alone, no matter how succinct are the descriptive powers of its author.

JUMP instructions

There are two types of JUMP instruction, conditional and unconditional. If an unconditional JUMP instruction is met, the program counter (PC) changes immediately to that value and JUMPs to it. If the jump is conditional, the JUMP is only carried out if that condition is met, for example if a register contains a particular number, or if it equals zero or is non-zero, etc. Some of the conditional JUMP instructions depend upon the contents of the flag register (or status register). This aspect of employing JUMP instructions is considered further in Chapter 10. 6502 CPU users should know that the equivalent of the Z80 'JP' instruction is 'JMP' and that 'relative jumps' are called 'branches'. For example 'BEQ label' means the same as 'JR Z,label' and 'BNE label' means the same as 'JR NZ,label'. Table 7.1 lists some of the JUMP instructions used in the Z80. 6502 users are referred to Table 7.2, Chapter 10 (The flag register) and Appendix VII (Some Z80/6502 equivalents).

It can be seen from Table 7.1 that both conditional and unconditional JUMPs can be direct or relative (in the Z80). If they're direct, program execution is continued from the address specified; if relative, the value given is added to the contents of the program counter and the program jumps to that address.

Table 7.1   JUMP instructions in the Z80 instruction set

Unconditional		Conditional	
Direct	Relative	Direct	Relative
JP nn	JR e	JP cc,nn	JR C,e
JP (HL)		JP C,nn	JR NC,e
JP (IX)		JP NC,nn	JR Z,e
JP (IY)		JP Z,nn	JR NZ,e
		JP NZ,nn	DJNZ,e
		JP PO,nn	
		JP PE,nn	
		JP P,nn	
		JP M,nn	

Table 7.2   The jump and branch instructions for the 6502

Unconditional		Conditional	
Direct	Relative	Direct	Relative
JMP label			BCC
JSR label			BCS
			BEQ
			BNE
			BMI
			BPL

As we have seen, relative jumps are useful because they require less code; their only disadvantage is that they have a limited range, but this is not usually a problem.

The format for direct JUMPs is:

[JUMP OPCODE], [JUMP ADDRESS lsb], [JUMP address msb]

The opcode for JP nn is C3 (0C3H), so if we wanted program execution to continue from that part of the program which is held in memory location 5C5D, then we should write: C3 5D 5C, noting that the address goes in backwards, i.e. lsb first, as before. The first byte specifies the opcode for the instruction and informs the CPU whether the instruction is conditional or unconditional. The second and third bytes are the address of the JUMP location, which directly replace the PC contents once the instruction has been processed. An example of this procedure is given in Figure 7.13.

In Figure 7.13 we join the program at address 5C04H where a JUMP instruction is encountered. The CPU collects the address to which the jump is to be performed, places it in the program counter and then jumps to that new

Memory Location	Opcode	Details	Action
5C04H	C3H	JP nn	EXECUTE
5C05H	50H	address lsb	
5C06H	5CH	address msb	
5C07H	C6H	ADD A, n	bypass
5C08H	03H	value 03H	
5C09H	C6H	ADD A, n	bypass
5C0AH	04H	value 04H	
5C50H	C6H	ADD A, n	EXECUTE
5C51H	08H	value 08H	
5C52H	XXH	next instruction	

Figure 7.13   *An example of a jump instruction being executed.*

Memory Location	Opcode	Details	Action
5C04H	18H	JR e instruction	EXECUTE
5C05H	03H	displacement	
5C06H	3AH	LD A, nn	bypass
5C07H	BAH	address lsb	
5C08H	5CH	address msb	
5C09H	C6H	ADD A, n	EXECUTE
5C0AH	05H	value 05H	
5CBAH	41H	value 41H	
5CBBH	42H	value 42H	
5CBCH	XXH	next instruction	

Figure 7.14   *An example of a relative jump instruction being executed.*

address, 5C50H in this case. The code in between is bypassed; the code shown (the two ADDs) has no special significance, it's simply a bit of code that would normally be processed if the JP 5C50H instruction had not been encountered.

Relative jumps

The great advantage of relative jumps is that only one byte is needed to specify the destination address. Since addresses are almost always two bytes (16 bits) this does not seem possible. In fact, all that happens is that a number is given in the instruction which is added to the contents of the PC. The 8-bit number is therefore limited to the range −128 to +127 relative to the start address. The number of memory locations you wish to bypass, backwards or forwards, is called the displacement. In hand assembly of programs, small displacements may be worked out quite easily, for more complicated displacements special calculators may be used and when an assembler is used, the program does it for you. You simply give the place you wish to jump to a label and use, for example, a JR NZ,LABEL instruction. A jump relative instruction requires two bytes altogether, so the PC is automatically incremented by two, hence the value of the displacement should be reduced by two, unless you're jumping backwards. Some examples of the procedure are given in Chapter 5.

Figure 7.14 shows a relative jump instruction being executed. After the JR 5C09H instruction is fetched (18H) the PC increments to 5C05H. After the next instruction is fetched (03H, the value of the displacement), the PC increments to 5C06H. When the instruction is executed, therefore, the displacement is added to the PC giving 5C09H which is what is required. It would not have been correct to put a displacement of 05H into memory location 5C05H. Again, the rest of the code in the example has no particular significance.

1) PC 5C04H

A	XX
B	XX
PC	5C05H
IR	3E

2) PC 5C05H

A	15
B	XX
PC	5C06H
IR	3E

3) PC 5C06H

A	15
B	XX
PC	5C07H
IR	C3

4) PC 5C07H

Fetch CD and store in temporary register.	
PC	5C08H

5) PC 5C08H

Fetch 5C and put into PC.	
PC	5CCDH
IR	C3

6) PC 5CCDH

A	15
B	XX
PC	5CCEH
IR	C6

7) PC 5CCEH

A	25
B	XX
PC	5CCFH
IR	C6

8) PC 5CCFH

A	25
B	XX
PC	5CD0H
IR	06

9) PC 5CD0H

A	25
B	18
PC	5CD1H
IR	06

10) PC 5CD1H

A	25
B	18
PC	5CD2H
IR	C3

11) PC 5CD2H

Fetch 09 and store in temporary register.	
PC	5CD3H

12) PC 5CD3H

Fetch 5C and put into PC.	
PC	5C09H
IR	47

13) PC 5C09H

A	3D
B	18
PC	5C0AH
IR	80

14) PC 5C0AH

A	3D
B	3D
PC	5C0BH
IR	47

Figure 7.15 *The full fetch–execute sequence for Example 5.*

Example 5

Work out the following program sequence and show its operation by illustrating the fetch–execute cycle for each program line:

1   load the accumulator with 15H
2   JUMP to memory location 5CCDH (which contains the instruction: ADD 10H to the accumulator)
3   load the B register with 18H
3   JUMP back to the next available location after the first (LD A,15H) and then add the contents of the accumulator to the B register.

The solution is given below:

Solution to Example 5

The program is as follows:

```
5C04H      LD  A,15H      3E
5C05H      15H            15
5C06H      JP,5CCDH       C3
5C07H      0CDH           CD
5C08H      5CH            5C
5C09H      ADD  A,B       80
5C10H      LD  B,A        47

5CCDH      ADD  A,10H     C6
5CCEH      10H            10
5CCFH      LD  B,18H      06
5CD0H      18H            18
5CD1H      JP,5C09H       C3
5CD2H      09H            09
5CD3H      5CH            5C
```

The fetch–execute sequence for the program is shown in Figure 7.15.

Questions

1 (Z80/6502) Draw up a flow chart to illustrate the fetch–execute sequence.

2 (Z80/6502) Where in the CPU are instructions held?

3 (Z80/6502) How does the CPU distinguish between hex codes which represent data, and hex codes which represent instructions?

4 Illustrate the fetch–execute sequence by showing the changing register contents for the following program:

```
(Z80)                    (6502)
5C00H      LD  A,56H      &0200      LDA  #&56
5C02H      LD  B,15H      &0202      ADC  #&15
5C04H      ADD  A,B       &0204      BRK
5C05H      HALT
```

5 Illustrate the fetch–execute sequence by showing the changing register contents for the following program:

```
(Z80)                    (6502)
5C00H      LD  B,16H      &0200      LDX  #&16
5C02H      LD  C,24H      &0202      LDY  #&24
5C04H      LD  A,C        &0204      TXA
5C05H      ADD  A,B       &0205      ADC  #&24
```

6 Describe the fetch–execute cycle for a program which fetches three numbers from memory, adds the numbers together by jumping to a subroutine and then outputs the answer to a different memory location.

7 Illustrate the fetch–execute sequence by showing the changing register contents for the following program:

```
(Z80)                     (6502)
5C00H      LD  A,0DFH      &0200      LDA  &32
5C02H      LD  HL,5DD0H    &0202      STA  &02E0
5C05H      LD  (HL),A      &0205      TAY
5C07H      INC  A          &0206      INY
5C08H      INC  HL         &0207      TYA
5C09H      LD  (HL),A      &020A      STA  &02E1
```

8 Illustrate the fetch–execute sequence by showing the changing register contents for the following program:

(Z80)		(6502)	
5C04H	JP 5DD0H	&0204	JMP &02E0
5C07H	LD B,42	&0207	LDX #42
5C0BH	AND B	&0209	AND &32
5DD0H	LD A,42	&02D0	LDA #42
5DD2H	JP 5C07H	&02d2	JMP &0207

9 Explain briefly, using the 'JP 5D00H' (Z80) or 'JMP &02D0' (6502) instruction, the sequence of events within the system and CPU during the fetch/decode/execute cycle.

10 Illustrate the fetch–execute sequence by showing the changing register contents for the following program:

(Z80)		(6502)	
5C00H	LD A,25H	&0200	LDA #&25
5C02H	LD HL,5DD0H	&0202	STA &34
5C05H	LD (HL),26H	&0204	TAX
5C08H	LD B,A	&0205	INX
5C09H	LD A,(HL)	&0206	TXA
5C0CH	ADD A,B	&0207	ADC &34

Chapter 8
Timing diagrams

Signals involved in the fetch–execute sequence

In the last chapter, we looked at the fetch–execute sequence in some detail in order to see how instructions and data are brought into the CPU and manipulated. We continue this discussion by looking at timing diagrams which effectively summarise bus activity and the action of the various control signals during the fetch–execute sequence.

There are some similarities between Z80 and 6502 timing diagrams and their associated theory, so we discuss both in the first part of this chapter. Examples of each are also given, with the 6502 section appearing towards the end.

CPU timing

All CPU timing is controlled by a crystal-controlled CLOCK oscillator, designated: Φ. (The 6502 has a two-phase clock which we look at later.) The use of a crystal ensures precise and constant frequency. The idealised clock oscillator output is a square wave as shown in Figure 8.1 and each clock period is called a 'T cycle', or 'T State'.

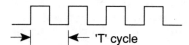
→| |← 'T' cycle

Figure 8.1 *The CPU clock square wave, indicating the period of the waveform, known as a 'T' cycle.*

The Z80 CPU (and the 6502) executes instructions by stepping through a very precise set of a few basic operations. These include:

Opcode fetch
Memory read or write
I/O device read or write
Interrupt acknowledge

All instructions are merely a series of these basic operations, each of which can take from three to six clock periods to complete (although they can be lengthened by using 'wait states' (T_w) used to synchronise the CPU to the speed of external devices). The basic operations, composed of T cycles, are called machine (M) cycles, and this is illustrated in Figure 8.2.

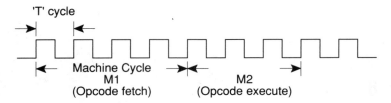

Figure 8.2 *Opcode fetch and execute machine cycles. The instruction set indicates the number of 'M' cycles for each instruction and how this is made up of 'T' cycles.*

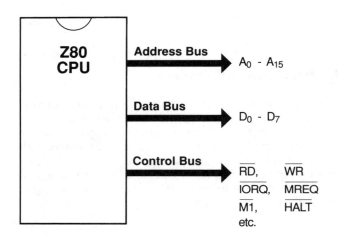

Figure 8.3 *Simple diagram of the Z80 CPU showing the important bus connections. It is the bus activity which must be shown on a timing diagram.*

Timing diagrams

A timing diagram shows the relationship between clock pulses and bus signals. Figure 8.3 shows a simplified diagram of the Z80 CPU indicating the address bus, data bus and some of the signals on the control bus.

Many of these changing signals will be shown on the timing diagram, but which are included will depend upon what the diagram is describing. For example, if we wish to show the timing diagram for an instruction opcode fetch, it will be necessary to show:

(a)   the clock pulses
(b)   the address lines
(c)   the data lines
(d)   relevant control lines: $\overline{\text{MREQ}}$, $\overline{\text{RD}}$ and $\overline{\text{M1}}$

It is important to understand that as well as address information and data, control signals are constantly appearing on certain pins of the CPU. A pin-out diagram of the Z80 is shown in Figure 8.4 and we list below the meaning of many of the control pins marked:

$\overline{\text{BUSACK}}$ Bus acknowledge (output, active low). Bus acknowledge indicates to the requesting device that the CPU address bus, data bus and control signals, $\overline{\text{MREQ}}$, $\overline{\text{IORQ}}$, $\overline{\text{RD}}$ and $\overline{\text{WR}}$, have entered their high impedance states. The external circuitry can now control these lines.

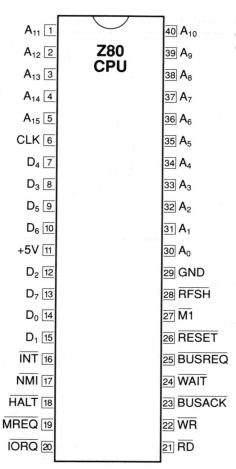

Figure 8.4 *The Z80 CPU and pin connections. The microprocessor is packaged in an industry standard 40-pin dual-in-line package.*

$\overline{\text{BUSREQ}}$	Bus request (input, active low). BUSREQ forces the CPU address bus, data bus and control signals, $\overline{\text{MREQ}}$, $\overline{\text{IORQ}}$, $\overline{\text{RD}}$ and WR, to go to a high impedance state so that other devices can control these lines.
$\overline{\text{HALT}}$	Halt state (output, active low). $\overline{\text{HALT}}$ indicates that the CPU has executed a halt instruction and is awaiting an interrupt before operation can resume. While halted, the CPU executes NOPs to maintain memory refresh.
$\overline{\text{INT}}$	Interrupt request (input, active low).
$\overline{\text{IORQ}}$	Input/output request (output, active low, 3-state). $\overline{\text{IORQ}}$ indicates that the lower half of the address bus holds a valid I/O read or write operation.
$\overline{\text{M1}}$	Machine cycle one (output, active low). $\overline{\text{M1}}$, together with $\overline{\text{MREQ}}$, indicates that the current machine cycle is the opcode fetch cycle of an instruction execution.
$\overline{\text{MREQ}}$	Memory request (output, active low, 3-state). $\overline{\text{MREQ}}$ indicates that the address bus holds a valid address for a memory read or memory write operation.

Table 8.1   Z80 and 6502 CPU terminology compared

	Z80		6502	
Signal	Signal	Pin no.	Signal	Pin no.
Clock (osc)	Φ	6	Φ₀	37
Reset	RESET	26	RES	40
Read	RD	21	R/W	34
Write	WR	22	R/W	34
Ready	WAIT	24	RDY	2
Interrupt	INT	16	IRQ	4
Non-maskable Interrupt	NMI	17	NMI	6
Input/output request	IORQ	20	n/a	–
Memory request	MREQ	19	n/a	–

Table 8.2   Categories of Z80 control signals

CPU control	Bus control	System control
INT	BUSREQ	RFSH
NMI	BUSACK	M1
HALT		WR
RESET		RD
WAIT		MREQ
		IORQ

> RD   Memory read (output, active low, 3-state). RD indicates that the CPU wants to read data from memory or an I/O device. The addressed I/O device or memory should use this signal to gate data onto the CPU data bus.

> WR   Memory write (output, active low, 3-state). WR indicates that the CPU data bus holds valid data to be stored at the addressed memory or I/O location.

Table 8.1 compares Z80 and 6502 terminology and lists the pin numbers for each CPU. The control signals are categorised in Table 8.2.

Before looking at a timing diagram, let's take a look at a more complete diagram of the Z80, as shown in Figure 8.5.

The clock signal and the three buses are shown entering the CPU (on the 6502, clock pulses are generated internally). Each bus has a buffer and their connections can be seen clearly. The data bus control logic allows data to be placed in the various registers; if a number is to be placed in the C register, for example, the data bus control operates the necessary tri-state logic to allow this to happen.

The external address bus is connected to the address control logic and the program counter (PC); this enables the current address held in the PC to be placed on the address bus at the appropriate time. The two-way control bus signals are taken to the timing and control logic which is closely associated with the instruction decoder.

The timing diagram for an instruction opcode fetch will show the clock pulses, the address and data lines and the relevant control lines. The CPU first

121

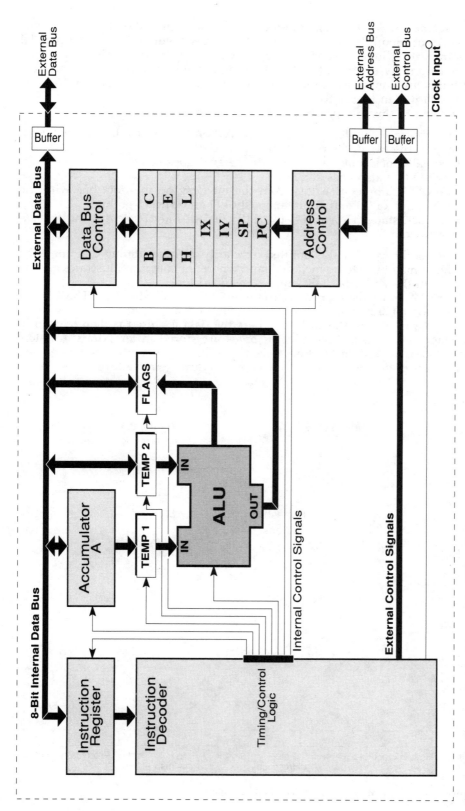

Figure 8.5 A more complete diagram of the Z80 architecture.

presents the address of the memory location being selected on the address bus wires (16 of them). It would be difficult to draw them all in, so two parallel lines indicating logic '0' and '1' are drawn which represent all the possible combinations. When the address may change, the logic levels on the address bus wires are shown to cross. A timing diagram for the instruction opcode fetch is shown in Figure 8.6.

Referring to this, the procedure is as follows:

1 The program counter contents are placed on the address bus at the beginning of the M1 cycle; the $\overline{M1}$ line goes low.
2 One half clock time later, the $\overline{MREQ}$ signal goes active.
3 The $\overline{RD}$ line also goes active.
4 The CPU samples the data from the memory on the data bus with the rising edge of T3 and simultaneously $\overline{RD}$ and $\overline{MREQ}$ go high.
5 On clock states T3 and T4, the CPU decodes and executes the fetched instruction.

Sometimes, the $\overline{M1}$ signal is omitted for clarity. In other timing diagrams, the $\overline{WR}$ signal may have to be included, and where the diagram illustrates an I/O transfer (in the Z80), the $\overline{IORQ}$ line is included too.

Figure 8.7 shows the four M cycles for an LD A,(nn) instruction. The instruction opcode fetch has been illustrated in Figure 8.6.

The lsb and msb of the address where the data is stored is then fetched in two stages as shown; these address bytes are stored in temporary registers associated with the instruction decoder/timing and control logic. At the appropriate time, this 16-bit address is put on the external address bus via the

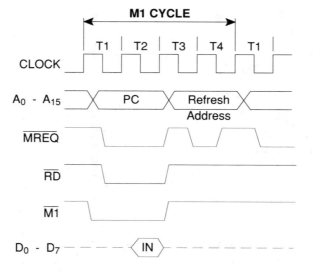

Figure 8.6 *Timing diagram for an instruction opcode fetch.*

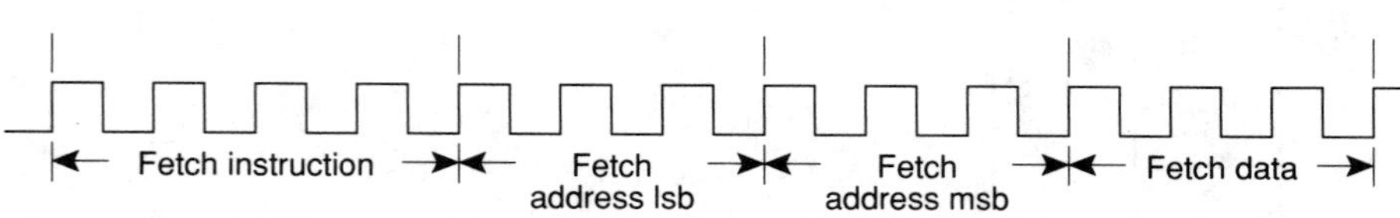

Figure 8.7 *The 4 M cycles for an LD A,(nn) instruction.*

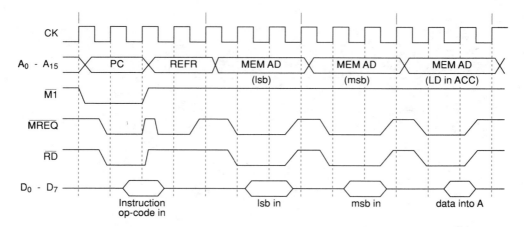

Figure 8.8 *Timing diagram for a LD A,(nn) instruction. M cycles: 4 T states: 13 (4, 3, 3, 3).*

address control logic. During the next 3 T-states, which make up the fourth M cycle, the data is fetched and placed in the accumulator. The complete timing diagram is shown in Figure 8.8.

Referring to this and the Z80 CPU diagram shown as Figure 8.5, let's take a step-by-step look at how the instruction LD A,(5D00H) is fetched and executed by the CPU.

We'll assume (as usual) that the program origin is 5C00H.

```
PC →   5C00H      3A      ;opcode for an LD A,(nn)
       5C01H      00      ;lsb
       5C02H      5D      ;msb
         .         .       .
         .         .       .
         .         .       .
       5D00H      99      ;data  (= 153D)
```

The sequence is as follows:

1   The control line causes the $\overline{M1}$ pin on the CPU to go low; address 5C00H then appears on the address bus; the PC is incremented.
2   The $\overline{MREQ}$ and $\overline{RD}$ lines go low.
3   After a short delay (the memory access time), the opcode at memory location 5C00H (3A) appears on the data bus and is read by the CPU.
4   The $\overline{M1}$ line goes high and remains so until the next machine cycle is initiated.
5   The control associated with the IR places the opcode into the IR.
6   The IR decodes the instruction; it 'realises' that a data value is required from a memory location; the PC contents are placed on the address bus and the $\overline{MREQ}$ and $\overline{RD}$ lines go low again.
7   The lsb of the target address is read from memory location 5C01H and stored in a temporary register associated with the IR. This takes three clock cycles.
8   The control logic takes $\overline{MREQ}$ and $\overline{RD}$ temporarily high, and, about one clock cycle later, they go low again; the PC has incremented to 5C02H and the msb is read into the CPU.

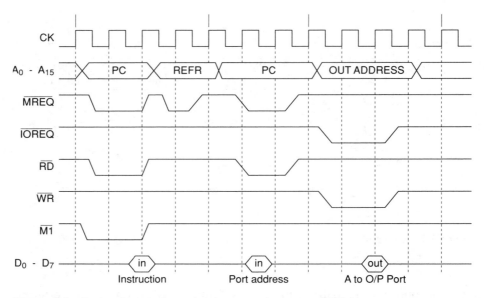

Figure 8.9 *Timing diagram for writing the accumulator to an I/O device.*

9   The target address (5D00H) then appears on the address bus; the data lines are connected to the accumulator by the control logic, and the data value (99H) is copied into the accumulator about two clock pulses later.

Figure 8.9 shows a timing diagram for writing the accumulator to an I/O device. This diagram includes the $\overline{\text{IORQ}}$ line and the $\overline{\text{WR}}$ line; they were absent in Figure 8.8, but they both become active during execution of this instruction. Examination of this diagram shows how the sequence takes place and the reader should have no difficulty in following it through.

6502 timing diagrams

A pin-out diagram of the 6502 is shown in Figure 8.10. We can identify five main groups of signals: power supplies; clock pulses; control, data and address lines.

The 6502 has a two-phase clock denoted Φ_1 and Φ_2 as shown in Figure 8.11. These waveforms are non-overlapping square waves, which means that they are never at logic '1' at the same time. Address changes take place during Φ_1 and data transfers during Φ_2.

6502 control pins

RDY   Disabled if held high. Taking this pin low halts the CPU on all cycles except write cycles.

$\overline{\text{IRQ}}$ Interrupt request – taking this pin low causes the processor to complete the current instruction being executed and then service the interrupt. Only occurs if the interrupt disable flag in the status register is at zero.

$\overline{\text{NMI}}$ A non-maskable interrupt is initialised by a negative-going transition on this pin.

SO   Set overflow flag – a negative-going transition causes the overflow flag in the status register to be set. Rarely required in practice, and is usually permanently tied to logic 1 in most systems.

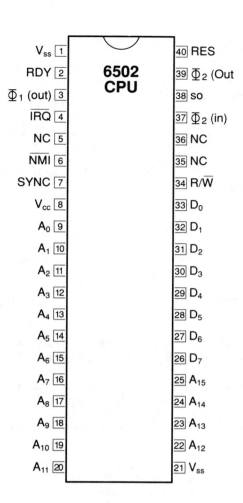

Figure 8.10 *The 6502 CPU and pin connections. The microprocessor is packaged in an industry standard 40-pin dual-in-line package.*

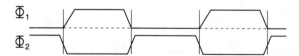

Figure 8.11 *Waveforms for the 6502 two-phase clock. Note that the Φ_2 waveform begins to fall, before the Φ_1 begins to rise, and does not reach its maximum before Φ_1 has fallen to zero.*

$\overline{\text{SYNC}}$ This pin goes high during Φ_1 of a fetch operation and stays high for the remainder of that cycle. (This is similar to the $\overline{\text{M1}}$ control signal on the Z80.)

$\overline{\text{RES}}$ RESET. A positive-going transition on this pin resets the CPU.

$\text{R}/\overline{\text{W}}$ Read/Write – when this pin is high, the 6502 reads in data; when it's low, data is written from the CPU.

Instruction timing

Because the 6502 index registers (X and Y) are only 8 bits wide (rather than the 16-bit Z80 index registers) a large number of addressing modes are

Table 8.3 Instruction timing for some 6502 addressing modes. The implied mode requires two cycles; one to fetch the instruction opcode and the other to execute it. Other addressing modes use extra cycles for loading operands and internal operations

Mode	Bytes	Cycles
Implied	1	2
Immediate	2	2
Zero page	2	3
Absolute	3	4
Indexed	2	4
Indirect	2	5
Indirect – indexed	2	5
Indexed – indirect	2	6
Relative	2	2/3

employed in order to achieve a full 64K indexed address range. Table 8.3 shows the number of bytes required for each mode and the number of cycles used.

The execution time for branch instructions is normally two cycles if no branch is made, but increases to three or four when a branch is executed. The higher value is when the branch is to a different page of memory.

Referring to Figures 8.12 and 8.13, let's examine an instruction opcode fetch (LDA #&55) in some detail:

1   The read operation begins when the R/$\overline{\text{W}}$ pin goes high. This is timed to coincide with the rising edge of the Φ_1 clock.
2   A short time later, the PC contents are placed on the address bus.
3   The SYNC pin goes high at this point and remains high for the remainder of the cycle.

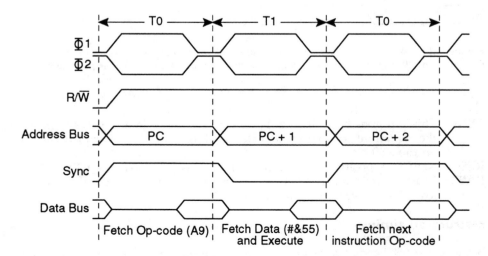

Figure 8.12. An instruction op-code fetch.

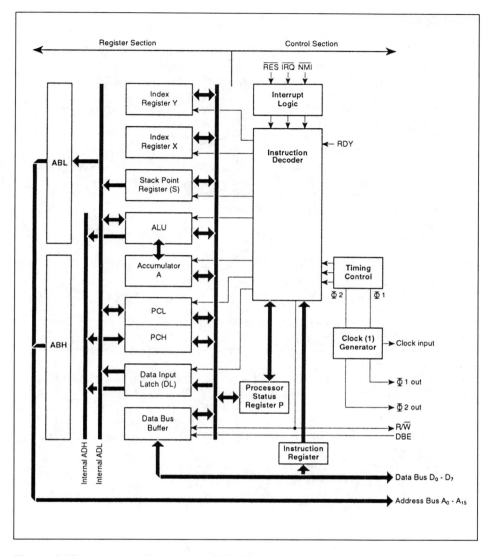

Figure 8.13   *Internal architecture of a 6502 CPU.*

4   The Φ_1 clock goes low and the Φ_2 goes high; shortly after the Φ_2 waveform reaches logic 1, the CPU reads the instruction opcode and latches it into the instruction register.

5   The first clock cycles complete as Φ_2 falls to zero and Φ_1 rises to a logic 1. When this is established, the address bus is cleared momentarily and the incremented PC contents are placed on it.

6   Just after the Φ_2 clock rises to a logic 1, the data is read into the accumulator. The value &55 is now in the accumulator.

7   The cycle repeats by putting the incremented PC contents onto the address bus and fetching the next opcode as before.

Questions

1 (a) Why is a crystal-controlled *clock* necessary in a CPU?
  (b) What sort of waveform represents the idealised clock oscillator output?
2 Name three basic operations which occur during the execution of a typical machine code program.
3 Define: (a) a clock cycle
           (b) a machine cycle
4 Draw a simple diagram of a CPU showing only three bus connections, and labelling the basic functions of each bus.
5 For the Z80 or 6502 CPU (or both if you like) state the pin numbers of the following signals or voltages:
  (a) V_{cc} (+5V)
  (b) GND (0V)
  (c) Clock input
  (d) Non-maskable interrupt
  (e) Start of an opcode fetch
6 (a) (Z80) Draw and label a simple diagram of a Z80 clock waveform
  (b) (6502) Draw and label a simple diagram of the 6502 two-phase clock waveforms
7 Name the control signals used in an opcode fetch, briefly stating their purpose.
8 In both the Z80 and 6502 CPUs, a varying number of clock cycles and machine cycles are required. State why this is so giving examples.
9 If the CPU clock runs at 1MHz, what is the length of one T-state?
10 Show diagrammatically the activity of all appropriate bus signals during the execution of:
  (a) (Z80) LD A,55H
  (b) (6502) LDA #&55

Assignment 8.1

(a) Show the relationship between clock pulses and bus signals on a timing diagram for the following:
  (i) Transfer the content of a memory location into the accumulator.
  (ii) Write the content of the accumulator to an I/O device.
(b) State two reasons why the timing diagram shown in Figure 8.14 will not give successful data transfer between CPU and memory.

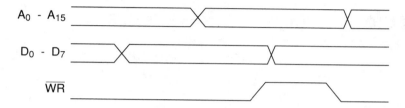

Figure 8.14 *Timing diagram.*

(c) A program in Z80 code contains the following instruction: (Address) 5C10H LD A,(0E059H). The instruction occupies four machine cycles as shown below:

	1	2	3	4
A_0–A_{15}				
D_0–D_7				

  (i) name the types of machine cycle, 1 to 4;
  (ii) state what the contents of A_0–A_{15} and D_0–D_7 are in each machine cycle.

Chapter 9
Organisation and use of the stack

Location of the stack

The stack is a reserved area of RAM; it therefore consists of several memory locations (frequently about 256, or 100H), used for the temporary storage of data. We introduced the concept in Chapter 4 whilst discussing memory maps. The RAM locations dedicated to stack use are no different from any others, so it's quite possible to store data there and retrieve it in the normal way. This isn't often done, of course, since it would obviously compromise stack action. Before going any further, let's look at some typical uses of the stack.

1   The contents of registers may be stored temporarily.
2   The contents of registers may be changed over via the stack.
3   The return address following a CALL to a subroutine may be saved automatically, using the stack.
4   Data and addresses may be saved during the servicing of an 'interrupt'.

The reader will be aware of the great speed at which microprocessors operate. Generally, they can deal with about half a million instructions a second; some processors work even faster. The Z80 instruction set quotes execution times for a clock frequency of 4MHz (the 'MARC' operates at 2MHz). Higher clock frequencies are common: 12MHz, 33MHz, 66MHz are already here and 100MHz is just becoming available. It is this great speed of operation that allows the computer to do several different things, apparently all at the same time. In fact, the illusion is created by the processor switching from one task to another, but doing it so rapidly that events appear to occur simultaneously. In this way, it's possible for a microelectronic system to execute a program, whilst 'at the same time' accepting external control signals and responding to them; this is done by allowing the external device to 'interrupt' normal program execution.

This is where the stack comes in. When the 'interrupt' is received, register contents and appropriate address information are quickly saved on the stack, the interrupting peripheral is serviced, data is retrieved from the stack and program execution continues, all as if nothing had happened.

The stack could theoretically be located anywhere in RAM and programmers are free to make their own choice in the matter. Generally, however, microelectronic systems will designate a suitable area and it's wise to stick to it.

Of course, data and addresses could be saved by the programmer, simply by storing the values in memory locations. But this would involve keeping track of those memory locations which could be used for the purpose, remembering which ones had been used so that the data can be retrieved and, finally, remembering to be careful not to overwrite the data with other values!

Setting up a stack in a reserved area of RAM is therefore very convenient; it eliminates all the problems just described and its use can be invoked with just a few simple instructions. You don't even have to know where the data has gone since the stack pointer does that for you.

The interrupt mechanism can access the stack facility automatically and we shall be discussing this and other advantages later.

Z80 and 6502 stacks

The organisation and use of the stack in each system is very similar, but there are sufficient differences between the two to justify treating them separately, Z80 first.

PUSH and POP

There are two instructions associated with stack action: PUSH and POP. Data is PUSHed onto the stack and POPped off it. Registers can't be pushed or popped on their own, the instruction refers only to register pairs, hence it's possible to store the register pairs AF, BC, DE, HL and the index registers IX and IY on the Stack.

The stack pointer (SP)

The last-used address (the 'top') of the stack is kept in a special CPU register called the stack pointer (SP). This has already been described for the Z80 CPU, where it is a 16-bit register; this means that it's large enough to hold any of the available RAM addresses in the system. In the 'MARC' micro, the stack is located at address 4300H and it builds downwards. This means that when the first data value is pushed onto the stack, it goes into memory location 42FFH, the next item goes into 42FEH and the third, into 42FDH, etc. as shown in Figure 9.1. The new address value of the top of the stack is kept in the stack pointer (SP register) so the CPU knows where to store any further values.

Last in, first out (LIFO)

The stack is always arranged as a LIFO file; last in, first out. In the example given above, if four data values have been pushed onto the stack, the stack pointer will be holding 42FCH. If you now want to retrieve data from the

Stack Memory location	Stack Pointer contents
* *	4300H
Value 1	42FFH
Value 2	42FEH
Value 3	42FDH
Value 4	42FCH
. .	42FBH

Figure 9.1 *Data values pushed onto the stack with appropriate stack pointer contents.*

Instruction	Register pair contents	Stack Memory location contents	Stack Pointer contents
PUSH BC	5E10H	＊ ＊	4300H
		5E	42FFH
PUSH DE	5F20H	10	42FEH
		5F	42FDH
		20	42FCH ←

Figure 9.2 *Pushing BC and DE register pairs. At the end of the pushes, the SP contains 42FCH.*

stack, it *must* come from location 42FCH first; the values located at 42FDH and 42FEH come out next, in that order. In this example, we're assuming that the BC register pair contains 5E10H and DE contains 5F20H. If we now 'push' the contents of the BC and DE register pairs onto the stack, the action is as shown in Figure 9.2.

Now you can have the CPU fill BC and DE register pairs with different numbers, perform an operation or two and then go back to the stack for the original values. If you want the BC values on the stack, you *must* 'pop' DE first. There is no way you can get at BC contents without disturbing DE, because of the LIFO principle. This may sound like something of a limitation, but it's not in practice.

If you decide against popping BC and DE for the moment, you could push another register pair onto the stack as well, HL for example. When the instruction is executed, the SP decrements (in this case to 42FBH) and the content of H goes into the stack at that point; this is followed by the content of L which goes into location 42FAH, which becomes the new SP value.

The action of pushing and popping these three register pairs is shown in Table 9.1. We assume that HL contains 5F99H.

Subroutines and calls

We have said very little about subroutines so far, mainly because the programming techniques already encountered have been sufficient for our modest needs. For any reasonably sized program, however, subroutines are essential and we have already mentioned their use in producing time delays. Very complicated programs, in fact, may consist of nothing but a succession of subroutines and we have noted that JSP is a particularly appropriate design in this respect. More of this later; for the moment, let's look at the part played by the stack when a 'call' to a subroutine is encountered.

CALL and RET

These are the two instructions involved in accessing a subroutine. To use the previous example, suppose we wish to use a time delay to slow down a video display. All we have to do is note the address of the first instruction in the delay routine (say 5F00H) and 'CALL' it. Provided the routine has the instruction RET (return) at the end of it, program execution will continue as normal

Table 9.1 Pushing and popping BC, DE and HL register pairs – last in, first out

Instruction	Register pair contents	Stack memory location contents	Stack pointer contents
PUSH BC	5E10H	**	4300H
		5E	42FFH
		10	42FEH
PUSH DE	5F20H	5F	42FDH
		20	42FCH
PUSH HL	5F99H	5F	42FBH
		99	42FAH
		Register contents	
POP HL		(L) 99	42FAH
		(H) 5F	42FBH
POP DE		(E) 20	42FCH
		(D) 5F	42FDH
POP BC		(C) 10	42FEH
		(B) 5F	42FFH
			4300H

as soon as the routine is completed. The following outline should make things clear:

```
ORG     5C00H
ENT     5C00H
LD      HL,0E02AH
LD      A,41H
LD      (HL),A
CALL    5F00H
INC     HL
INC     A
LD      (HL),A
HALT
```

The program is a trimmed down version of the famous 'Print ABCD' program, first introduced in Chapter 1. At that time, we only wanted to show the possible comparable lengths between machine code programs and those written in a high-level language such as BASIC.

Since then, we delayed detailed explanation of the program until we got to the appropriate chapters; we've examined the use of mnemonics, various different instructions and the use of program loops, etc. The explanation remained incomplete because we didn't want to jump into the use of an automatic assembler just yet. It's felt that the reader *must* have a good working knowledge of machine code, gained to a large extent by hand assembly, before enjoying the luxury of an automatic assembler! We'll get to this soon, of course, but as far as BTEC courses are concerned, we prefer to work at the nuts and bolts of the system in Level II and leave automatic assembly until Level III.

However, for the sake of clarity, it's probably worthwhile mentioning a few points now, not least because we need some more information about utilising

CALLs. Looking again at 'Print ABCD' (see page 3), we note the use of labels. The first one 'VIDEO:' (note the colon at the end) is equated (EQU) with 0E02AH (a video RAM location). This means that in any other part of the program, we can use the label 'VIDEO' rather than the number 0E02AH, the former being much easier to remember. It's only used once in this simple program (line 6) but in larger programs it could be used any number of times.

The label 'CLS:' is equated with 01B9H which is a memory address in the MARC micro operating system. When the processor meets the instruction 'CALL CLS' it pushes the program counter (PC) contents onto the stack. Then it rushes off to 01B9H which contains a clear screen routine, kindly inserted there by the manufacturers for our great convenience. This 'user routine' as it's called, has a RET statement at the end of it, which has the effect of popping the address stored on the stack and putting it back into the PC. Program execution then continues, in this case with the instruction 'LD HL,VIDEO'.

Another important point arises here. The CLS routine uses the HL register pair, so every time it's used, anything we've put into HL gets corrupted. It doesn't matter in this case, because we're loading HL immediately after the CALL but if we'd used a different order, we'd be looking at a program bug (see next chapter). To explain what we mean, look at part of the revised version of 'Print ABCD':

```
VIDEO:   EQU       0E02AH
CLS:     EQU       01B9H
         LD        HL,VIDEO
         CALL      CLS
         LD        A,41H
         etc.
```

Can you see what's going to happen? We put 0E02AH into HL using a label, then we CALLed the CLS routine. The routine uses HL, so 0E02AH is lost and the program doesn't run properly. Solution number 1 is to use the correct order; solution number 2 is to use the stack. OK, we don't need to use it in this instance, it's only an example, but if we had no simple alternative, this is what we'd do:

```
VIDEO:   EQU       0E02AH
CLS:     EQU       0B19H
         LD        HL,VIDEO
         PUSH      HL
         CALL      CLS
         POP       HL
         LD        A,41H
         etc.
```

Remembering the LIFO principle, we save HL contents using a PUSH, the CALL causes automatic stack action, and the PC contents are saved; the RET at the end of the routine automatically POPs the return address back into the PC and our own POP HL restores 0E02AH as required. All registers can be saved and restored in this way; memory locations too, by first loading them into a register pair.

If we use 'PUSH AF' this saves accumulator and flag register contents; it is therefore the only PUSH/POP instruction which affects the flag register.

Table 9.2 lists a typical Z80 program which uses the stack for CALLs and saving register pair contents. Careful examination of this program will show how the stack's used to exchange register pair contents. We've made use of labels in this example, and note that, on this occasion, the ORG and ENT values are not the same. This is a good illustration of how these instructions are used.

Table 9.2 Z80 program showing examples of stack action

Address	Machine code	Label	Mnemonics	Comments
			ORG 5C00H	;origin
			ENT 5C1BH	;entry point
5C00H	01 00 5D	FILL:	LD BC,5D00H	;load register pair
5C03H	11 10 5D		LD DE,5D10H	;load register pair
5C06H	21 20 5D		LD HL,5D20H	;load register pair
5C09H	78		LD A,B	;load A with B
5C0BH	32 30 5D		LD (5D30H),A	;put A into memory
5C0CH	1F		RRA	;rotate right A
5C0DH	32 31 5D		LD (5D31H),A	;put A into memory
5C0EH	3C		INC A	;increment accumulator
5C0FH	C9		RET	;return from subroutine
5C10H	3A 31 5D	LOAD:	LD A,(5D31H)	;load A from memory
5C13H	C9		RET	;return from subroutine
5C14H	C5	CHANGE:	PUSH BC	;BC onto stack
5C15H	D5		PUSH DE	;DE onto stack
5C16H	E5		PUSH HL	;HL onto stack
5C17H	D1		POP DE	;HL goes into DE
5C18H	C1		POP BC	;DE goes into BC
5C19H	E1		POP HL	;BC goes into HL
5C1AH	C9		RET	;return from subroutine
5C1BH	3A 99	MAIN:	LD A,99H	;99H into accumulator
5C1DH	32 32 5D		LD (5D32H),A	;load memory with A
5C20H	3A 00		LD A,0	;clear accumulator
5C22H	CD 00 5C		CALL FILL	;call subroutine
5C25H	CD 10 5C		CALL LOAD	;call subroutine
5C28H	CD 14 5C		CALL CHANGE	;call subroutine
5C2BH	76		HALT	;halt processor

The program's origin is 5C00H, but execution is meant to start at 5C1BH, labelled 'MAIN'. When you're happy with it, have a look at Example 1 below.

Example 1
Let's write yet another version of 'Print ABCD'. In this re-write, include a time-delay routine to slow down the 'printing' of each letter. We'd like you to use A and B registers for the delay routine so you'll have to save these register contents on the stack before CALLing your delay. You probably won't need it, but we've included a typical solution in Table 9.3.

Before leaving this section on CALLing subroutines and consequent stack action, we present two possible methods of displaying the information.

Example 2

```
5C21H   00                      NOP
5C22H   CD  00  55              CALL  CHANGE
5C25H   00                      NOP

5500H   E5           CHANGE:    PUSH  HL  ;HL onto stack
5501H   D5                      PUSH  DE  ;DE onto stack
5502H   E1                      POP   HL  ;DE into HL
5503H   D1                      POP   DE  ;HL into DE
5504H   C9                      RET       ;return from
                                           subroutine
```

Table 9.3 Z80 program showing more examples of stack action

Address	Machine code	Label	Mnemonics	Comments
			ORG 5C00H	;origin
			ENT 5C00H	;entry point
				;
		VIDEO:	EQU 0E02AH	;assign a label
		CLS:	EQU 01B9H	;MARC subroutine
5C00H	CD B9 01		CALL CLS	;
5C03H	21 2A E0		LD HL,VIDEO	;load video address
5C06H	3A 41		LD A,41H	;load A with ASCII 'A'
5C08H	06 04		LD B,4	;load B with counter
5C0AH	77	LOOP:	LD (HL),A	;put A onto screen
5C0BH	F5		PUSH AF	;save A contents
5C0CH	C5		PUSH BC	;save B contents
5C0DH	CD 00 5C		CALL DELAY	;call delay subroutine
5C10H	C1		POP BC	;retrieve B contents
5C11H	F1		POP AF	;retrieve A contents
5C12H	23		INC HL	;increment HL
5C23H	3C		INC A	;increment A
5C24H	05		DEC B	;
5C25H	20 F3		JR NZ,LOOP	;
				;
5C27H	76		HALT	;halt processor
5C28H	3A AA	DELAY:	LD A,0AAH	;
5C2AH	06 AA	LOOP2:	LD B,0AAH	;
5C0CH	05	LOOP1:	DEC B	;
5C0DH	C2 0C 5C		JP NZ,LOOP1	;
5C10H	3D		DEC A	;
5C11H	C2 2A 5C		JP NZ,LOOP2	;
5C14H	C9		RET	;return from subroutine

Assume that we have previously loaded HL with 2040H and DE with 4020H. The program enables HL and DE contents to be exchanged via the stack. A suitable layout for recording the action is shown in Table 9.4.

Table 9.4 A suitable format for recording stack action during program execution

Before CALL		During CALL		After CALL	
Memory	Stack	Memory	Stack	Memory	Stack
5C21H 00	4300H **	5C22H **	4300H **	5C25H **	4300H **
5C22H CD			42FFH 5C	5C26H	
5C23H 00			42FEH 25	etc.	
5C24H 55		5500H E5	42FDH 20		
5C25H 00			42FCH 40		
		5501H D5	42FBH 40		
			42FAH 20		
		5502H E1	42FAH 20		
			42FBH 40		
		5503H D1	42FCH 40		
			42FDH 20		
		5504H C9	42FEH 25		
			42FFH 5C		

Example 3

The following program demonstrates stack action just for CALL and RETurn commands:

```
                   ORG      5C00H
                   ENT      5C16H
5C00H    TRACE:    LD       BC,8080H
5C03H              LD       DE,5D00H
5C06H              LD       HL,5E00H
5C09H              LD       SP,4300H
5C0CH              LD       B,A
5C0DH              INC      A
5C0EH              INC      A
5C0FH              INC      H
5C10H              RET
5C11H    LOAD:     LD       A,(5CFFH)
5C14H              RET
5C16H    MAIN:     CALL     LOAD
5C19H              CALL     TRACE
5C1CH              HALT
```

Table 9.5 A simple illustration of stack action during CALL and RETurn instructions

Before load CALL		After load CALL		After RETurn	
PC=5C16H		PC=5C11H		PC=5C19H	
	stack		stack		stack
SP=4300H	**	SP=4300H	**	SP=4300H	**
	**	42FFH	5C		**
	**	42FEH	19		

Table 9.5 shows very simply how the stack stores the subroutine return address.

Parameter passing

Subroutines need data to act upon, so there must be some way of getting that data into the routine. The values could be held in appropriate memory locations of course, but this limits the flexibility of the routine. If there are only two values, they can be placed in the B and C registers for example, and the routine written to act upon them. You may sometimes require a routine which multiplies two numbers together. If the routine is designed to have the numbers in the B and C registers, then it becomes a generalised block of code which can be slotted into any program at any time. Whatever method is used, this is what is known as parameter passing; loading the data to be operated on into the subroutine.

Parameters can, of course, be passed via the stack. There may be many occasions where programmers find it useful to do just this; but then there's the problem of keeping track of the data. It's easy to imagine the BC register pair contents being PUSHed onto the stack; a subroutine is called and in it BC is POPped off the stack. NO! Remembering the LIFO principle, we see that the first POP would result in obtaining the PC contents and not BC at all, causing all sorts of problems. The stack may be used for passing parameters, but it's not always the best way. In fact, the chances of errors arising are so high, it's generally best to avoid the method altogether!

Interrupts

At the beginning of this chapter we noted the great speed at which the micro-processor executes instructions and introduced the concept of 'interrupts'. In terms of BTEC work, the use of interrupts is a Level III topic; we say more about interrupts later whilst discussing the Input/Output (I/O) ports (Chapter 12). For the moment, a few stack-oriented comments only are given.

All of the programs we have written and examined so far merely perform mathematical operations or shift data around the CPU and memory. Where appropriate we've put messages on the screen, and we've seen how time delays may be generated. However, when it comes down to serious micro-electronic systems applications, the majority of the work will be dealing with the servicing of peripherals. The word 'peripheral' is meant to mean all electronic or electromechanical devices which are external to the main computer. We're already familiar with obvious peripherals such as: keyboard, VDU, printer, etc. but there are many others including: light displays, solenoids, stepper motors, robot arms, temperature sensors, etc. and the peripherals will probably not only accept signals but supply them as well. Bearing in mind the great speed with which the CPU operates, it's convenient to have it 'interrupted' only when a peripheral needs to be 'serviced'; it can then be doing other important work in the meantime.

As far as we're concerned, we can assume that an interrupt is rather like a CALL instruction, although, as we shall see later, certain types of interrupt may be inhibited by software instructions, whilst others may not. In each case, a peripheral supplies a signal (normally a transition to a logic 0), indicating that it requires attention. The signal is taken to the interrupt pin on the CPU (INT or NMI), directly if there's only one peripheral, or via some sort of priori-tising logic, if there's more than one. It's at this point that a special interrupt servicing routine is accessed; the details need not concern us here, but this is when the stack is used to save the return address.

As with CALLs, the interrupt service routine has a RET statement at the end of it, 'RETI' in this case (RETurn from Interrupt); this causes the return address to be POPped off the stack and program execution continues as normal.

Note that PUSHes and POPs must be nested, and that in doing this the programmer needs to be aware of the CPU also possibly using the stack in CALLs and interrupts.

To summarise, the steps are as follows:

1 After executing each instruction in a program, the CPU checks to see if an interrupt signal has been received.
2 If an NMI or INT signal is received, it has the effect of a CALL instruc-tion. The current PC contents are PUSHed onto the stack and the CPU jumps to the interrupt service routine.
3 Where CPU registers are used by the interrupt service routine, their contents are also saved on the stack.
4 The interrupt is serviced.
5 Original CPU register contents are restored.
6 The original PC contents are POPped off the stack.

The 6502 stack

In the 'EMMA' micro system, the stack is located in Page 1 with 256 memory locations. It therefore starts at memory location 01FF and builds down. Special instructions are used which operate only on the stack. Generally, only a few of the memory locations dedicated to stack use will be required, since it's only possible to 'push' three registers onto the stack: accumulator, processor status register and program counter. However, the stack soon starts to fill up when subroutines are 'nested' as we shall see later, and there may be occasions when it's necessary to save most of the zero page addresses, to provide space for an interrupt service routine, for example.

PUSH and PULL

Readers who looked through the Z80 section will have noted that the equivalent instructions are PUSH and POP. There are other differences; for example, whilst the Z80 stack instructions operate on all the register pairs, the 6502 operates only on the three registers mentioned earlier on their own. There are four important instructions associated with this process and they are listed in Table 9.6.

Table 9.6   6502 stack operation instructions

Instruction	Operation
PHA	Push accumulator onto stack
PLA	Pull accumulator from stack
PHP	Push processor status on stack
PLP	Pull processor status from stack

Unlike the Z80 instructions, the X and Y register contents cannot be PUSHed or PULLed directly. The easiest way of saving these register contents on the stack is to do it via the accumulator.

The stack pointer (SP)

This is a special register in the CPU, 16 bits wide, which holds the current top of the stack. Initially, therefore, the SP contains 01FF. When the first value is PUSHed onto the stack, the SP decrements to 01FE. Subsequent PUSHes cause the SP to decrement further; in this way the CPU always knows where the last value has been placed.

Last in, first out (LIFO)

The stack is arranged as a LIFO file; the last data item to be PUSHed onto the stack is always the first to be PULLed from it. To demonstrate this, assume that we have the hex numbers &45, &99, &AF and &F1 at consecutive memory locations starting at &02E0 which we wish to save on the stack, and subsequently retrieve them; the action would be as shown in Table 9.7.

Subroutines and calls

A subroutine may be regarded as a small, self-contained program which can be CALLed from any part of the main program. In 6502 programming, the instruction JSR (jump to subroutine) is used. We have mentioned the use of time delays, which are frequently contained within a subroutine. The programs we have discussed so far have been relatively simple, used in the main to demonstrate the basic principles of machine code programming, but larger ones will inevitably contain many subroutines and some programs

Table 9.7 Illustrating stack action and the LIFO principle. The first data value is loaded from memory into accumulator; the SP points to 01FF. When the instruction PHA is executed, accumulator contents are pushed onto the top of the stack and the SP decrements (initially to 01FE)

Instruction	Memory location contents	Stack memory location contents	Stack pointer contents
LDA &02E0	&45		(&01FF)
PHA		&45	&01FE
LDA &02E1	&99		
PHA		&99	&01FD
LDA &02E2	&AF		
PHA		&AF	&01FC
LDA &02E3	&F1		
PHA		&F1	&01FB
.	.	.	.
.	.	.	.
.	.	.	.
PLA		&F1	&01FC
STA &02E3	&F1		
PLA		&AF	&01FD
STA &02E2	&AF		
PLA		&99	&01FE
STA &02E1	&99		
PLA		&45	&01FF
STA &02E0	&45		

consist entirely of them, CALLed one after another. (The JSP methodology leads logically to this kind of program design.)

As an example, suppose we want a program which will output a message on a series of seven-segment displays. The message is to 'move' along from one display to the next, with a time delay to slow down the 'printing'; there's a means of interrupting the message should we wish to, and an electronic 'beep' sounds when it's completed. There are many ways of doing this (and we'll be considering the implications later) but even though this will be quite a complicated program, perhaps a bit beyond us at the moment, we could still write down the basic structure. We'd list it quite simply as a series of sub-routines, the actual code in each routine being supplied later.

One possible listing is shown below:

```
          ;MESSAGE DISPLAY PROGRAM
          CALL MAIN
.MAIN     CALL MESSAGE
          CALL MOVE
          CALL DELAY
          CALL INTERRUPT
          CALL BEEP
```

Each subroutine is then written as a separate program, the main program calling each subroutine as it's required.

When the instruction JSR is encountered, the current PC contents are PUSHed onto the stack. The subroutine is terminated with the instruction RTS (return from subroutine) at which point the address is PULLed off the stack and put back into the PC. Program execution then continues as normal.

Table 9.8   6502 program showing examples of stack action. The program is written in true 6502 code, but was originally used within a BBC BASIC program; this explains the addresses used

Address	Machine code	Label	Mnemonics	Comments
			ORG &1500	;origin
			ENT &1500	;entry point
&1500	A9 FF	.START	LDA #&FF	;immediate load
&1502	8D 62 FE		STA &FE62	;code for output
&1505	A0 00	.AGAIN	LDY #0	;start count
&1507	D0	.NXT	STY &FE60	;Y contents are output
&150A	20 13 15		JSR DELAY	;CALL delay routine
&150D	C8		INY	;increment Y
&150E	D0 F7		BNE NXT	;inner loop
&1510	4C 05 15		JMP AGAIN	;outer loop
&1513	98	.DELAY	TYA	;transfer Y to A
&1514	48		PHA	;PUSH A onto stack
&1515	A0 FF		LDY #255	;load Y with 255D
&1517	A2 FF	.LOOP1	LDX #255	;load X with 255D
&1519	CA	.LOOP2	DEX	;decrement X
&151A	D0 FD		BNE LOOP2	;
&151C	88		DEY	;decrement Y
&151D	D0 F8		BNE LOOP1	;
&151F	68		PLA	;PULL A from stack
&1520	A8		TAY	;transfer A to Y
&1521	60		RTS	;return from subroutine

If the X and Y registers, for example, are being used in the subroutine, and were also being used in the main program, as soon as the subroutine is processed the previous X and Y contents would be lost. The stack can therefore also be used to save X and Y contents (by first transferring them to the accumulator) and they can be PULLed back afterwards so that normal program execution may continue. An example appears in Table 9.8 showing how the process could work.

This program, written in 6502 machine code for the BBC Micro, allows a bank of LEDs connected to the output port to count up in binary. A delay routine slows down the 'flash' rate so that it can be observed. The Y register is used in both the main program and the delay routine, so Y contents are pushed onto the stack before DELAY is entered, and pulled off the stack afterwards. The programmer must always remember to save registers in this way when necessary. PC contents are automatically pushed and pulled when a JSR instruction is executed.

Parameter passing

The subroutine will need to have data to act upon, so there must be some way of getting that data into the routine. The values could be held in appropriate memory locations of course, but this limits the flexibility of the routine. If there are only two values, they can be placed in the X and Y registers, and the routine written to act upon them. A common example may be where the routine multiplies two numbers together. If the routine is designed to have the numbers in the X and Y registers, then it becomes a generalised block of code which can be slotted into any program at any time. Whatever method is used, this is what is known as parameter passing; loading the data to be operated on into the subroutine.

Parameters can, of course, be passed via the stack. There may be many occasions where programmers find it useful to do just this; but then there's the problem of keeping track of the data. It's easy to imagine X and Y register contents being PUSHed onto the stack; a subroutine is called, inside which we PULL the X and Y values off the stack. NO! To show why, a typical program is given below:

```
0200        A2      02                      LDX   #&02
0202        A0      D0                      LDY   #&D0
0204        8A                              TXA
0205        48                              PHA
0206        98                              TYA
0207        48                              PHA
0208        20      E0      02              JSR   .STORE
020B

02D1        00                              BRK

02E0        68              .STORE          PLA
02E1        85      30                      STA   &30
02E3        68                              PLA
02E4        85      31                      STA   &31
02E6        60                              RTS
```

Now, looking at Table 9.9 and remembering the LIFO principle, we see that the first PULL would result in obtaining the PC contents, and not the register contents at all, causing all sorts of problems. When the JSR .STORE instruction is executed, the last address (020A) is put on the stack. (When it's retrieved at the end of the subroutine, the PC increments so that it would then contain &020B and the CPU fetches the instruction from that location.)

Meanwhile, the PLA and STA &30 puts &0A into zero page memory location &30, when what we wanted was &D0. Finally, RTS PULLS &02D0 off the stack (instead of &020A) and program execution continues from &02D1.

The stack may be used for passing parameters, but it has to be used with care and it's not always the best way; in fact, the chances of errors arising when doing this are so high, it's generally best to avoid the method altogether.

Table 9.9 Illustrating stack action before, during and after a subroutine CALL (JSR .STORE). At the end of this routine, the value &02D0 goes into the PC instead of the correct address, &020A. When the PC has incremented, program execution continues from &02D1

Instruction	Stack		Stack pointer
	Location	Contents	
TXA, PHA	&01FF	&02	&01FE
TYA, PHA	&01FE	&D0	&01FD
JSR .STORE	&01FD	&02	&01FC
	&01FC	&0A	&01FB
PLA, STA &30	&01FC	&0A	&01FB
PLA, STA &31	&01FD	&02	&01FC
RTS	&01FE	&D0	&01FD
	&01FF	&02	&01FE
			&01FF

Interrupts

We mentioned interrupts at the beginning of this chapter and some relevant comments are made in the Z80 section. The 6502 introduces a slight difference in nomenclature and operation, which we examine now.

There are four types of interrupt available to the 6502 programmer, three 'hardware' interrupts: IRQ, NMI and RES (interrupt request, non-maskable interrupt and reset) and one 'software' interrupt: BRK (break). The hardware interrupts are activated by signals applied to the CPU pins bearing the names. When the IRQ pin is taken low (the line above IRQ means it's active low), the CPU completes its present instruction and then looks at the interrupt disable flag (IDF, Flag I or bit 2 in the status register); if it's clear then the CPU implements the interrupt procedure as follows:

1   The high byte of the PC (PCH) is pushed onto the stack;
2   The low byte of the PC (PCL) is pushed onto the stack;
3   The status register contents are pushed onto the stack;
4   PCL is loaded from address 0EFE;
5   PCH is loaded from address 0EFF;
6   The interrupt disable flag (IDF) is set;
7   Program execution then continues from the address found in 0EFF/0EFE.

The first three procedures save PC and flag registers on the stack. The next two obtain the address of the interrupt service routine; it's up to the programmer to put this address (which is called the interrupt vector) into 0EFF/0EFE. These addresses relate to the 'EMMA'; in most other 6502 machines, the vectors may be found at FFFE and FFFF so you do need to check! The interrupt disable flag is now set (becomes a logic 1) so that no further interrupts can take place until the present one has been serviced.

The routine itself may enable an external sensor to supply data on, for example, the operating temperature of some device or system. If it's too high, the CPU might typically generate and output suitable control signals to reduce it. The routine is usually completed so quickly, you wouldn't even notice it; almost immediately, registers are being pulled off the stack and everything's back to normal. When the status register is pulled, it will contain a zero in the IDF, (otherwise, it wouldn't have been interrupted in the first place), restoring the status quo ready for another interrupt.

NMI is a non-maskable interrupt. This ignores the state of the IDF in the status register, i.e. it can't be 'masked' off, hence the name. The rest of the procedure is the same as for an ordinary IRQ except that the vectors are taken from 0EFC (low byte) and 0EFD (high byte). Again, in 6502 machines other than the EMMA, the vectors may be found elsewhere, typically at FFFA and FFFB.

RES indicates the reset pin on the CPU. Taking this from a logical 0 to 1 causes the CPU to begin the reset sequence. After a system initialisation period of six clock cycles, the interrupt disable flag is set and the CPU loads the PC with the reset vectors FEE0 (in the EMMA; it's probably different in other machines). The same effect can be achieved by using JMP FEE0 (4C E0 FE in machine code); this instruction causes the CPU to jump to the reset condition in the monitor program.

BRK is an instruction which halts the CPU. It's a software IRQ interrupt and the programmer can insert it at any suitable point in a program. It's most often used to halt CPU execution for debugging (see Chapter 11) but can have a variety of other uses too. Setting the IDF will have no effect on the BRK instruction, but execution of the BRK instruction causes the break command flag (BCF) to be set in the flag register. This enables the programmer to identify the source of an interrupt; if the BCF is at '1' it was a break, if it's at '0' then it was an interrupt.

The small program shown in Table 9.10 shows a method of doing this. The first four instructions use the stack to transfer the flag register contents to the accumulator, and then restore the original flag register contents.

Table 9.10 Short 6502 program showing how the stack may be used to examine the BRK bit in the flag register

Address	Machine code	Label	Mnemonics	Comments
&0100	06	.CHECK	PHP	;push flag register ;onto stack
&0101	68		PLA	;pull into accumulator
&0102	48		PHA	;push flag register ;back onto stack
&0103	28		PLP	;restore flag register
&0104	29 10		AND #&10	;test bit 4; is it '1'?
&0106	D0		BNE	;branch to break sub-routine ;rest of program

ANDing the accumulator with &10 has the effect of identifying whether bit 4 (the break command bit) is a 1 or not. For readers who may find this confusing, here are a few words of explanation. A logical AND operation returns a logic 1 result *only* if all the inputs are logic 1. In this case, there are only two inputs (as with all logic operations with the CPU accumulator) so if the AND instruction 'sees' two ones, then it will return a 1; otherwise it returns a zero. This is shown in Figure 9.3. It doesn't matter what the other bits in the flag register hold, because they're all ANDed with zero; so only if the bit we're interested in is a 1, will a 1 be returned. If it is, the BNE instruction (branch not equal – which actually means branch not zero – explained more fully in the next chapter) causes a branch to the BRK sub-routine.

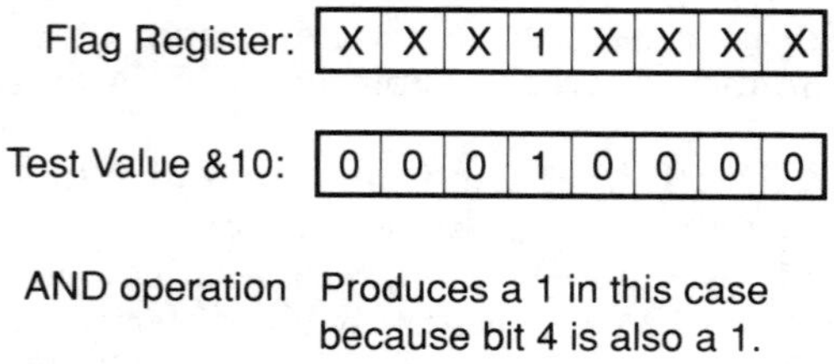

Flag Register: | X | X | X | 1 | X | X | X | X |

Test Value &10: | 0 | 0 | 0 | 1 | 0 | 0 | 0 | 0 |

AND operation Produces a 1 in this case because bit 4 is also a 1.

Figure 9.3 *A logical AND used to test the BRK command bit.*

It should be emphasised that PUSHes and PULLs must be nested and that in doing this the programmer needs to be aware that the CPU may also use the stack in subroutine calls and interrupts. There's more to say about interrupts, but we leave that until Chapter 12.

Multiple choice questions

1 In a micro system the stack is a:
   (a)   reserved area of memory ☐
   (b)   special register in the CPU ☐
   (c)   location in ROM ☐
   (d)   special EPROM ☐
2 The stack pointer (SP):
   (a)   tells the CPU where the stack is located ☐
   (b)   points to the last stack address used ☐
   (c)   points to the next stack address to be used ☐
   (d)   tells the control logic where the stack is located ☐
3 A typical memory size for a stack is:
   (a)   256 bytes ☐
   (b)   2560 bytes ☐
   (c)   25K ☐
   (d)   2.5Mb ☐
4 The kind of interrupt known as $\overline{\text{NMI}}$:
   (a)   cannot be software inhibited ☐
   (b)   does not affect the stack ☐
   (c)   has less priority than other interrupts ☐
   (d)   is the same as $\overline{\text{IRQ}}$ and $\overline{\text{INT}}$ ☐
5 The Z80 stack instructions are:
   (a)   push and pull ☐
   (b)   push and pop ☐
   (c)   store and save ☐
   (d)   pick and place ☐
6 Stack action is:
   (a)   always LIFO ☐
   (b)   sometimes LIFO ☐
   (c)   always FIFO ☐
   (d)   sometimes FIFO ☐
7 The 6502 stack instructions are:
   (a)   push and pull ☐
   (b)   push and pop ☐
   (c)   store and save ☐
   (d)   pick and place ☐
8 When stack action is initiated, the top of the stack is located by inspecting the:
   (a)   flag register ☐
   (b)   processor status register ☐
   (c)   program counter ☐
   (d)   stack pointer ☐
9 Data is loaded into a subroutine by a technique known as:
   (a)   register pointing ☐
   (b)   register addressing ☐
   (c)   parameter passing ☐
   (d)   parametric storing ☐
10 BRK (6502) or HALT (Z80) are examples of:
   (a) $\overline{\text{NMI}}$ ☐
   (b) $\overline{\text{IRQ}}$ or $\overline{\text{INT}}$ ☐
   (c)   RESETS ☐
   (d)   software interrupts ☐

Assignment 9.1

(a) With the aid of a suitable diagram, explain the meaning of the term 'STACK' and the method used by the CPU to organise the stack.

(b) Describe the part played by the stack and the CPU program counter during the execution of the line of a Z80 program shown below:

Address	Machine code	Mnemonic	Comment
5C20H	CD 00 5D	CALL STORE	;call subroutine STORE

(c) Describe the part played by the stack and the CPU program counter during the execution of the line of a 6502 program shown below:

Address	Machine code	Mnemonic	Comment
&0220	20 E0 02	JSR STORE	;jump to subroutine STORE

Chapter 10
The flag register

Meaning and use of flags

A flag may be described as a vivid, visual indication of the status of an object or situation. In a CPU it is useful to know the status of operations in the ALU, such as whether or not a previous operation produced a zero, non-zero, negative or positive result or if there were a carry or not. The *flag register* carries out this function. As such, it is also frequently described as the 'status register' or condition code register. In a Z80 CPU, the flag register is the register that stores the bits which keep track of the results of ALU operations within the CPU; to a large extent, the 6502 processor status register performs exactly the same function.

The Z80 flag register has 8 bits, but only 6 of them are used. They all function quite independently so their positions in the register are unimportant, except for the carry flag. We'll see later that it's convenient to have this bit in position 0. Each flag bit can be either set (a logic 1) or reset (a logic 0) and this frequently occurs after ALU operations.

Reference to the flag register enables program flow to be controlled. For example, if the result of a previous ALU operation were zero, execution could be diverted to a different part of the program. Examples of this action have been given already (in Chapter 5, for example) where it was pointed out that if the contents of the flag register are being used to determine a JUMP, CALL or BRANCH, etc. reference *must* be made to the previous ALU operation, i.e. the one that occurred *immediately* before the JUMP or BRANCH instruction.

Some ALU operations do not affect the flag register. In the Z80 instruction set, for example, the 8-bit loads LD A,I and LD A,R do not affect the flags. Nor do any of the 16-bit (register pair) increments or decrements.

Z80 and 6502 flag registers

Although the basic layout, operation and use of flag registers is the same for most processors, there are sufficient differences to justify dealing with the Z80 and 6502 separately. However, details which are common to both are not always duplicated here, so where the Z80 description covers the 6502 as well, readers are referred to it in the text.

Like many other registers in the CPU, the flag register is 8 bits wide. The Z80 flag register is illustrated in Figure 10.1 (the 6502 version is shown in Figure 10.6).

7	6	5	4	3	2	1	0

S Z - H - P/V N C

Figure 10.1 *The Z80 flag register.*

S = Sign flag
Z = Zero flag
H = Half carry flag (BCD ops)
P = Parity/overflow flag
N = Add/subtract flag (BCD ops)
C = Carry flag

The Z80 flag register

A brief description of the operation of each of the flags is given in Table 10.1.

Table 10.1 Functions and meanings of the bits in the Z80 flag register

Symbol	Flag	Description
S	Sign	Bit 7 of a result is copied into the sign flag. If signed arithmetic is being used, this indicates the sign of a result. (1 = neg, 0 = pos)
Z	Zero	Indicates whether or not the last instruction gave a result of zero. (1 = zero, 0 = not zero)
H	Half carry	Indicates when a carry out of the lowest 4 bits of a byte occurs (only useful in BCD arithmetic). (1 = half-carry, 0 = no half-carry)
P/V	Parity/overflow	See below for function.
N	Negative operation	Indicates whether the last instruction involved a subtraction operation or an addition operation (1 = sub, 0 = add) (only useful in BCD arithmetic).
C	Carry	Indicates whether the result of an arithmetic operation exceeds the storage capacity of the register holding the result (1 = carry, 0 = no carry resulted)

Function of the parity/overflow flag (P/V)

P – The flag is affected as a parity flag

For example, if an instruction gives a result byte with an even number of 1 bits, this is described as being even parity. If the number of 1 bits in the result is odd then this is described as being odd parity.

(1 = even parity, 0 = odd parity)

V – The flag is affected as an overflow flag

(This is only useful when performing arithmetic on *signed* numbers.)
For example, when two numbers of the *same* sign are *added* or *different* signs *subtracted*, the true result may be too large to hold in the number of bytes used for the result. The result will then appear to have the wrong sign (when considered as a signed number). This situation is called arithmetic overflow.

(1 = overflow occurred, 0 = no overflow occurred)

Flag	State	Meaning
S	0	The sign of the result is positive
	1	The sign of the result is negative
Z	0	The result is non-zero
	1	The result is zero
P/V	0	The parity of the result is odd; No overflow
	1	The parity of the result is even; Overflow
C	0	A carry or borrow did not occur
	1	A carry or borrow did occur

Figure 10.2 *Summary of the effect of the Z80 flags.*

H and N flags

Generally, we will not be concerned with these flags, so for convenience and clarity, the effect of the other flags is summarised in Figure 10.2.

Carry flag

This flag is set if an ALU operation produces a carry, i.e. if the result is greater than 255D. If 200D is added to 300D, for example, the result is 500D. This would leave 244D in the accumulator with a carry of 1 (representing 256 (2^8)); the carry flag is often regarded as the ninth bit of the accumulator. Figure 10.3 shows how this would look. Without this facility, the carry would be lost.

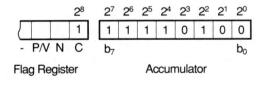

Flag Register   Accumulator

Figure 10.3 *Between the two registers, 500D is being stored: 244 in the accumulator and 256 in the carry flag. The binary weightings for each digit are shown above the appropriate bit for clarity.*

Example 1 (using the carry flag)
```
LD   A,25H
LD   C,16H
ADD  A,C
```

The addition of 25H and 16H produces 3BH (11 1011B) which will fit into the accumulator easily. The carry flag is therefore set to 0.

Example 2
```
LD   A,0AAH
LD   B,07BH
ADD  A,B
```

The addition of 0AAH and 07BH gives 0125H (1 0010 0101B). This will not fit into the accumulator, so after the program is executed, the A register will contain 25H and the carry flag will be set to '1'.

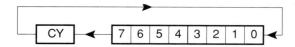

Figure 10.4 *The rotate left instruction. After execution of RLA, bit 7 is placed in the carry flag and the carry flag is placed into bit 0 of the accumulator.*

Example 3
There are several ways of utilising the carry flag; here are two.

The RLA instruction (see Figure 10.4)

This instruction rotates the contents of the accumulator to the left. The content of bit 7 of the accumulator is copied into the carry flag and the previous content of the carry flag is copied into bit 0. In this way, the accumulator contains the carry status of an ALU operation. For this to work, the accumulator must of course have been previously set to zero itself.

For example add 63H and F5H and put the result in the HL register pair. (The result should be: 99D + 25D = 344D which is 158H.)

```
LD      A,63H
LD      B,0F5H
ADD     A,B
LD      L,A
LD      A,00H
RLA
LD      H,A
HALT
```

After execution of the program, the HL register pair should contain 0158H. We have used two 8-bit registers to store a result which is larger than 8 bits.

Use of JUMP

Use one of the following JUMP instructions:

```
JP cc,nn
JP NC,nn   or   JR NC,LABEL
JP C,nn
```

using the same numbers as previously a suitable program may look like this:

```
        LD      A,63H
        LD      B,0F5H
        ADD     A,B
        JR      NC,ZERO
        LD      L,A
        LD      H,01H
        JR      NEXT
ZERO:   LD      L,A
        LD      H,00H
NEXT:   LD      A,45H
        etc.
```

Obviously, many other methods of utilising this may be devised by the programmer. Try Examples 4 and 5 (below) for yourself:

Example 4

Using the RLA instruction, add 0C8H and 0E1H and put the result in the HL register pair. (HL should contain 01A9H after execution.)

Example 5

Using a JUMP instruction, add 91H to 0A9H putting the result in the HL register pair. (HL should contain 013AH after execution.)

Zero flag

The zero flag is set whenever the result of a previous ALU operation produces all zeros in the accumulator. If the result were non-zero (any number but zero) then this flag is reset (to a logic 0).

Example 6

```
        LD    B,10H
LP1:    INC   HL
        LD    (HL),A
        DEC   B
        JR    NZ,LP1
        HALT
```

When the program is executed the B register contains 16 (decimal); on the first pass, this drops to 0FH (15D), a non-zero number, which sets the zero flag to '0', diverting program flow back to line 2, INC HL. The loop continues until the contents of the B register have decremented to zero. At this point, the zero flag is set to '1', program execution is no longer diverted and execution ceases on 'HALT'. For this to happen, every time the CPU comes across JR NZ,LP1 it 'looks' at the Z bit in the flag register to find out what it should do, then acts upon it.

The compare (CP) instruction

This instruction affects both carry and zero flags. When a compare instruction is used, a number (using immediate, register, indirect or indexed addressing) is subtracted from the contents of the accumulator. The great difference between CP and SUB is that the result is *not* placed in the accumulator, but the appropriate flags (zero and carry) are affected.

Figure 10.5 shows how the flags are affected for the three possibilities, using a compare instruction.

Possibility	Z	C
Accumulator < test value	0	1
Accumulator = test value	1	0
Accumulator > test value	0	0

Figure 10.5 *The effect on the zero and carry flags when a compare instruction is used.*

As an example let's compare accumulator contents with B register contents as follows:

```
        CP   B
        JP   Z,EQUAL   ;A=B  goto  EQUAL
        JP   C,LESS    ;A<B  goto  LESS
        JP   MORE      ;A>B  goto  MORE
EQUAL:
           .

LESS:
           .

MORE:
```

The JP Z,EQUAL instruction tests for equality; if the zero flag is set, then the accumulator and B register contents are equal. If A is less than B, the carry flag is set, and 'JP C,LESS' will detect this. If neither condition is true, then accumulator contents must be greater than B and program execution will jump to label 'MORE' or can continue as normal as program requirements dictate. In either case, it's not necessary to look at the flag register again.

If only the A>B condition is required, then a JP NC,MORE could be used on its own. The reader can easily use these instructions in a variety of programming requirements and test them with simple values.

A real-world application for the short program given might be some sort of temperature control. A digital value representing a controlled temperature may be placed in the accumulator (using IN A,(n)), whilst the required temperature value may be held in the B register. If the external temperature falls below the standard, a heater may be turned on; if it's greater, the heater's turned off and if the temperature is just right, nothing occurs.

The rest of the flags

At this level, and some way beyond, we'll only really use the zero and carry flags. A brief description of the others now follows; you'll probably not need to use them.

The sign flag – S

When two's complement arithmetic is used, bit 7 is the sign bit (see Chapter 2). It is this bit which is copied into the S flag consequent upon a two's complement ALU operation; 0 for a positive result, 1 for a negative result.

The half carry flag – H

We've already indicated that this flag is only useful in BCD arithmetic. It won't normally be used by a programmer since it cannot be utilised in a jump instruction although the DAA (decimal adjust accumulator) instruction makes use of it. BCD numbers were discussed in Chapter 2 where it was noted that 1001 is the highest, group-of-four binaries which is legal (1010 = 10D and only single digit decimal values (0 to 9) are allowed).

Consider a register as consisting of two halves (two nibbles). The right-hand nibble contains the lsb, the left-hand nibble the msb. Each nibble can contain a BCD number. As an example, consider this addition:

```
0001 1001      (19)
0001 1001 +    (19)
-------------------
  11 0010      (38)
```

The half carry flag would be set because the addition of bit 3 causes a carry from the lower half to the upper half. The BCD number 0001 1001 is equal to 19D. The addition should therefore produce 38D but it doesn't. It appears to be 32; the DAA instruction adjusts this result so that the correct BCD representation is obtained. (11001 in decimal is actually 25 and the result 110010 (50D) would be correct.)

The parity/overflow flag P/V

The parity flag is set consequent upon a *logical* ALU operation and the overflow flag following an *arithmetic* ALU operation. Parity is the term used when the number of logic ones in a binary value is even; if it is, the parity flag is set. Hence, 1100 1100 would set the parity flag to a '1' (four logic ones are even) whereas 1110 1100 would not (five logic ones are odd).

The overflow flag is set or reset according to the value remaining in bit 7 of the accumulator after an arithmetic ALU operation. It operates assuming two's complement arithmetic which only allows values in the range +127 to −128.

Hence, 45D + 63D would result in the overflow flag being cleared (a logic 0); the decimal addition 85 + 92, however, would set the overflow flag to a logic 1, since the sum exceeds those values allowed in two's complement arithmetic. In normal programming, however, these flags look after themselves and are of no concern to the programmer.

The add/subtract flag, N

This flag differs from all the others in that it is set or reset according to the type of instruction last executed by the ALU and not its result. For example, an ADD or INC instruction would cause the N flag to be reset to a logic 0; SUB and DEC cause it to be set to a logic 1. Like the H flag, N is used by the DAA instruction to adjust for decimal arithmetic, but it is rarely of any concern to the programmer.

The 6502 processor status register

Although we are dealing with the 6502 'flag register' separately, there are some similarities to the Z80 functions and use and we refer the reader to those relevant sections. The function of each bit (except bit 5 which isn't used) is shown in Figure 10.6.

A brief description of each flag is given in Table 10.2. The carry (C), zero (Z) and overflow (V) flags operate in exactly the same way as their Z80 equivalents. The interrupt disable (I) and break (B) flags were described in the last

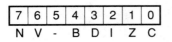

7	6	5	4	3	2	1	0
N	V	-	B	D	I	Z	C

Figure 10.6 *The 6502 processor status register (P).*

C = Carry
Z = Zero
I = Interrupt disable
D = Decimal mode
B = Break
V = Overflow
N = Negative

Table 10.2 Functions and meanings of the bits in the 6502 processor status register (P)

Symbol	Flag	Description
N	Neg	Bit 7 of a result is copied into the negative flag. If signed arithmetic is being used, this indicates the sign of a result. (1 = neg, 0 = pos)
V	Overflow	Indicates overflow in two's complement arithmetic. (1 = overflow occurred, 0 = overflow did not occur)
B	Break	This flag is set when the processor meets a BRK instruction. (1 = break, 0 = no break)
D	Decimal	Processor performs decimal arithmetic (as opposed to binary) when set. (1 = decimal, 0 = binary arithmetic)
I	Interrupt disable	Indicates IRQ interrupts disabled when set. (1 = disabled, 0 = enabled)
Z	Zero	Indicates whether or not the last instruction gave a result of zero. (1 = zero, 0 = not zero)
C	Carry	Indicates whether the result of an arithmetic operation exceeds the storage capacity of the register holding the result. (1 = carry, 0 = no carry resulted)

chapter (the stack). The negative (N) flag operates in the same way as the Z80 sign (S) flag.

As with the Z80, the most important flags (from the point of view of their being utilised by programmers at this level) are the carry (C) and zero (Z) flags. To clarify their operation, some small programs follow:

Example 7 (carry flag)

(a)	LDA #&25	(b)	LDA #&AA
	CLC		CLC
	ADC #&16		ADC #&7B

(a) &25 + &16 = 3BH (0011 1011) which is less than &FF (1111 1111) and the carry flag is not set
(b) &AA + &7B = &125. This is greater than &FF so the carry flag is set.

    1 0010 0101

The msb is held in the carry flag, the rest of the digits are in the accumulator; this is similar to the representation shown in Figure 10.4.

Example 8

```
0250    A9    B6                      LDA   #&B6
0252    18                            CLC
0253    69    A7                      ADC   #&A7
0255    85    30                      STA   &30
0257    B0    07                      BCS   CARRY   ; (+7)
0259    A9    00                      LDA   #0
025B    85    31                      STA   &31
025D    4C    64    02                JMP   &0264
0260    A9    01          .CARRY      LDA   #01
0262    85    31                      STA   &31
0264    00                            BRK
```

The program adds &B6 to &A7 producing &15D (349D). This obviously exceeds accumulator capacity. Loading and running this program will place 5D into memory location &30 and a carry of 1 into memory location &31. If the value at 0251 (B6) is changed to 11, then, after program execution, &30 will contain the sum (&B8) and &31 will contain zero (no carry). Note that the '(+7)' which appears after the 'BCS CARRY' instruction does not form part of the program. It's a comment to remind us that the required displacement is 7 forward. Remember that we're counting from 025A (00) because the PC has incremented by now, and we arrive at 0260 (A9) on the count of seven.

The program uses the BCS instruction to check for a carry and branch to a new address if there is one. Another method of checking for a carry following an ALU operation is to use the ROL A instruction as shown in Example 3.

Example 9

```
0250   A9  B6      LDA    #&B6
0252   18          CLC
0253   69  A7      ADC    #&A7
0255   85  30      STA    &30
0257   A9  00      LDA    #0
0259   2A          ROL    A
025A   85  31      STA    &31
025C   00          BRK
```

The ROL A instruction rotates the contents of the accumulator to the left. The content of bit 7 of the accumulator is copied into the carry flag and the previous content of the carry flag is copied into bit 0. In this way, the accumulator contains the carry status of an ALU operation. For this to work, the accumulator must of course have been previously set to zero itself.

The reader should load and run these programs using different values and observe the results.

Example 10 (zero flag)

Now let's examine the use of the zero flag in order to determine program branches. Refer to the following program:

```
0200   A9  00            LDA    #&0
0202   A2  10            LDX    #&10
0204   A0  00            LDY    #&0
0206   85  30            STA    &30
0208   C8        .LOOP   INY
0209   95  30            STA    &30,Y
020B   CA                DEX
020C   D0  FA            BNE    LOOP
020E   00                BRK
```

This program clears a section of zero-page memory.

The X register acts as the loop counter. When the program is executed the X register contains 16 (decimal); on the first pass, this drops to 0FH (15D), a non-zero number, which sets the zero flag to '0', diverting program flow back to the label 'LOOP'. The loop continues until the contents of the X register have decremented to zero. At this point, the zero flag is set to '1', program execution is no longer diverted and execution ceases on 'BRK'. For this to happen, every time the CPU comes across BNE LOOP, it 'looks' at the Z bit in the flag register to find out what it should do, then acts upon it.

Possibility	N	Z	C
Accumulator < test value	1	0	0
Accumulator = test value	0	1	1
Accumulator > test value	0	0	1

Figure 10.7 *The effect of CMP on the N, Z and C flags.*

The compare (CMP) instruction

This instruction affects the zero, carry and negative flags. When a compare instruction is used, a number is subtracted from the contents of the accumulator. The great difference between CMP and SBC is that the result is *not* placed in the accumulator, but the appropriate flags (zero, carry and negative) are affected. Figure 10.7 shows how the flags are affected for the three possibilities using a compare instruction.

Example 11

As an example, let's compare the contents of a zero-page memory location (&30) with the contents of the accumulator. If we want a branch *only* if the values are equal, then we'd use BEQ; a branch only if the accumulator holds a value *greater* than the test value would use BCS and where the accumulator holds a value *less* than the test value, we'd use BMI. To test for all the possibilities and branch accordingly, the following program may be used:

```
        CMP  &30
        BEQ  EQUAL    ;A = memory, goto EQUAL
        BMI  LESS     ;A < memory, goto LESS
        JMP  MORE     ;A > memory, goto MORE
              .    .
.EQUAL
              .    .
.LESS
              .    .
.MORE
              .    .
```

'JMP MORE' could be replaced by 'BCS MORE' although it's not necessary in this case. The relevant instructions are listed in Table 10.3.

Table 10.3 Branch instructions which can be used in conjunction with the CMP instruction

Instruction mnemonic	Operation	Meaning	Machine code
BCC	Branch on C=0	Branch if there's no carry	90
BCS	Branch on C=1	Branch if there is a carry	B0
BNE	Branch on Z=0	Branch if result is non-zero	D0
BEQ	Branch on Z=1	Branch if result is zero	F0
BPL	Branch on N=0	Branch if result is positive	10
BMI	Branch on N=1	Branch if result is negative	30

Questions

1 (a) Draw a diagram of the Z80 flag register indicating the positions of each flag.
   (b) Draw a diagram of the 6502 processor status register indicating the positions of each flag.
2 (a) Draw up a table which gives a brief description of the Z80 flags.
   (b) Draw up a table which gives a brief description of the 6502 flags.
3 (a) Which of the Z80 flags is set or reset according to the type of instruction last executed by the ALU and not its result?
   (b) Which of the 6502 flags is set or reset consequent upon executing a software interrupt instruction?
4 In relation to the accumulator, what is the carry flag sometimes referred to as?
5 If the result of a previous ALU operation is zero, is the zero flag set to a zero?
6 (Z80) Write a program which will add together the numbers 91H and A9H putting the result in the HL register pair. Use (a) the RLA instruction and (b) a JUMP instruction.
7 (6502) Write a program which will add together the numbers &84 and &D3 putting the result in consecutive zero-page memory locations. Use (a) the ROL A instruction and (b) a BRANCH instruction.
8 Using the CP (Z80) or CMP (6502) instruction, write a program which will sort and store numbers in different memory locations, according to whether they are equal, less than or greater than 10D.
9 Make a list of the instructions which operate according to the setting of the carry flag.
10 List the flags which operate identically in both the Z80 and 6502 flag registers.

Multiple choice questions

1 A flag is said to be set when:
   (a) power is applied   ☐
   (b) power is removed   ☐
   (c) it goes to a logic 1   ☐
   (d) it goes to a logic 0   ☐
2 Even parity occurs in a byte when:
   (a) it contains all ones   ☐
   (b) it contains all zeros   ☐
   (c) there is an odd number of zeros   ☐
   (d) there is an even number of zeros   ☐
3 The Z80 sign flag is set when the previous ALU operation:
   (a) produced a positive result   ☐
   (b) produced a negative result   ☐
   (c) involved addition   ☐
   (d) involved subtraction   ☐

4 The carry flag is set:
  (a) only when BCD arithmetic is being used □
  (b) only when signed arithmetic is being used □
  (c) when the storage capacity of the accumulator is
      exceeded □
  (d) when bit 7 of the accumulator becomes a 1 □

5 The zero flag becomes a one when the last ALU operation
  produced a:
  (a) zero □
  (b) one □
  (c) a value > zero □
  (d) a value < zero □

6 The decimal addition 159 + 297 sets the carry flag. The decimal
  value left in the accumulator is:
  (a) 99 □
  (b) 100 □
  (c) 199 □
  (d) 200 □

7 The hexadecimal addition A7 + B2 sets the carry flag. The binary
  value left in the accumulator is:
  (a) 0110 0001 □
  (b) 0101 1001 □
  (c) 0110 1001 □
  (d) 0101 0001 □

8 The addition which causes the overflow flag to be set is:
  (a) 7EH + 01 □
  (b) 7EH + 02 □
  (c) 80H + 01 □
  (d) 80H + 02 □

9 The Z80 half carry flag (H) is utilised by the DAA instruction to:
  (a) adjust ALU results to correct BCD representation □
  (b) adjust ALU results to correct decimal representation □
  (c) adjust ALU results to correct hexadecimal
      representation □
  (d) adjust ALU results to correct binary representation □

10 When a Z80 CP instruction is executed, the flags affected are:
  (a) zero and sign □
  (b) zero and carry □
  (c) carry and half carry □
  (d) all of them □

11 The Z80 flag which is set or reset according to the *type* of
   instruction last executed by the ALU and not its result is:
  (a) sign (S) flag □
  (b) zero (Z) flag □
  (c) parity/overflow (P/V) flag □
  (d) add/subtract (N) flag □

12 The number of flags in the 6502 processor status register is:
  (a) 5 □
  (b) 6 □
  (c) 7 □
  (d) 8 □

13 When the 6502 negative (N) flag is set, the sign of the result is:
   (a)   negative ☐
   (b)   positive ☐
   (c)   zero ☐
   (d)   indeterminate ☐

14 The carry flag is set following the addition of the hexadecimal numbers:
   (a)   A9 and 52 ☐
   (b)   AA and 57 ☐
   (c)   52 and A2 ☐
   (d)   57 and A3 ☐

15 The 6502 instruction which clears the carry flag is:
   (a)   CCF ☐
   (b)   CLL ☐
   (c)   CLC ☐
   (d)   CLS ☐

16 Which one of the following addition operations would cause the carry flag to be set?
   (a)   7EH + 01 ☐
   (b)   7EH + 02 ☐
   (c)   FEH + 01 ☐
   (d)   FEH + 02 ☐

17 Which of the additions below would cause the zero flag to be set?
   (a)   &FE + 01 ☐
   (b)   &FE + 02 ☐
   (c)   &FE + 03 ☐
   (d)   &FE + 04 ☐

18 When the instruction JR M,label (Z80) is executed, program flow will be diverted if:
   (a)   the sign flag is set ☐
   (b)   the sign flag is reset ☐
   (c)   the parity flag is set ☐
   (d)   the parity flag is reset ☐

19 The only invalid Z80 JUMP RELATIVE instruction below is:
   (a)   JR P,label ☐
   (b)   JR PB,label ☐
   (c)   JR PE,label ☐
   (d)   JR P0,label ☐

20 The only invalid 6502 BRANCH instruction given below is:
   (a)   BCS ☐
   (b)   BPL ☐
   (c)   BMI ☐
   (d)   BWW ☐

Assignment 10.1 – the flag register

(Z80) Write a program which displays the flag register on the screen, showing each bit correctly set or reset after execution of the following program:

```
LD    B,1
LD    A,0BBH
ADD   A,0AAH
ADC   22H
DEC   B
```

A typical display (with all the bits set to zero) is shown in Figure 10.8.

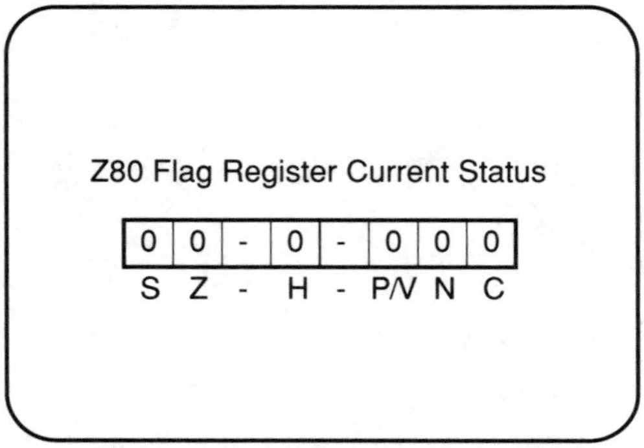

Z80 Flag Register Current Status

```
| 0 | 0 | - | 0 | - | 0 | 0 | 0 |
  S   Z   -   H   -  P/V  N   C
```

Figure 10.8 *Display for a possible solution to Assignment 10.1.*

Assignment 10.2 – arithmetic, logic and test using flags

(Z80) Using Z80 mnemonics and identifying the addressing modes used, produce a flow chart, write and run a program to:

(a) put the numbers 80H and BAH into memory locations 5500H and 5501H respectively;
(b) add the content of memory location 5500H to 5501H returning the result to memory locations 5502H and 5503H (carry);
(c) Load the B register with the number D2H and the C register with 2DH;
(d) Logical AND the content of the B register with that of the C register storing the result in the D register. If the result is zero, the E register should be set to 01, if not, the E register should be set to 00.

Chapter 11
Debugging programs

In line with the rest of this book, we consider the Z80 CPU first and look at some basic principles. Most of the information is applicable to all processors, only the code differs really, and we discuss the 6502 in some detail towards the end of the chapter.

Introduction

The secret of writing successful machine code programs lies in using a methodical approach: algorithm, JSP/flow chart, source code, object code; in a word *documentation*. Experience, as in all things, is also a valuable contributor and as experience increases, software problems tend to decrease (in theory anyway!). Nevertheless, problems can and do occur, with infuriating regularity at first; the Americans named these problems 'bugs', hence the need for 'debugging'. Generally, program bugs are of two types; a basic misconception in programming or a careless error made whilst typing in the code. It goes without saying that the computer only does what you tell it to do, not what you *think* you've told it to do.

The Z80 system

Fortunately, many software aids to debugging exist. For example, when using the assembler in the MARC system, errors are highlighted as the code is listed on the screen and the number of errors is given at the end. One of the most common mistakes made lies in carelessness over hex numbers. The hex number 5000 *must* have a trailing 'H' otherwise the computer will think you mean a decimal number. Such a bug may not be highlighted in a listing; but it will obviously affect the way your program will run. The correct format of 5000H must always be used.

The hex number E02A must have the trailing 'H' but also a leading zero; this is to distinguish the number from a label. Hence, the correct format is 0E02AH.

When creating program loops, using 10H as a loop counter will produce 16 passes – not 10. Quite obvious when you think about it, but I've seen this error a 1000 times (that's 1000 decimal of course!), and indeed the use of the letter O instead of the number 0.

Program loops themselves also cause problems owing to the careless use of labels. It is wise to list the labels you use *on paper* for easy reference. Watch out for the repeated use of the label 'LOOP'; if you want to 'JR NZ,LOOP1' then you must say so. The obvious danger is that 'JR NZ,LOOP' is used instead so that the program flow is diverted to a different subroutine.

Another problem may occur if you loop back to the wrong part of the program. For example, in Chapter 5 we described a program which would clear a VDU screen by putting ASCII blanks into every video RAM location (see Example 5, p. 78). If the unconditional jump (or relative jump) caused program execution to loop back to line 5C08H (instead of the correct 5C06H), the screen would be filled with graphics characters and not be blanked out at all. You may care to try this out to see the effect.

Using the process of single-stepping (which we describe later) careful analysis of the program would reveal that the accumulator at line 5C08H does not contain 20H (the code for an ASCII blank) which should then lead you to the cause of the problem.

Sometimes, single-stepping through loops can be a tedious process because of their great length. In the 'CLS' example just quoted, the bug would be discovered on the first pass, but don't be tempted to single-step through some MARC user routines – unless you've got a lot of time on your hands!

Use of the 'HALT' instruction

In this case (and many similar ones), it is useful to insert a 'HALT' statement after a certain section of code, enabling that section to be checked independently. Faulty program segments can often easily be identified using this technique.

This discussion could be continued for some time and you will doubtless have come across many similar examples already. For really stubborn program bugs, you may have to resort to writing out the code for yourself, working out what *should* happen and then comparing it with the program as it is being executed to see what *is* happening. Two aids to this process exist: trace tables and single-stepping and a discussion of this now follows.

Trace tables

A trace table shows the contents of relevant registers in the CPU and specific memory locations (where necessary) for each step in a program. Table 11.1 shows a suitable layout. When using such a table, you need only include the registers and memory locations (where applicable) used in your program. Consider the following program which uses the registers A, B, C and D only:

Example 1

```
LD      A,0F0H
LD      B,75H
LD      C,11H
SUB     C
LD      D,A
AND     B
LD      C,A
OR      92H
HALT
```

Obvious causes of error lie in the omission of the 'H' in any of the numbers. Omission of the H in 0F0H may throw up an error message in automatic assembly, but putting 75, 11 and 92 for 75H, 11H and 92H would not; program execution would continue with the CPU assuming the numbers were in decimal.

The trace table is built up by *working out* by hand what should be in the registers after each line of program execution then comparing this with what

Table 11.1 Typical trace table format. You only need to list the registers used in your program. Where memory locations are needed, insert the appropriate address; 5D00H is shown as an example

| Line number | Registers | | | | | | | Memory locations |
	A	B	C	D	E	H	L	5D00H
0	00	00	00	00	00	00	00	
1								
2								
3								
4								
5								
6								
.								
.								

actually happens by stepping through the program one line at a time. Many micros allow you to do this, the MARC monitor system, for example, provides the single-stepping facility.

The trace table for Example 1 is shown in Table 11.2, but before looking at it, try and work out its contents for yourself and then compare it with what is given.

Single-stepping

Brief instructions for using the single-step facility in the MARC system now follow. The 6502 'EMMA' comes later. For other systems, the reader is referred to the instruction manual for that system.

First, enter your program and then press ESC at the end. Press CTRL Z. This displays the CPU registers A, B, C, D, E, H, L, IX, IY, the flag register (F), the stack pointer (SP) and the program counter (PC). Note that the PC points to the next memory location, *not* the current one.

These registers can then be edited by positioning the cursor (SPACE BAR = cursor right; DELETE = cursor left) over the data to be changed and enter-

Table 11.2 A typical trace table format, using Example 1

| Line number | Register | | | |
	A	B	C	D
	00	00	00	00
1	F0	00	00	00
2	F0	75	00	00
3	F0	75	11	00
4	DF	75	11	00
5	DF	75	11	DF
6	55	75	11	DF
7	55	75	55	DF
8	D7	75	55	DF
9	D7	75	55	DF

ing the new data. Normally, the registers will be set to zero with the exception of the stack pointer which will usually be set to 4300H at the beginning. The PC is then edited to the start address of the program to be debugged. Any of the other displayed registers can be edited if preset start conditions are required.

Press ESC (or RETURN) to exit edit mode. Press CTRL N. This executes the first instruction and shows the contents of the registers. Repeatedly pressing CTRL N single-steps through the entire program. This procedure can be tried out using Example 1 and the trace table given.

Example 2
Now try drawing up a trace table for the following program:

```
5C00    LD      B,16
5C02    LD      C,23
5C04    LD      A,B
5C05    ADD     A,C
5C06    LD      D,A
5C07    HALT
```

Notes: The on-screen display shows the PC pointing to the next memory location, *not* the present one. As a further exercise you could try including the contents of the flag register, using the format shown in Table 11.3.

Notes on Example 1

Table 11.4 gives program comments for information, starting with a listing of the object code.

Notes on Example 2

In this example a format similar to the one seen on the screen of the MARC micro is used. (It doesn't much matter which format you use as long as it's clear and accurate.) You will note from Table 11.5 that the registers are listed in pairs. Note also that the system always displays hexadecimal values. Since hexadecimal numbers were not specified in the program (16 not 16H), decimal values are assumed. In the on-screen trace table, 16(D) has been converted to 10(H). Note again that the PC always points to the address of the *next* instruction to be executed. So when the result of the instruction LD B,16 is being recorded, the PC has incremented to 5C02H.

Example 3
Work out the trace table for the following example before consulting the solution given in Table 11.6.

```
5C00    LD      A,62H
5C02    LD      B,33H
5C04    LD      C,B
5C05    ADD     A,C
5C06    ADD     A,C
5C07    ADD     A,C
5C08    LD      D,A
5C09    HALT
```

Table 11.3 Slightly different format resembling the on-screen display of the MARC which includes the addresses (rather than line numbers)

Address	AF	BC	DE	HL
5C00H	0000	0000	0000	0000
5C02H	0000	1000	0000	0000
5C04H				
.				etc.
.				

Table 11.4 Object code and program notes on Example 1

5C00	3E	F0	06	75	0E	11	91	57
5C08	A0	4F	F6	92	76	**	**	**

LD	A,0F0H	;F0H = 240D	
LD	B,75H	;75H = 117D	
LD	C,11H	;11H = 17D	
SUB	C	;F0H − 11H which is 240 − 17 = 223D or 0DFH	
LD	D,A		
AND	B	;AND 75H with 0DFH	1101 1111
			0111 0101
	;	giving:	0101 0101
	;	which is:	85D or 55H
LD	C,A	;55H goes into C register	
OR	92H	;OR 92H with 55H	0101 0101
			1001 0010
	;	giving:	1101 0111
	;	which is:	215D or 0D7H

Table 11.5 Trace table for Example 2. The flag register contents are also included

Address	AF	BC	DE	HL
5C00H	0000	0000	0000	0000
5C02H	0000	1000	0000	0000
5C04H	0000	1017	0000	0000
5C05H	1000	1017	0000	0000
5C06H	2720	1017	0000	0000
5C07H	2720	1017	2700	0000

Table 11.6 Trace table for Example 3

	Registers			
Line	A	B	C	D
	00	00	00	00
1	62	00	00	00
2	62	33	00	00
3	62	33	33	00
4	95	33	33	00
5	C8	33	33	00
6	FB	33	33	00
7	FB	33	33	FB

Example 4

We now present a program containing a 'JR NZ,LOOP' instruction. The associated trace table demonstrates the operation of the loop:

```
5C00            LD    A,22H
5C02            LD    B,22H
5C04            LD    C,03H
5C06   LOOP:    INC   A
5C07            DEC   B
5C08            DEC   C
5C09            JR    NZ,LOOP
5C0B            INC   A
5C0C            INC   B
5C0D            INC   C
```

The trace table is given in Table 11.7.

Table 11.7 Trace table for Example 4

Address	Registers		
	A	B	C
5C00	00	00	00
5C02	22	00	00
5C04	22	22	00
5C06	22	22	03
5C07	23	22	03
5C08	23	21	03
5C09	23	21	02
5C06	23	21	02
5C07	24	21	02
5C08	24	20	02
5C09	24	20	01
5C06	24	20	01
5C07	25	20	01
5C08	25	1F	01
5C09	25	1F	00
5C0B	25	1F	00
5C0C	26	1F	00
5C0D	26	20	00
5C0E	26	20	01
5C0F	26	20	01

Example 5

Finally, in this section, we present the program listed in Table 11.8. This is useful for demonstrating trace table operation and use; using 'HALT'; checking a section of RAM and fulfilling the requirements of C&G 224 Part II Electronics Systems (Control Systems) microelectronics assignment. Other readers should also find the program instructive.

Note: only the section of code from 5C13H (START:) need be entered in the assignment book. Only this part of the program need be single-stepped through, and only the resulting changes in register contents for this part of the program need be entered.

The first part of the program is a loop which is designed simply to put '11' in the appropriate RAM locations. Insert a 'HALT' statement after this section

Table 11.8 Suitable (Z80) C&G 224 assignment program

Address	Label	Assembler instructions	Machine code	Comments
		ORG 5C00H		
		ENT 5C00H		
				;RAMTEST
5C00		LD HL,5E00H	21 00 5E	;start of RAM
5C03		LD A,11H	3E 11	;
5C05		LD B,0AH	06 0A	;counter
5C07	LOOP:	LD (HL),A	77	;
5C08		INC HL	23	;next location
5C09		DEC B	05	;dec counter
5C0A		JR NZ,LOOP	20 FB	;
				;
5C0C		LD HL,5E00H	21 00 5E	;reload HL
5C0F		LD B,05H	06 05	;counter
5C11		LD A,33H	3E 33	;test number
5C13	START:	LD HL),A	77	;put in RAM
5C14		INC A	3C	;inc test number
5C15		INC HL	23	;next location
5C16		DEC B	05	;dec counter
5C17		JR NZ,START	20 FA	;relative jump
5C19		PUSH AF	F5	;save acc conts
5C1A		LD A,99H	3E 99	;
5C1C		LD (HL),A	77	;
5C1D		POP AF	F1	;
5C1E		LD B,4	06 04	;new counter
5C20		INC HL	23	;
5C21	AGAIN:	LD (HL),A	77	;
5C22		INC A	3C	;
5C23		INC HL	23	;
5C24		DEC B	05	;
5C25		JR NZ,AGAIN	20 FA	;
5C27		HALT	76	;

(at address 5C0CH) and then run it. If this proves satisfactory, remove the 'HALT', input the rest of the program code and run the whole program.

Finally, single-step through the program from 5C0CH and note down the RAM contents in the appropriate section of the assignment book.

If all the code has been entered, a HALT placed at address 5C0CH would enable a programmer to run only the first part of the program and check that all the appropriate locations contain 11H before proceeding with the rest of the program.

It is a simple matter for the reader to write his or her own programs, draw up trace tables and check them with single-stepping.

6502 code and the EMMA

Much of the first part of this chapter is relevant to the 6502 CPU so the reader is urged to read through it. In summary, we note that although good documentation will not guarantee a trouble-free program which works perfectly first time, bad documentation inevitably results in bad programs that do *not* run perfectly first time. A well-documented program, if it does not perform as expected, is much easier to debug than a badly documented program.

Hexadecimal numbers must always be identified; in the 6502 system we have seen that this is by the use of the prefix '&' (the ampersand – a common abbreviation for 'and' in normal English) so that 64 in hex is shown as &64 (decimal 100). This is important when creating program loops. Using &64 as a loop counter produces 100 passes – not 64. Quite obvious when you think about it, but it is a common error in programming.

For really stubborn program bugs, two aids to debugging exist: trace tables and single-stepping, and we discuss these now with particular reference to the 6502 and the EMMA micro.

Trace tables

A trace table shows the contents of relevant registers in the CPU and specific memory locations (where necessary) for each step in the program. The idea is to write down the coded program line by line in the table, working out what should be in the relevant registers and memory locations.

Then you step through the program one line at a time and examine each location. In this way, program bugs can usually be detected. As an example, consider the following program:

Example 1

```
0200   A9  F0    LDA  #&F0
0202   A2  75    LDX  #&75
0204   A0  11    LDY  #&11
0206   86  35    STX  &35
0208   84  36    STY  &36
020A   E5  35    SBC  &35
020C   85  37    STA  &37
020E   C6  36    DEC  &36
0210   E6  37    INC  &37
0212   00        BRK
```

The trace table is built up by *working out* by hand what should be in the registers and memory locations after each line of program execution and comparing this with what actually happens by stepping through the program one line at a time. Many micros allow you to do this; the EMMA, for example, provides the single-stepping facility.

Table 11.9 Trace table for Example 1

Line	Registers			Memory locations		
	A	X	Y	&35	&36	&37
	00	00	00	00	00	00
1	F0	00	00	00	00	00
2	F0	75	00	00	00	00
3	F0	75	11	00	00	00
4	F0	75	11	75	00	00
5	F0	75	11	75	11	00
6	7B	75	11	75	11	00
7	7B	75	11	75	11	7B
8	7B	75	11	75	10	7B
9	7B	75	11	75	10	7C

The trace table for Example 1 is shown in Table 11.9. Glance at it to see the format if you like, but try and work out its contents for yourself before studying the solution.

Single-stepping

Full details of single stepping on the EMMA appear in the manual. It is sufficient to give only a brief description here. First of all, consider Example 2 which follows:

Example 2

```
0200    A9  30      LDA  #48
0202    AA          TAX
0203    E8          INX
0204    A9  58      LDA  #88
0206    A8          TAY
0207    C8          INY
0208    48          PHA
0209    A9  57      LDA  #&57
020B    85  30      STA  &30
020D    68          PLA
020E    85  31      STA  &31
0210    98          TYA
0211    85  32      STA  &32
0213    00          BRK
```

First of all, set memory locations &30, &31 and &32 to zero. With the program loaded, put the single step switch to ON. Press R and you should see the 'GO' address on the display (if not, enter 0200). Pressing R again should produce the number: 30 02 00 25. (Don't worry if it doesn't – but read on.)

The first two digits show the accumulator contents (30), followed by X and Y contents (02 and 00). The final two digits represent the contents of the flag register. This is shown more clearly in Figure 11.1.

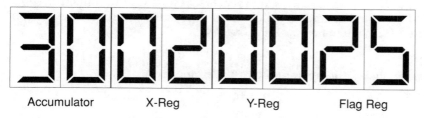

| Accumulator | X-Reg | Y-Reg | Flag Reg |

Figure 11.1 *EMMA display showing register contents during single-stepping.*

PC contents   Stack pointer

Figure 11.2 *Display showing PC and SP contents.*

Table 11.10   Trace table for Example 2

	Registers			Memory locations			
Address	A	X	Y	&30	&31	&32	SP
0200	00	00	00	00	00	00	01FC
0202	30	00	00	00	00	00	01FC
0203	30	30	00	00	00	00	01FC
0204	30	31	00	00	00	00	01FC
0206	58	31	00	00	00	00	01FC
0207	58	31	58	00	00	00	01FC
0208	58	31	59	00	00	00	01FC
0209	58	31	59	00	00	00	01FB
020B	57	31	59	00	00	00	01FB
020D	57	31	59	57	00	00	01FB
020E	58	31	59	57	00	00	01FC
0210	58	31	59	57	58	00	01FC
0211	59	31	59	57	58	00	01FC
0213	59	31	59	57	58	59	01FC

The X and Y registers could contain any number – it depends upon what they were last used for, so don't be concerned if your display doesn't show 02 and 00. It *must*, however, show &30 in the accumulator. Pressing the R button again will give a display like that shown in Figure 11.2. The PC contents are on the left, the SP is on the right.

Repeatedly pressing R takes you through the single-step sequence, displaying register contents and then PC/SP contents alternately, as described. In this program, when the PC is pointing to the program origin, it contains &0200. At this point, register contents could be anything, even though in Table 11.10 we have shown the initial contents as all zeros.

After the first press of 'R' the first instruction is executed, and &30 appears in the accumulator. Pressing 'R' again reveals that the PC now contains &0202; you should see that the trace table contains the results of program execution and updated PC contents *on the same line*.

Example 3
This program shows the operation of a 'BNE' loop:

```
0200   A2  03              LDX   #&03
0202   A9  22              LDA   #&22
0204   A0  22              LDY   #&22
0206   18         .LOOP    CLC
0207   69  01              ADC   #&01
0209   88                  DEY
020A   CA                  DEX
020B   D0  F9              BNE   LOOP
020D   E8                  INX
020E   C8                  INY
020F   A9  FF              LDA   #&FF
0211   ..  ..               .  ..
```

The trace table is given in Table 11.11.

Table 11.11 Trace table for Example 3

| Address | Registers | | |
	A	X	Y
0200	00	00	00
0202	00	03	00
0204	22	03	00
0206	22	03	22
0207	22	03	22
0209	23	03	22
020A	23	03	21
020B	23	02	21
0206	23	02	21
0207	23	02	21
0209	24	02	21
020A	24	02	20
020B	24	01	20
0206	24	01	20
0207	24	01	20
0209	25	01	20
020A	25	01	1F
020B	25	00	1F
020D	25	00	1F
020E	25	01	1F
020F	25	01	20
0211	FF	01	20

The break point

Sometimes, it's useful to insert a break point (BRK) in a program to test program execution up to that point and isolate the rest. This technique will also enable you to check that PC contents are correct for the point at which you have inserted the BRK. In other cases, inserting a NOP (hex code EA) to substitute for suspect code is a useful device. You can't just eliminate a few bits of code in a micro system like the EMMA; even if it were possible, any JUMPs or BRANCHes may be upset.

As with everything else, experience is the most useful debugging aid. However, some ability at producing and using trace tables will be found to have great value.

Finally in this chapter, we present the program listed in Table 11.12. This is suitable for demonstrating trace table operation and use, using BRK, checking a section of RAM and fulfilling the requirements of C&G 224 Part II Electronics Systems (Control Systems) microelectronics assignment. Other readers may also find it of interest, however.

Note: only the section of code from 020D (START) need be entered in the assignment book. Only this part of the program need be single-stepped through, and only the resulting changes in register contents for this part of the program need be entered.

Table 11.12   Suitable C&G 224 assignment program. Note that the final 'JUMP' simply resets the computer and needn't be included in the assignment book documentation

Address	Label	Assembler instructions	Machine code	Comments
		ORG 0200		
		ENT 0200		
0200		LDX #&0A	A2 0A	;counter
0202		LDY #&00	A0 00	;pointer
0204		LDA #&11	A9 11	;data
0206	.LOOP	STA &02E0,Y	99 E0 02	;store '11'
0209		INY	C8	;inc pointer
020A		DEX	CA	;dec counter
020B		BNE LOOP	D0 F9	;branch
020D	.START	LDA #&33	A9 33	;new number
020F		LDY #&A5	A0 A5	;
0211		STY &02E0	8C E0 02	;store in
0214		STA &02E1	8D E1 02	;locations
0217		LDA #&00	A9 00	;new number
0219		STA &02E2	8D E2 02	;store here
021C		STA &02E3	8D E3 02	;and here
021F		LDX #&03	A2 03	;new number
0221		STX &02E4	8E E4 02	;store
0224		LDA #&44	A9 44	;new number
0226		STA &02E5	8D E5 02	;store
0229		INY	C8	
022A		STY &02E6	8C E6 02	;store
022D		INX	E8	
022E		STX &02E7	8E E7 02	;store
0231		LDA #&99	A9 99	;new number
0233		STA &02E8	8D E8 02	;store
0236		PHA	48	;push on stack
0237		LDA #&66	A9 66	;new number
0239		STA &02E9	8D E9 02	;store
023C		PLA	68	;pull from stack
023D		STA &02EA	8D EA 02	;store
0240	.JUMP	JMP &FEE0	4C E0 FE	;reset computer

The first part of the program is a loop which is designed simply to put '11' in the appropriate RAM locations. When this has been 'RUN' satisfactorily, input the rest of the program code and run the program.

Finally, single-step through the program from 020D (START) and note down the RAM contents in the appropriate section of the assignment book.

If all the code has been entered, a BRK placed at address &020D would halt program execution at that point. This would enable a programmer to run only the first part of the program and check that all the appropriate locations contain &11 before proceeding with the rest of the program.

It is a simple matter for the reader to write his or her own programs, draw up trace tables and check them with single-stepping.

Questions

Work out the machine code, draw up trace tables, then load, run and check the following programs:

1 (Z80)
```
LD      A,0DFH
LD      B,22H
LD      C,15H
LD      D,44H
ADD     A,C
LD      A,E
ADD     A,B
LD      A,90H
ADD     A,E
LD      (5D00H),A
LD      A,00H
RLA
LD      (5D01H),A
```

2 (Z80)
```
LD      A,0FFH
LD      B,63H
SUB     B
LD      A,44H
SCF
RRA
LD      C,A
ADD     C
LD      D,A
SUB     C
```

3 (6502)
```
LDA     #&DF
LDX     #&22
LDY     #&15
CLC
ADC     #&15
STA     &30
TAY
TXA
STA     &31
ROL     A
STA     &32
```

4 (6502)
```
LDA     #&63
STA     &30
LDA     #&FF
SBC     &30
STA     &31
ROR     A
TAY
STA     &32
```

5 Use debugging techniques to find out what's wrong with the following programs:

174   Microprocessor Technology

(a) (Z80) This program should display '123' on the screen:

```
5C03H                LD    A,31H
5C05H     START:     LD    HL,E224H
5C08H                LD    (HL),A
5C09H                INC   A
5C0AH                INC   HL
5C0BH                LD    (HL),A
5C0CH                INC   A
5C0DH                INC   HL
5C0EH                LD    (HL),A
```

(b) (6502) This program should add two numbers together. The result is held in memory location &30 and the content of the carry flag in memory location &31.

```
0250   A9   B6              LDA   #&B6
0252   18                   CLC
0253   69   A7              ADC   #&A7
0255   80   09              BCS   CARRY
0257   85   30              STA   &30
0259   A9   00              LDA   #0
025B   85   32              STA   &31
025D   4C   65  02          JMP   &0265
0260   A9   00      .CARRY  LDA   #01
0262   85   31              STA   &31
0264   00                   BRK
0265   E0                   NOP
```

(c) (6502) This is a subroutine that should multiply a number (loaded into the accumulator) by 4. 'LODE' is used because in some systems 'LOAD' is a reserved word and cannot be used for a label:

```
0270   A9   06      .LODE   LDA   #3
0272   A2   04      .MULT4  LDX   #4
0274   85   30      .STORE  STA   &30
0276   18           .LOOP1  CLC
0277   65   30              ADC   &30
0279   85   31              STA   &31
027B   CA                   DEX
027C   D0   F8              BNE   LOOP1
```

(d) (Z80) A number loaded into the accumulator should be multiplied by four, using this routine:

```
5C00H                LD    A,6        ;multiplicand
5C02H     MULT4:     LD    B,4        ;counter
5C04H                LD    HL,5F00H
5C07H                LD    (HL),0     ;zero total
5C09H     LOOP:      ADD   A,(HL)     ;add to total
5C0AH                LD    HL,A       ;save total
5C0BH                DEC   B          ;dec counter
5C0CH                JR    NZ,LOOP
5C0FH                RET
```

Assignment 11.1

(a) Explain how 'single-stepping' and a 'trace table' can be useful in 'debugging' a program.

(b) Refer to the subroutine TRACE shown in Table 11.13.

Table 11.13 Subroutine 'TRACE'

Address	Machine code	Label	Assembler instructions	Line number
			ORG 5C00H	
			ENT 5C00H	
5C00	3E 00	TRACE:	LD A,05H	1
5C02	0E 02		LD C,02H	2
5C04	06 1C		LD B,1CH	3
5C06	80	LOOP:	ADD A,B	4
5C07	0D		DEC C	5
5C08	C2 06 5C		JP NZ,LOOP	6
5C0B	C9		RET	7

Construct a trace table showing how the contents of the registers PC, A, B and C and the zero flag Z change during program execution. The layout and first line of the program is given below:

PC	CPU registers			Flag	Next line
	A	B	C	Z	
5C00	00	00	00	0	1

Chapter 12
The input/output (I/O) ports

Introduction

Computer I/O (input/output) ports allow the CPU to control devices in the outside world. Signals may be output to control lights, motors and heaters for example. Conversely, signals from the outside world, provided by sensors such as simple switches, thermistors, opto diodes (light sensors) and pressure sensitive devices, may be input to the CPU. These signals can be analysed and processed, with the CPU outputting new signals in consequence. So, like sea ports along the coast which allow goods to be brought into and out of a country, I/O ports allow signals to pass into and out of a computer.

The word 'signal' in this context is meant to imply an electrical or electronic value, or variation in value, which carries information. Signals have no power; they only carry information. You can't simply connect devices to the ports and expect them to work. Obviously (at least I hope it's obvious), you can't connect ordinary, mains light bulbs to output ports. A 60W mains lamp needs 240V at 250mA; the output ports of 6502 and Z80 systems can only manage a maximum voltage of 5V, at a current some hundred times less than the 250mA quoted for the lamp. The signals could not even light up a small LED, but they can be amplified of course.

The television signals appearing at the end of the aerial cable are of the order of microvolts and they can't drive anything either. But all the complex information about the sound, vision and colour, together with synchronising pulses and teletext, are carried in the signal. It has only to be processed and amplified by the television receiver.

Turning the low voltage, low current computer signals into something which can drive lights, motors or heaters requires 'interfacing'. The interface is an electronic block which accepts low level signals, and conditions them so that they are suitable in amplitude (both voltage and current) for driving the peripheral to which the interface is connected. We discuss interfacing methods in Chapter 14, but since we'll need some minimal equipment to test the operation of the programs about to be described, we discuss some simple concepts here.

The EMMA and MARC systems have I/O monitors which can be used to indicate port activity, and similar systems will have their own versions. In the absence of such a facility, it's a simple matter to build up a small circuit that will do the job; in order to make a start, let's consider the circuit of Figure 12.1.

A simple interface circuit

Almost any npn transistor would work in this circuit. It is TTL compatible, so the power may be taken from the microcomputer supply. A wire is then

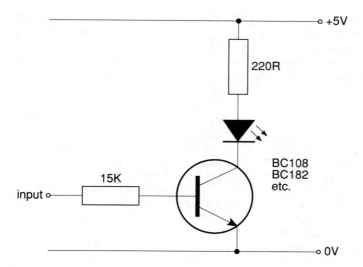

Figure 12.1 *Simple, single transistor interface circuit, suitable for monitoring a single bit of a computer output port.*

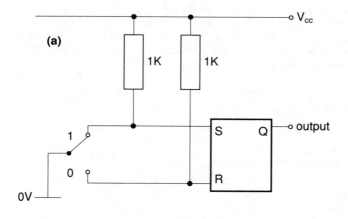

Figure 12.2 *(a) Using a simple S-R bistable to debounce a switch; (b) how the circuit may be implemented using NAND gates.*

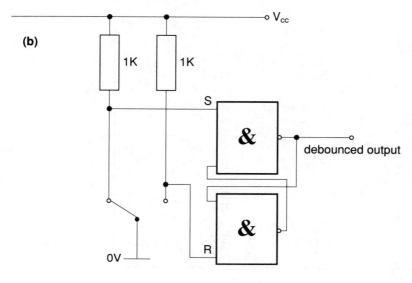

connected between the required bit of the output port and the input of the circuit. If bit 0 of port A is being monitored, then the LED lights if bit 0 is set to a 1 and it doesn't light if bit 0 is zero. It's as simple as that.

The circuit can be constructed in a few seconds using 'locktronics' or any breadboarding technique, and in a few minutes using stripboard, as the reader prefers.

Inputting data

Any data appearing on the ports can be read into the accumulator and processed just like any other data. This kind of input could derive quite simply from a bank of eight switches, each providing either a logic 1 or a logic 0. However, switches suffer from something known as contact bounce. Mechanical switches do not tend to connect 'cleanly' and the contacts bounce for a while after the first 'make'. Generally, this effect is of no importance, but in electronic circuits, designed to respond to maybe a million variations a second, the problems can become serious, distorting the required input and producing false data. The solution is to use some sort of 'debounce' circuit and one possible design is shown in Figure 12.2.

Figure 12.3 shows the mode of action and a description of this now follows.

At first, the switch is being thrown from the S position, to the R. When contact is made, R = 0 and the output takes up a 0. When the contacts bounce, R becomes a 1 again; however, the other input to the lower NAND gate is at zero, so its output remains at a logic 1 and the debounced output stays at 0. So, no matter how often the switch bounces, the output takes up a logic 0 on the first contact and remains there.

A similar analysis can be made when the switch is returned to the S position. The output takes up a logic 1 immediately and stays there. The switch has been debounced.

Switch Positions	Comments	Bistable State	Output
S o o R	Moving to make at R	S = 1 R = 1	Q = 1
S o o R	Circuit made at R	S = 1 R = 0	Q = 0
S o o R	Bounce from R	S = 1 R = 1	Q = 0
S o o R	Circuit made at R again	S = 1 R = 0	Q = 0
S o o R	Circuit made at S	S = 0 R = 1	Q = 1
S o o R	Bounce from S	S = 1 R = 1	Q = 1
S o o R	Circuit made at S again	S = 0 R = 1	Q = 1

Figure 12.3
Operation of the S-R bistable debounce circuit.

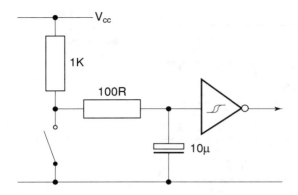

Figure 12.4 *A switch debouncing circuit using a Schmitt trigger with an inverting output.*

Another method of debouncing a switch is shown in Figure 12.4. This circuit uses a Schmitt trigger which gives an output only when it reaches a critical level. Its effect in the circuit is to produce debounced outputs of logic 0 when the switch is open and logic 1 when it's closed. (The device is an inverter.)

We'll be referring to these simple circuits later and programming examples will be given to show how they can be used.

The I/O ports

The ports themselves simply consist of registers or memory locations; in the systems we are considering, they will be 8 bits wide just like all the others. In 6502 systems, the ports are memory mapped (this was mentioned in Chapter 4), so, in order to output a signal, you simply load it (or store it) in the appropriate memory location and away you go.

The EMMA and MARC systems (and many others) have two ports, designated port A and port B. In each case, utilising the ports consists of two steps:

(1) set up the port to indicate how you want it to operate: input or output; all bits or just some of them; which ones?
(2) output or input the actual data.

Once the ports have been set up at the beginning of the program, no further action is required unless you need to change the configuration. The only commands then required are those associated with inputting or outputting the actual data; this is done via the data registers, designated DRA and DRB for the data registers of ports A and B respectively. The same nomenclature is used for both 6502 and Z80 systems.

The control registers

These are used only for the setting up procedure which normally occurs once at the beginning of a program. In Z80 systems, the registers are usually designated CRA and CRB, the control registers for ports A and B. In 6502 systems, they are normally designated DDRA and DDRB, the data direction registers for the two ports.

These registers are 8 bits wide, just like the ports themselves. In the 6502 system, if you want a particular bit to be output, you set it to a logic 1; if you want it to be input, you set it to a logic 0. Figure 12.5 shows some of the many possibilities.

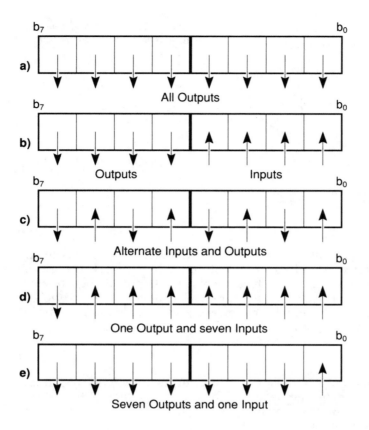

Figure 12.5 *Some possible I/O configurations.*

In Z80 systems, it's the other way round. Since the actual process differs somewhat between Z80 and 6502 systems, we consider them separately. The latter is simpler in many respects so we deal with the 6502 first.

6502 I/O ports

To use the ports then, you first need to select the mode required and send the control word for it to the data direction register (DDRA or DDRB). When this has been done, data may be stored in or loaded from the relevant memory location in the usual way. The port addresses for the EMMA are as follows:

```
DDRA    0903
DDRB    0902
DRA     0901
DRB     0900
```

In order to set up or 'initialise' a port, all that's necessary is to send the appropriate code to it. This consists of storing a logic 1 in each of the bits required for outputs, and a logic 0 in each of the bits required for inputs. To take a simple example, if we want port A to output on every bit, then we'd simply send the code &FF to DDRA. This is because, if we want output, then all bits are set to a logic 1, and all bits set to a 1 produce the hexadecimal number &FF. This is shown in Figure 12.6.

If we wanted port A to input, then the code sent to DDRA would be &00. Figure 12.7 shows some other possible combinations and the resulting hex

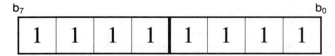

Figure 12.6 *All bits are set to a logic 1. The DDR therefore contains 1111 1111B which is &FF.*

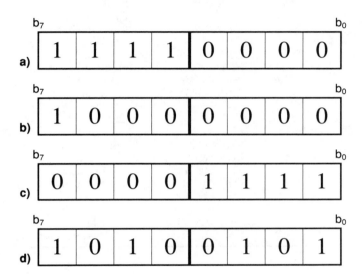

Figure 12.7 *Some of the many possible combinations of input/output settings in the data direction registers, a) &F0 b) &80 c) &0F d) &A5.*

codes. Each of the 8 bits in both ports A and B can be independently set or reset for any combination of inputs and outputs.

Example 1

Table 12.1 shows a simple program with appropriate comments, which will turn bit 0 of port A on and off continuously. You will see that DDRA is first configured so that bit 0 is output. (In fact, in this case, we've set all the bits to output by storing &FF in DDRA.) A 1 is then sent to DRA (at address 0901) making bit 0 a 1 and lighting the LED in the interface circuit, assuming it's correctly connected. The program then calls a delay, and then sends a zero to bit 0, turning the LED off. A jump instruction causes the sequence to continue indefinitely.

In summary, the program shown in Table 12.1 does the following:

1   port A is set up for outputs by sending &FF to DDRA;
2   &01 is sent to bit 0 of DRA – the lamp lights;
3   a software delay is used;
4   &00 is sent to bit 0 of DRA – the lamp goes out;
5   the delay routine is jumped to again;
6   program execution jumps back to (2); &0205 in the program.

The reader can speed up the 'flash' rate by changing the value in the accumulator in the delay routine at address 0221 from &88 to, say, &22. Different bits can be turned on and off simply by sending the appropriate

Table 12.1 6502 I/O program with comments

Address	Machine code	Label	Mnemonics	Comments
			ORG &0200	;origin
			ENT &0200	;entry point
0200	A9 FF	.START	LDA #&FF	;immediate load
0202	8D 03 09		STA &0903	;DDRA to output
0205	A9 01	.AGAIN	LDA &01	;bit 0 'ON'
0207	8D 01 09		STA &0901	;DRA (output)
020A	20 20 02		JSR &0220	;Jump to subroutine
				;DELAY
020D	A9 00		LDA #00	;bit 0 'OFF'
020F	8D 01 09		STA &0901	;output to DRA
0212	20 20 02		JSR DELAY	;Jump to DELAY
0215	4C 05 02		JMP AGAIN	;do again
				;DELAY
0220	A9 88	.DELAY	LDA #&88	;immediate load
0222	8D 20 00		STA &0020	;store in memory
0225	A9 FF	.LOOP1	LDA #&FF	;fill accumulator
0227	8D 21 00		STA &0021	;store in memory
022A	CE 21 00	.LOOP2	DEC &0021	;decrement in
022D	D0 FB		BNE LOOP2	;inner loop
022F	CE 20 00		DEC &0020	;decrement in
0232	D0 F1		BNE LOOP1	;outer loop
0234	60		RTS	;return from routine

codes to DRA. Port B is configured and operated in exactly the same way; set up DDRB for the required I/O sequence and then send the data out to DRB. A bank of LEDs can be made to count up in binary simply by repeatedly adding 1 to the contents of the accumulator and storing the new value in the data register. A suitable delay can be incorporated between successive additions. Given the necessary hardware, a bank of 8 LEDs would be seen to count up in binary.

Example 2
Assuming suitable equipment to be available then, let's take a look at some programming examples. Table 12.2 shows a revised version of the program given earlier which flashes a lamp on and off. The main program is identical to the one shown in Table 12.1 (addresses 0200 to 0215), only the delay routine has been changed. In this case, the setting of the switch affects the rate at which the lamp flashes. In one position the rate is fast, in the other position it is slow.

To make the program work, all you need to do is ground bit 0 of port B with a fly-lead. Removing the short to ground allows bit 0 to 'float' back up to a logic 1 quite naturally. A two-way, debounced switch arrangement is not necessary in this particular application.

These are very simple programs of course, but they should serve to demonstrate how a user can interact with the CPU via the I/O ports. By building up suitable interfacing circuitry (see Chapter 14), the flash rate of the lamp in the last example could have at least eight different values; by more sophisticated programming, it could have 256 different values. Think about it.

It wasn't necessary to use both ports in the last example; the same effect could have been achieved using any two of the bits in port A (or port B), one

Table 12.2 The delay routine has been modified so that the accumulator takes a 'speed value' from port B. If bit 0 of that port is set to a 0, the flash rate is slow; if it's set to a 1 then the flash rate is fast

Address	Machine code	Label	Mnemonics	Comments
0215	4C 95 02		JMP &0205	;do again
.	.		.	.
.	.		.	.
.	.		.	.
				;DELAY
				;
0220	A9 00		LDA #&00	;load A with zero
0222	8D 02 09		STA &0902	;store in DDRB
				;makes port B input
0225	AD 00 09		LDA &0900	;load A with port B
				;contents
0228	29 01		AND &01	;a 1 or a zero?
022A	D0 05		BNE FAST	;branch on a 1
022C	A9 88	.SLOW	LDA #&88	;slow speed value
022E	4C 33 02		JMP &0233	;bypass fast speed
0231	A9 22	.FAST	LDA #&22	;fast speed value
0233	8D 20 00		STA &0020	;store in memory
0236	A9 FF	.LOOP1	LDA #&FF	;fill accumulator
0238	8D 21 00		STA &0021	;store in memory
023B	CE 21 00	.LOOP2	DEC &0021	;decrement in
023E	D0 FB		BNE LOOP2	;inner loop
0240	CE 20 00		DEC &0020	;decrement in
0243	D0 F1		BNE LOOP1	;outer loop
0245	60		RTS	;return from routine

configured for input, the other for output. As a final I/O exercise, see if you can write and run a program which will achieve this.

Interrupts

The process of interrupting the execution of a program was introduced in Chapter 9 (The stack). We cover the subject in much more detail here since interrupts are closely related to the use of I/O ports.

You will recall that we described two particular types of interrupt: NMI and IRQ. The latter is an interrupt request which can be disabled in software by a simple instruction (DI). The former (a non-maskable interrupt) cannot be disabled and *always* causes an interrupt when initiated. The NMI is sensitive to a transition from a logic 1 to logic 0. The IRQ interrupt is logic low enabled and we will see soon, how these different interrupts can be achieved.

Starting with an NMI interrupt, let's modify the program given in Table 12.1. We're going to use the technique of flashing the light on and off using port B. At the same time, we'll use the on/off (logic 1 to logic 0) *transition* to cause the NMI interrupt. The interrupt routine counts the number of flashes and outputs the running total to a bank of LEDs connected to port A.

Before describing the process in detail, let's take an overall look at what we're trying to achieve. The interrupt is initiated by applying the appropriate transition to the NMI pin on the 6502. On the EMMA panel, this socket is made available alongside the I/O port connections. Once the transition is received, the CPU finishes the current instruction and then goes along to service the interrupt.

Table 12.3 Interrupt vector memory locations

0EFC	low byte	$\overline{\text{NMI}}$ Vector
0EFD	high byte	
0EFE	low byte	$\overline{\text{IRQ}}$ Vector
0EFF	high byte	

To do this, the CPU looks for the interrupt vectors. Unless you're familiar with the phrase, it might sound like something out of *Star Trek*; in fact the interrupt vectors are simply the memory address locations which contain the interrupt service routine. You will write the routine, you can get it do what you like and you can put it where you like, subject to the usual constraints. To tell the CPU where you've put your interrupt service routine, you simply load the start addresses in the appropriate memory locations. These may well vary from one system to another but for the EMMA the memory locations are as given in Table 12.3.

In summary, the process is as follows:

1   Write the main program (in this case it'll be a bit like the flashing light program in Table 12.1).
2   Write the interrupt service routine (in this case it will contain a simple increment to count how many times the light flashes).
3   Load in the interrupt vectors: this means, put the start address of the routine in the appropriate memory location. For example, if you write your routine starting at memory location 02E0, then E0 goes into 0EFC and 02 goes into 0EFD (see Table 12.3).

The program, suitably annotated for clarity, is shown in Table 12.4. A look at this will show that the interrupt service routine occupies addresses from &0220 to &0223 and (in this case) only contains two instructions. It is this start address (&0220) which must be loaded into &0EFC and &0EFD using the usual format of lsb first and msb second. This is shown in Table 12.5.

To test and demonstrate the program connect an interfaced indicator to bit 0 of port B. The count will appear at port A so whilst a bank of eight LEDs would be ideal, four would do. We only want to see the effect after all. Finally, connect a lead from bit 0 of port B to the NMI socket. The negative going transition from logic 1 to logic 0, as the indicator turns off, acts as the interrupt signal. Normal CPU execution ceases after the current instruction, the interrupt service routine is accessed and port A data register (DRA) is incremented.

To demonstrate the use of an $\overline{\text{IRQ}}$ interrupt, we'll need to modify the software as the interrupting signal is slightly different in this case. One of the first things we must do is reset the interrupt flag in the processor status register using the CLI instruction. The $\overline{\text{IRQ}}$ itself is level low enabled; this means that we need to keep the level high until the interrupt is required. We must also remember to return to the high state afterwards, otherwise the interrupt signal will be present all the time. Studying the program carefully (see Table 12.6) will show how this has been achieved.

The program itself sets bit 7 of port A to a logic 1; this is connected to the $\overline{\text{IRQ}}$ input so that, when bit 7 goes low, the interrupt is initiated. A binary count appears on port A so that eventually all bits (including bit 7) will be

Table 12.4 6502 I/O program using $\overline{\text{NMI}}$ interrupts

Address	Machine code	Label	Mnemonics	Comments
			ORG &0200	;origin
			ENT &0200	;entry point
0200	A9 FF	.START	LDA #&FF	
0202	8D 03 09		STA &0903	;ports A and B
0205	8D 02 09		STA &0902	;to output
0208	A9 00		LDA #&00	;
020A	8D 01 09		STA &0901	;DRA set to zero
020D	8D 00 09	.LOOP	STA &0900	;DRB set to zero
0210	20 30 02		JSR DELAY	;delay subroutine
0213	A9 01		LDA #&01	;put light on
0215	8D 00 09		STA &0900	;store in DRB
0218	20 30 02		JSR DELAY	;delay subroutine
021B	A9 00		LDA #&00	;turn light off
021D	4C 0D 02		JMP LOOP	;repeat from &020D
				;INTERRUPT SERVICE
				;ROUTINE
0220	EE 01 09		INC DRA	;increment port A
0223	40		RTI	;return from ISR
				;DELAY
0230	A9 AA	.DELAY	LDA #&AA	;immediate load
0232	85 30		STA &30	;zero-page memory
0234	A9 FF	.LOOP1	LDA #&FF	;fill accumulator
0236	85 31		STA &31	;store in memory
0238	C6 31	.LOOP2	DEC &31	;decrement in
023A	D0 FC		BNE LOOP2	;inner loop
023C	C6 30		DEC &30	;decrement in
023E	D0 F4		BNE LOOP1	;outer loop
0240	60		RTS	;return from
				;subroutine

Table 12.5 Loading the interrupt service routine vectors

&0EFC 20	$\overline{\text{NMI}}$ vectors	&0EFE 20	$\overline{\text{IRQ}}$ vectors
&0EFD 02		&0EFF 02	

set to zero. The interrupt service routine is then accessed and this causes two things: one, the count on port B is incremented, and two, bit 7 is reset, thus inhibiting further interrupts until the binary count on port A returns to zero. The lights on port B, therefore, give an indication of how many times port A has cycled through its binary count.

To make things easier, we list the changes that have been made to the last program so that if it's still in memory a good deal of it can be retained.

Changes made to $\overline{\text{NMI}}$ interrupt program

1   Addresses &020A onwards in main program are different.
2   Interrupt service routine is at same address as previously but is changed.

Table 12.6 6502 I/O program using IRQ interrupts

Address	Machine code	Label	Mnemonics	Comments
			ORG &0200	;origin
			ENT &0200	;entry point
0200	A9 FF	.START	LDA #&FF	
0202	8D 03 09		STA &0903	;ports A and B
0205	8D 02 09		STA &0902	;to output
0208	A9 00		LDA #&00	;
020A	8D 00 09		STA &0901	;DRB set to zero
020D	A9 80		LDA #&80	;bit 7 on
020F	8D 01 09		STA DRA	;in Port A
0212	58		CLI	;clear flag
0213	EE 01 09	.LOOP	INC DRA	;increment count
0216	20 30 02		JSR &0230	;delay subroutine
0219	4C 13 02		JMP LOOP	;do again
				;INTERRUPT SERVICE
				;ROUTINE
0220	A9 80	.ISR	LDA #&80	;load accumulator
0222	8D 01 09		STA DRA	;turn bit 7 back on
0225	EE 00 09		INC DRB	;record count
0228	40		RTI	;return from
				;interrupt
				;DELAY
0230	A9 11		LDA #&11	;immediate load
0232	85 30		STA &30	;zero-page memory
0234	A9 FF	.LOOP1	LDA #&FF	;fill accumulator
0236	85 31		STA &31	;store in memory
0238	C6 31	.LOOP2	DEC &31	;decrement in
023A	D0 FC		BNE LOOP2	;inner loop
023C	C6 30		DEC &30	;decrement in
023E	D0 F4		BNE LOOP1	;outer loop
0240	60		RTS	;return from
				;subroutine

3 In the delay routine, only the value at &0231 has been changed (to 11); this makes the count cycle through more quickly. In general, the smaller the number, the smaller the delay. However, you can't use zero as a minimum; if you do, the value decrements to &FF and so on back down to zero making the count even longer!

Hardware

We used a bank of eight LEDs on port A and a bank of four on port B. This requirement can be modified to suit individual needs of course.

Once again, we are using very modest programs to demonstrate the use of interrupts. With only a little thought, it's possible to write some very interesting routines. It is a good idea to change the interrupt service routine, and change its location (not forgetting to modify the interrupt service routine vectors) so that you can get a feel for the way in which the programming is implemented. Even more fascinating things can be explored after looking at the next two chapters.

Z80 I/O ports

The MARC, in common with many similar systems, has two ports which are designated A and B. Associated with each port are two registers, a control register (CRA in the case of port A) and a data register (DRA). The former enables the port to be configured for either input or output or a combination of the two. This is achieved by sending the appropriate control word to the CRA. These different modes are summarised in Table 12.7.

Unlike the 6502 system which is memory mapped, special instructions exist for Z80 I/O operations; you can't just load the data registers. Two of them (the instructions 'IN' and 'OUT') inform the processor that I/O operations are required, so that all the usual memory locations (in ROM, RAM, EPROM, etc.) are ignored, and the address bus is now connected to the external I/O chips.

The I/O section has its own little memory map as shown in Table 12.8.

DRA and DRB are the data registers. Data written to or from these registers appears in, or is read from, the A or B ports. CRA and CRB are control registers. A special control word written to either of these registers indicates the mode in which it is operating as shown in Table 12.7.

Depending on the code sent to the control register, each of the ports can be configured as byte output (all 8 bits of the port configured as output) or byte input. In another mode, each of the bits can be set independently to be either input or output. Figure 12.7 shows some possibilities already referred to in the 6502 section.

Table 12.7 I/O modes used in the MARC Z80 system

Mode	Description	Code
* Mode 0	Byte output mode All 8 bits of the port are used for output	(00)
* Mode 1	Byte input mode All 8 bits of the port are used for input	(01)
* Mode 2	Byte input/output mode This is a bi-directional mode which uses port A for data transfer	(10)
* Mode 3	Bit input/output mode Any of the bits can be used for either input or output	(11)

Table 12.8 The MARC Z80 I/O port memory map

Port	Register	Address
PORT A	DRA CRA	00 02
PORT B	DRB CRB	01 03

Working out the control codes

The control registers are 8 bits wide and they perform a similar function to the 6502 data direction registers. In the case of the MARC Z80 system, *all*

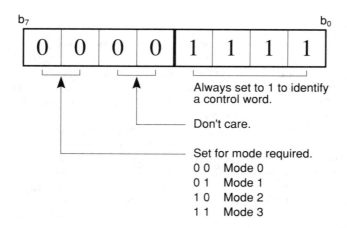

Figure 12.8 *How the appropriate control word value is calculated.*

control words used for selecting data direction will have the four least significant bits set to a 1. Bits 4 and 5 are of no consequence so it doesn't matter whether they're 1 or 0, so we'll put them to 0. The two most significant bits will be set according to the mode required, the value of which may be obtained from Table 12.7. For example, if the byte output mode is required, then bits 6 and 7 will be set to 00. This is shown in Figure 12.8.

Hence, to configure port A for output, the code would be as follows:

```
LD  A,0FH      ;0000 1111
OUT (02H),A    ;send code to CRA
```

To configure port B for input, the code would be like this:

```
LD  A,4FH      ;0100 1111
OUT (03H),A    ;send code to CRB
```

Having configured the ports, data is sent out to either DRA or DRB using the OUT instruction again. So, to send 88H to port A use:

```
LD  A,88H      ;1000 1000
OUT (00H),A    ;send pattern to DRA
```

To input data from port B, the simple IN instruction is used:

```
IN  A,(01H)    ;put the data at port B into the
                accumulator
```

Table 12.9 shows a simple program with appropriate comments, which will turn bit 0 of port A on and off continuously. You will see that CRA is first configured so that bit 0 is output. (In fact in this case, we've used mode 0 to set all the bits to output by storing 0FH in CRA.) A 1 is then sent to DRA (at address 00) making bit 0 a 1 and lighting the LED in the interface circuit, assuming it's correctly connected. The program calls a delay, and then sends a zero to bit 0, turning the LED off. A jump instruction causes the sequence to continue indefinitely, and the LED appears to blink.

In summary, the program shown in Table 12.9 does the following:

1 Port A is set up for outputs by sending 0FH to CRA;
2 A logic 1 is sent to bit 0 of DRA – the lamp lights;
3 A software delay is used;
4 A logic 0 is sent to bit 0 of DRA – the lamp goes out;
5 The delay routine is CALLed again;
6 Program execution jumps back to (2); 5C04H in the program.

Table 12.9   Z80 I/O program with comments

Address	Machine code	Label	Mnemonics	Comments
			ORG 5C00H	;origin
			ENT 5C00H	;entry point
5C00	3E 0F		LD A,0FH	;immediate load
5C02	D3 02		OUT (02H),A	;CRA to output
5C04	3E 01	START:	LD A,01H	;bit 0 'ON'
5C06	D3 00		OUT (00H),A	;DRA (output)
5C08	CD 16 5C		CALL DELAY	;slow down effect
5C0B	3E 00		LD A,00H	;bit 0 'OFF'
5C0D	D3 00		OUT (00H),A	;output to DRA
5C0F	CD 16 5C		CALL DELAY	;
5C12	C3 04 5C		JP START	;do again
5C15	76		HALT	
				;DELAY
5C16	3E FF	DELAY:	LD A,0FFH	;immediate load
5C18	06 FF	LOOP2:	LD B,0FFH	;immediate load
5C1A	05	LOOP1:	DEC B	;decrement B
5C1B	20 FD		JR NZ,LOOP1	;
5C1D	3D		DEC A	;decrement A
5C1E	20 F8		JR NZ,LOOP2	;
5C20	C9		RET	;return

The reader can speed up the 'flash' rate by changing the value in the accumulator in the delay routine at address 5C17H to, say, 55H. Different bits can be turned on and off simply by sending the appropriate codes to DRA. Port B is configured and operated in exactly the same way; set up CRB for the required I/O format and then send the data out to DRB. A bank of LEDs can be made to count up in binary simply by incrementing the contents of the accumulator and storing the new value in the data register. A suitable delay can be incorporated between increments. Given the necessary hardware a bank of eight LEDs would be seen to count up in binary.

Using port B for input

To configure port B for input, first of all send the code for the required mode to CRB. This is done, as before, by loading the accumulator with the control byte, and then sending it to CRB using the OUT instruction. Once this has been done, data appearing at DRB may be read into the accumulator using the IN instruction:

```
LD    A,4FH
OUT   (03),A
IN    A,(01H)
```

In the program (see Table 12.10) this is followed by an AND instruction which will test the bit and find out if it's a 1 or a 0. Study the program until you understand it then load and run it. Put a logic 0 on bit 0 of port B, simply by shorting it down to the 0V line (a debounced switch is not necessary in this case). When the short is removed, the input automatically rises to a logic 1.

It wasn't necessary to use both ports in the last example; the same effect could have been achieved using any two of the bits in port A (or port B), one

Table 12.10 The delay routine has been modified so that the accumulator takes a 'speed value' from port B. If bit 0 of that port is set to a 0, the flash rate is slow; if it's set to a 1 then the flash rate is fast

Address	Machine code	Label	Mnemonics	Comments
5C12	C3 04 5C		JP START	;do again
5C15	76		HALT	
.	.		.	.
.	.		.	.
.	.		.	.
				;DELAY
				;
5C16	3E 4F	DELAY:	LD A,4FH	;set port B to
5C18	D3 03		OUT (03H),A	;input
5C1A	DB 01		IN A,(01H)	;load A with port B
				;contents
5C1C	E6 01		AND 01	;a 1 or a zero?
5C1E	20 04		JR NZ,FAST	;branch on a 1
5C20	3E FF	SLOW:	LD A,0FFH	;slow speed value
5C22	18 02		JR LOOP2	;bypass fast speed
5C24	3E 44	FAST:	LD A,44H	;fast speed value
5C26	06 FF	LOOP2:	LD B,0FFH	;
5C28	05	LOOP1:	DEC B	;decrement B
5C29	20 FD		JR NZ,LOOP1	;
5C2B	3D		DEC A	;
5C2C	20 F8		JR NZ,LOOP2	;
5C2E	C9		RET	;return from subroutine

configured for input, the other for output. As a final exercise, see if you can write and run a program which will achieve this.

Z80 interrupts

The Z80 interrupt structure is more complex than that in the 6502. However, as a starting point, we'll describe Z80 versions of the simple 6502 routines already covered.

The process of interrupting the execution of a program was introduced in Chapter 9 (The stack). You will recall that we described two types of interrupt: NMI and INT. The former (a non-maskable interrupt) is sensitive to a transition from logic 1 to logic 0. The INT interrupt is level enabled, so utilising each kind of interrupt may require different programming techniques.

Starting with an NMI interrupt, we'll modify the program given in Table 12.9. This contains a routine to flash a light on and off and we can use the on/off (logic 1 to logic 0) *transition* to cause the NMI interrupt. The interrupt routine itself counts the number of flashes on port A and outputs the running total to a bank of LEDs connected to port B.

Before describing the process in detail, let's take an overall look at what we're trying to achieve. The interrupt is initiated by applying the appropriate transition to the NMI pin on the Z80. On the MARC panel, this socket is made available alongside the I/O port connections. Once the transition is received, the CPU finishes the current instruction and then goes off to service the interrupt.

To do this, the CPU looks for the interrupt vectors. As was noted earlier, the interrupt vectors are simply the memory address locations which contain

Table 12.11   Interrupt vector memory locations

403BH	low byte	$\overline{\text{NMI}}$ vector
403CH	high byte	
4050H	low byte	$\overline{\text{INT}}$ vector
4051H	high byte	(mode 1)

Table 12.13   Z80 I/O program using $\overline{\text{NMI}}$ interrupts

Address	Machine code	Label	Mnemonics	Comments
			;************** NMIINT *********************	
			;Simple program to flash a light and also	
			;to demonstrate NMI interrupts. JSA 1990	
			;***	
			;	
			ORG 5C00H	;origin
			ENT 5C00H	;entry point
5C00	0E 00		LD C,0	;counter in ISR
5C02	3E 00		LD A,0	
5C04	32 3B 40		LD (403BH),A	;ISR vectors
5C07	3E 5D		LD A,5DH	
5C09	32 3C 40		LD (403CH),A	
5C0C	3E 0F		LD A,0FH	;port A to mode 0
5C0E	D3 02		OUT (02H),A	;CRA to output
5C10	3E 01	START:	LD A,01H	;bit 0 'ON'
5C12	D3 00		OUT (00H),A	;DRA (output)
5C14	CD 22 5C		CALL DELAY	;slow down effect
5C17	3E 00		LD A,00H	;bit 0 'OFF'
5C19	D3 00		OUT (00H),A	;output to DRA
5C1B	CD 22 5C		CALL DELAY	
5C1E	C3 10 5C		JP START	;do again
5C21	76		HALT	
				;DELAY
5C22	3E FF	DELAY:	LD A,0FFH	;immediate load
5C24	06 FF	LOOP2:	LD B,0FFH	;immediate load
5C26	05	LOOP1:	DEC B	;decrement B
5C27	20 FD		JR NZ,LOOP1	;
5C29	3D		DEC A	;decrement A
5C2A	20 F8		JR NZ,LOOP2	;
5C2C	C9		RET	;return
			ORG 5D00H	;new origin
5D00	3E 0F	ISR:	LD A,0FH	;ISR start
5D02	D3 03		OUT (03),A	
5D04	0C		INC C	;ISR counter
5D05	79		LD A,C	;to accumulator
5D06	D3 01		OUT (01),A	;OUT to port B
5D08	ED 45		RETN	;RETurn from NMI

the interrupt service routine. You will write the routine, you can get it to do what you like and you can put it where you like, subject to the usual constraints. To tell the CPU where you've put your interrupt service routine, you simply load the start addresses in the appropriate memory locations.

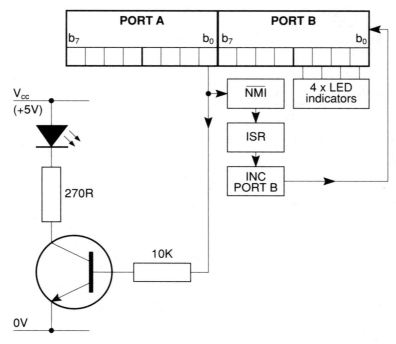

Figure 12.9   *General practical layout of $\overline{NMI}$ demonstration. Port A, b_0, is connected to the $\overline{NMI}$ socket.*

These may well vary from one system to another but for the MARC, the memory locations are as given in Table 12.11.

In summary, the process is as follows:

1   Write the main program (flash light on and off, port A).
2   Write the interrupt service routine to increment port B.
3   Load the interrupt vectors, i.e. put the start address of the routine in the appropriate memory locations. If the routine starts at 5D00H, then 00 goes into 403BH and 5D goes into 403CH, as shown in Table 12.11.

The program, suitably annotated for clarity, is shown in Table 12.13.

A glance at Table 12.13 will show that the interrupt service routine (ISR) begins at 5D00H (we used a second ORG statement to achieve this). It is this start address which must be loaded into 403BH and 403CH using the usual format of lsb (low byte) first and msb (high byte) second.

To test and demonstrate the program, refer to Figure 12.9. An interfaced indicator is connected to bit 0 of port A. Another lead is also connected to bit 0 and taken to the $\overline{NMI}$ socket. The negative going transition from logic 1 to logic 0 as the indicator turns off acts as the interrupt signal. Normal CPU execution ceases after the current instruction, the ISR is accessed and port B data register (DRB) is incremented as shown by the interfaced LED indicators connected to it.

$\overline{INT}$ interrupts

Unlike the 6502 $\overline{IRQ}$ interrupt, the Z80 has three software maskable interrupt modes, designated IM 0, IM 1 and IM 2.

Table 12.14 Interrupt control instructions. IFF indicates the interrupt enable flip-flop

Mnemonic	Description	Hex
DI	Disable interrupts IFF — 0	F3
EI	Enable interrupts IFF — 1	FB
IM 0	Set interrupt mode 0	ED 46
IM 1	Set interrupt mode 1	ED 56
IM 2	Set interrupt mode 2	ED 5E
RETI	RETurn from Interrupt	ED 4D
RETN	RETurn from non-maskable interrupt	ED 45
LD A,I	Contents of interrupt register to accumulator	ED 47
LD I,A	Contents of accumulator into interrupt register	ED 57
EX AF,AF'	Exchange contents of acc and flag register with alternate pair	08
EXX	Exchange contents of all general purpose register pairs with alternate pairs (BC, DE, HL)	D9

Eleven of the instructions associated with Z80 interrupts are given in Table 12.14.

The Z80 microprocessor has three modes of interrupt which can be set by using one of the three instructions which set the interrupt mode to 0, 1 or 2, respectively. The DI and EI instructions allow software controlled disabling and enabling.

Interrupt service routines are normal program sections which deal with a particular interrupt. The last instruction in such a routine is either a RETI (RETurn from Interrupt) or RETN (RETurn from Non-maskable interrupt) instruction which causes a return to the main program.

Z80 mode I $\overline{\text{INT}}$ interrupts (IM I)

These are probably the easiest to use, so we'll discuss them first. The procedure is as follows:

1   Set up the I/O ports as before, both to mode 0.
2   Enable interrupts using the EI instruction.
3   Tell the CPU that we want mode 1 interrupts using the IM 1 instruction.
4   Tell the CPU where the interrupt service routine (ISR) is by loading the addresses (the 'vectors') into memory locations 4050H (lsb) and 4051H (msb). We are using address 5D00H for the ISR.
5   Write the ISR starting at 5D00H, remembering to re-enable the interrupts (EI) and finally, return, using the RETI instruction.

The difference between $\overline{\text{NMI}}$ and $\overline{\text{INT}}$

In order to demonstrate the $\overline{\text{INT}}$ interrupt we'll use a similar program to the one used for the 6502 $\overline{\text{IRQ}}$. First of all, note the great difference between $\overline{\text{INT}}$ (and the 6502 $\overline{\text{IRQ}}$ which is more or less the same) and $\overline{\text{NMI}}$, in addition to the fact that the former cannot be masked off. $\overline{\text{NMI}}$ interrupts are initiated using a falling edge square wave, i.e. a transition from logic 1 to logic 0. Once the interrupt has occurred, and while logic 0 is maintained, further interrupts are inhibited – until the next logic 1 to logic 0 comes along.

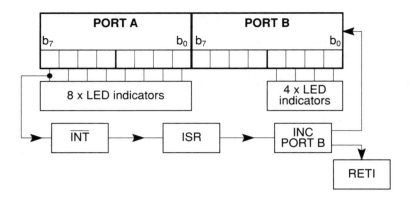

Figure 12.10   *General practical layout of $\overline{INT}$ demonstration. Port A, b_7, is connected to the $\overline{INT}$ socket.*

   With $\overline{INT}$ interrupts, provided they are properly enabled, a logic 0 on the $\overline{INT}$ CPU pin will cause repeated interrupts. So, once the ISR is completed, if logic 0 is maintained, another interrupt would take place – whether you wanted it or not. Fortunately, the system is designed to eliminate this problem; as soon as the $\overline{INT}$ is detected, IFF1 (interrupt flip-flop 1) assumes a logic 0, disabling further interrupts. This happens quite automatically, so your ISR must contain an EI before returning from interrupt (RETI). By this time, you must also have ensured that the logic 0 level which caused the $\overline{INT}$ in the first place has been removed. We'll look at the consequences of not doing this later.

IM 1 demonstration program description

Figure 12.10 shows the general layout. The ports are both set to mode 0 (byte output) as previously. The byte 80H is sent to port A, turning bit 7 on. The $\overline{INT}$ socket is connected to this, so that, when bit 7 becomes a zero, $\overline{INT}$ is activated. Meanwhile, port B is connected to a small display (doesn't need to be four) and this increments when the interrupt occurs.
   Bit 7 will go to zero because port A is made to go through a binary count from 80H. Once FFH has been reached, the next pulse turns all the LEDs off, including bit 7, of course, and $\overline{INT}$ is activated. The ISR increments port B, then program execution continues from where it left off.
   The complete program is given in Table 12.15. Load it and run it with the appropriate hardware connections and note the effect. We've reduced the time delay so that the port A display cycles through quite quickly and the effect of the ISR on port B can be seen.

Some $\overline{INT}$ (IM 1) program notes

1   C is loaded with 80H; later on, by the label START: this value is trans-ferred to the accumulator, incremented and sent out to port A. 80H is 1000 0000 in binary, so once incremented, the pattern 1000 0001 appears at port A.
2   The instructions from 5C04H set up ports A and B in mode 0.
3   Address 5C14H contains the enable interrupt instruction and this is followed by IM 1 which sets interrupt mode 1.

Table 12.15 Z80 I/O program using mode 1 $\overline{\text{INT}}$ interrupts

Address	Machine code	Label	Mnemonics	Comments
			ORG 5C00H	;origin 1
			ENT 5C00H	;entry point
5C00	0E 80		LD C,80H	;main counter
5C02	16 00		LD D,0	;ISR counter
5C04	3E 0F		LD A,0FH	;port A to mode 0
5C06	D3 02		OUT (02H),A	;CRA to output
5C08	3E 80		LD A,80H	;bit 7 'ON'
5C0A	D3 00		OUT (00H),A	;DRA (output)
5C0C	3E 0F		LD A,0FH	
5C0E	D3 03		OUT (03H),A	;port B to mode 0
5C10	3E 00		LD A,00H	;clear DRB
5C12	D3 01		OUT (01),A	
5C14	FB		EI	;enable interrupt
5C15	ED 56		IM 1	;mode 1 interrupt
5C17	3E 00		LD A,00	;ISR vectors
5C19	32 50 40		LD (4050H),A	
5C1C	3E 5D		LD A,5DH	
5C1E	32 51 40		LD (4051H),A	
5C21	79	START:	LD A,C	;get count
5C22	3C		INC A	
5C23	D3 00		OUT (00H),A	
5C25	4F		LD C,A	
5C26	FE 00		CP 0	;check for a zero
5C28	28 06		JR Z,LEDON	;put LED on if 0
5C2A	CD 34 5C		CALL DELAY	;slow down effect
5C2D	C3 21 5C		JP START	
5C30	0E 80	LEDON:	LD C,80H	
5C32	18 ED		JR START	;go again
				;DELAY
5C34	3E 05	DELAY:	LD A,05H	;immediate load
5C36	06 FF	LOOP2:	LD B,0FFH	;immediate load
5C38	05	LOOP1:	DEC B	;decrement B
5C39	20 FD		JR NZ,LOOP1	;
5C3B	3D		DEC A	;decrement A
5C3C	20 F8		JR NZ,LOOP2	;
5C3E	C9		RET	;return
			ORG 5D00H	;new origin
5D00	14	ISR:	INC D	;start of ISR
5D01	7A		LD A,D	
5D02	D3 01		OUT (01),A	;OUT to port B
5D04	57		LD D,A	
5D05	0E 80		LD C,80H	
5D07	79		LD A,C	
5D08	D3 00		OUT (00),A	
5D0A	FB		EI	
5D0B	ED 4D		RETI	;RETurn from $\overline{\text{INT}}$

4   Address 5C17H has the first of several instructions which load the interrupt service routine (ISR) vectors.

5   The main program (to produce a binary count from 80H) begins at 5C21H. The count is saved in C because it would otherwise be lost in the delay routine which also uses the accumulator.

6   We'll explain the CP 0 and LEDON instructions later. The delay routine comes next and this is already a familiar bit of code.

7   Note that we've got a second origin (ORG) statement next. This will ensure that the ISR starts at 5D00H. The D register was set to zero at 5C02H. When the ISR is entered, interrupts are automatically disabled by setting a flip-flop (IFF1) to zero, and then D is incremented. After the transfer to the accumulator, this count is output to port B, then saved in the D register.

8   At this point, all the LEDs are still out. If interrupts are enabled, then an interrupt would occur and program execution would continue from 5D00H in an endless loop. To prevent this, 80H is sent out to port A before EI and RETI are executed.

9   Program execution then continues from the point at which it left. This was when the binary count reached zero (turning bit 7 off). So the next instruction is at address 5C25H, i.e. LD C,A. A must have been zero, as we've just noted, so LD C,A puts 0 into the C register. If we left it at that (and didn't put in CP 0 and the 'LEDON' instructions) we'd end up back at label 'START:'; LD A,C puts zero into the accumulator, INC A makes it a 1 and when this is output at 5C23H, an interrupt occurs immediately. This is because the INT socket is connected to bit 7 of port A which is at zero.

   The CP 0 instruction detects the zero situation, causing a relative jump to 'LEDON:' which puts 80H into C, restoring bit 7 to a logic 1. Program execution then continues in the way we want.

There are some very important programming concepts here, quite apart from discovering the nature of INT interrupts. It is essential that you read through these nine points until you understand the meaning and use of every line of code in the program.

Other Z80 INT interrupts

The INT mode is chosen by using the appropriate instruction IM 0, IM 1 or IM 2. These set the interrupt mode flip-flops as shown in Figure 12.11.

   The interrupt flip-flops also shown can be set or reset by the enable or disable instructions, EI and DI. Interrupts are automatically disabled in mode 1, following an INT interrupt as we have seen. Figure 12.12 summarises the ISR vector locations. We've dealt with one example of mode 1 INT interrupts, so now lets look at mode 2. At the same time, we'll also look at mode 3 I/O, the bit input/output mode. This mode is most frequently used for control applications because it's simple to use and does not require handshake signals.

Z80 mode 3 I/O

To set a port to mode 3, you need to do two things: tell the control register that mode 3 is required and then tell it which bits should input, and which should output. Each instruction takes the form of a control word, the latter being formed by setting each bit to 0 for output, and to 1 for input (the opposite format to that used for 6502 data direction registers).

IFF1

IFF2

0 = disabled
1 = enabled

Stores IFF1
During an $\overline{\text{NMI}}$ service

Figure 12.11   *The interrupt flip-flops and interrupt mode flip-flops.*

Interrupt Mode Flip-Flops

IMF$_a$	IMF$_b$	Mode
0	0	0
0	1	not used
1	0	1
1	1	2

Z80 Interrupts	ISR Vectors
$\overline{\text{NMI}}$ (Negative edge, (1→ 0) triggered)	403B - Low byte 403C - High byte
$\overline{\text{INT}}$ Mode 0 (IM0)	The interrupting device places a Z80 instruction on the data bus which is executed by the CPU. This is normally a RESTART (RST) instruction which initiates a call to the selected one of 8 RST locations in zero-page memory.
$\overline{\text{INT}}$ Vectors Mode 1 (IM1)	4050 - Low byte 4051 - High byte
$\overline{\text{INT}}$ Vectors Mode 2 (IM2)	Peripheral supplies low byte. I register supplies high byte (programmer uses LD I, A instruction). The peripheral device can select any of a large number of ISRs located in memory.

Figure 12.12   *Summary of the $\overline{\text{NMI}}$ and $\overline{\text{INT}}$ interrupt service routine vectors.*

b_7	b_6	b_5	b_4	b_3	b_2	b_1	b_0

a) The I/O register control word for input or output. Easy to remember because 0 looks like the letter O standing for Output; 1 is like the letter I for Input.

1	1	1	1	0	0	0	0

b) The control word is F0.
This sets bits 4 - 7 as Input and bits 0 - 3 as Outputs.

Figure 12.13   *Mode 3, I/O control words.*

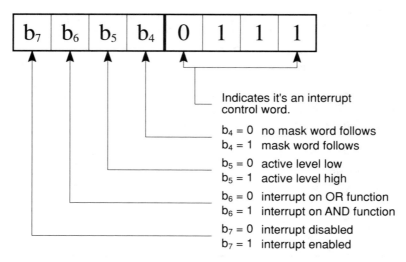

Figure 12.14 *Functions of an interrupt control word.*

Referring back to Table 12.7, we see that mode 3 requires bits 6 and 7 to be set to a 1. This is similar to setting them to 0 for mode 0 as shown in Figure 12.8. To set the individual bits, the programmer constructs the control word and this is demonstrated in Figure 12.13.

Z80 mode 2 $\overline{\text{INT}}$ interrupts (IM 2)

Setting up the interrupt mode follows a similar pattern. We've already noted that when bits 0 to 3 are all set to 1, the control register recognises an I/O control word. If bits 0 to 2 are set to 1, whilst bit 3 is a zero, then the control register expects an interrupt mode control word. This is shown in Figure 12.14, where the function of the four most significant bits is also shown.

The demonstration program that follows will require a bank of eight debounced switches to be connected to port A. Setting bits 4 to 7 to a 1 produces an interrupt control word of 0F7H and has the following effects:

1 b_7 enables interrupts;
2 b_6 causes an interrupt on the AND function;
3 b_5 sets active high mode;
4 b_4 tells the control register that a mask word follows.

Number 1 (above) is obvious, but what do the others mean? To understand them, let's look at the basic nature of mode 2 interrupts. A bank of switches is connected to one of the ports, let's say port A. This enables the user to set any of the bits to either a 1 or a zero. In a real application, the interrupting devices themselves will probably supply the appropriate logic signals. If bit 6 is set (number 2 above), then an interrupt occurs on an AND function. This works in conjunction with bits 4 and 5 so we'll look at them too.

The 'mask' word tells the CPU which bits to monitor, so if interrupt signals may appear on bits 4 to 7 of port A, then the mask would be: 0000 1111 in binary, or 0F in hex. A zero makes a particular bit active, so in this case the interrupt logic will only monitor the top four bits of port A. Bit 5 determines whether the interrupting signal should be active high or active low. Setting this bit to a 1 produces the active high mode, so that interrupts are generated only when logic ones are received. Putting all this together, an interrupt is

Table 12.16 Program to demonstrate $\overline{INT}$ mode 2 interrupts

Address	Machine code	Label	Mnemonics	Comments
			ORG 5C00H	;origin 1
			ENT 5C00H	;entry point
5C00	3E 0F		LD C,0FH	;port B to
5C02	D3 03		OUT (03),A	;output
5C04	3E 00		LD A,0	;
5C06	D3 01		OUT (01H),A	;
5C08	FB		EI	;enable interrupt
5C09	ED 5E		IM 2	;interrupt mode 2
5C0B	3E 5E		LD A,5EH	;intvec HB
5C0D	ED 47		LD I,A	
5C0F	3E CF		LD A,0CFH	;I/O mode 3
5C11	D3 02		OUT (02),A	
5C13	3E FF		LD A,0FFH	;all inputs at
5C15	D3 02		OUT (02),A	;port A
5C17	3E FE		LD A,0FEH	;intvec LB
5C19	D3 02		OUT (02),A	
5C1B	3E F7		LD A,0F7H	;interrupt control
5C1D	D3 02		OUT (02),A	;EI, AND, HIGH
5C1F	3E 0F		LD A,0FH	;mask
5C21	D3 02		OUT (02),A	
5C23	3A 00 00	LOOP:	LD A,(0)	
5C26	18 FB		JR LOOP	
			ORG 5EFEH	;new origin
5EFE	00 5E		DEFW INTVEC	
			ORG 5E00H	;new origin
5E00	CD 06 5E	INTVEC:	CALL LEDON	;interrupt
5E03	FB		EI	;routine
5E04	ED 4D		RETI	
5E06	1E 0F	LEDON:	LD E,0FH	;LED display on
5E08	3E FF		LD A,0FFH	;port B
5E0A	D3 01		OUT (01),A	
5E0C	CD 16 5E	LOOP3:	CALL DELAY	
5E0F	3D		DEC A	
5E10	D3 01		OUT (01),A	
5E12	1D		DEC E	
5E13	20 F7		JR NZ,LOOP3	
5E15	C9		RET	
				;DELAY
5E16	06 3F	DELAY:	LD B,3FH	;immediate load
5E18	0E FF	LOOP2:	LD C,0FFH	;immediate load
5E1A	0D	LOOP1:	DEC C	;decrement C
5E1B	20 FD		JR NZ,LOOP1	;
5E1D	05		DEC B	;decrement B
5E1E	20 F8		JR NZ,LOOP2	;
5E20	C9		RET	;return

therefore *only* generated if all of the chosen four bits on port A go to a logic 1 – that's where the AND function comes in. If we'd set bit 6 to a zero, then an interrupt would occur if any *one* of the top four bits went to a logic 1, by using the OR function.

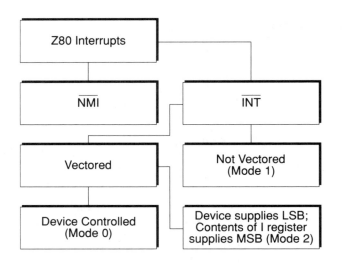

Figure 12.15 *Summary of the Z80 interrupts.*

Now let's look at the demonstration program given in Table 12.16. The first instructions set up port B for output and send zero to its data register to turn the LEDs off. Interrupts are then enabled and IM 2 sets up for mode 2 interrupts. The high byte of the interrupt vector comes next and this is loaded into the interrupt register via the accumulator.

Port A is then set up in mode 3, with all bits used for input. The interrupt vector low byte is then sent to the control register. We are now ready to load the interrupt commands. 0F7H will enable interrupts, using the AND function, and tells the CPU that the appropriate mask word follows. The mask (0FH) sets the upper four bits of port A active by loading in zeros there.

Port A data register is continuously monitored in the next instruction, with its contents being placed in the accumulator. Now, if all four most significant bits in the accumulator become 1, an interrupt occurs. In this case, the ISR causes the LEDs on port B to count down from 15 to zero. You can change this simple routine to anything you like.

More on Z80 interrupts

The $\overline{\text{NMI}}$ interrupt is relatively simple to program; the $\overline{\text{INT}}$ interrupts have three different modes of operation, 0, 1 and 2. We've suggested means of utilising modes 1 and 2, so we'll now summarise this information and describe the use of interrupt mode 0. Figure 12.15 shows the family of Z80 interrupts quite clearly.

Mode 0 interrupts

This mode utilises the Z80 RST (restart) instructions. These are special one-byte instructions like a fixed CALL. When the CPU encounters the instruction, the current PC contents are pushed onto the stack and one of eight zero-page memory locations is loaded into the PC. Program execution is hence diverted to that address and the opcode found there is fetched and executed.

Exactly what the one-byte instruction is will depend upon the system you are using or indeed the system you have designed yourself. Table 12.17 gives the restart instructions and their machine codes, together with the secondary instructions used in the MARC Z80 micro at those addresses. For example, the instruction RST 18H will cause a JumP to address 4043H. It is up to the

Table 12.17 The eight Z80 RST instructions. The final column shows what secondary instructions will be found at the zero-page addresses in the MARC microcomputer system

RST instruction	Machine code	JUMP instruction in MARC
RST 00H	C7	JP 0EC6
RST 08H	CF	DI JP 403D
RST 10H	D7	JP 4040
RST 18H	DF	JP 4043
RST 20H	E7	JP 4046
RST 28H	EF	JP 4049
RST 30H	F7	JP 404C
RST 38H	FF	JP 404F

programmer to insert the required interrupt service routine (or interrupt vectors) at that point.

The interrupting peripheral will supply the one-byte restart address; in its simplest form, this can be done by setting up the binary pattern using the 5V supply via a 1K resistor for '1' and grounding the rest for '0' as shown in Figure 12.16.

There are several I/O chips available which can provide the necessary functions for interrupt logic, notably the Z80 PIO and the 8255 PIO. The latter has programmable interrupt control logic; the Z80 PIO also has daisy chain (see next section) priority interrupt logic in order to provide automatic interrupt vectoring without external logic. The Z80 PIO is shown in Figure 12.17.

Daisy chain connections

Z80 family PIOs can be connected one after another in what is called a 'daisy chain'. The relevant pins are shown in Figure 12.17 and correct connection of these pins allows priority to be established in a set of PIOs. A simple representation of a typical system consisting of four Z80 PIOs connected to a single CPU is shown in Figure 12.18.

The PIO can only initiate an interrupt when its IEI pin is at a logic 1. PIO1, therefore, has that pin connected permanently to the 5V line, giving it the highest priority; its interrupts will *always* be serviced immediately by the CPU, regardless of what the other PIOs are doing (assuming of course that interrupts have been previously correctly enabled).

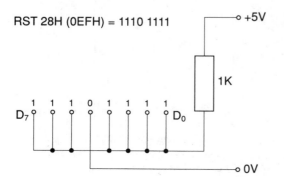

RST 28H (0EFH) = 1110 1111

Figure 12.16 *A method of holding the restart instruction (RST 28H) in binary. The interrupt logic places the pattern on the data bus at the appropriate time.*

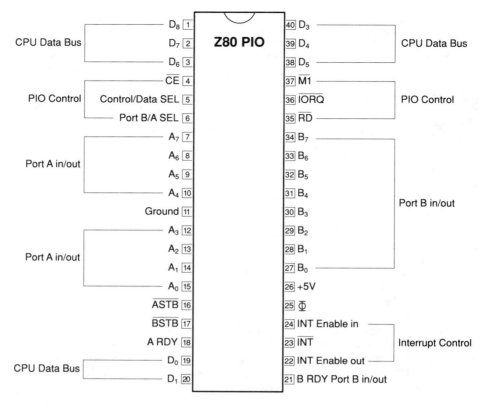

Figure 12.17 *The Z80 PIO showing connections. The interrupt control pins are 22, 23 and 24.*

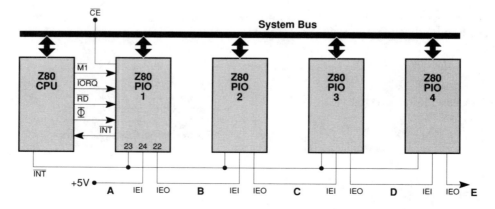

Figure 12.18 *Illustrating multiple PIO connections to a single Z80 CPU. Some control lines shown between CPU and PIOI also go to all other PIOs and are omitted only for clarity. The interrupt enable in (IEI) and interrupt enable out (IEO) connections are shown at the bottom.*

Table 12.18 Logic states during a multiple interrupt sequence; column headings relate to labels in Figure 12.18

State	A	B	C	D	E
1	1	1	1	1	1
2	1	1	1	0	0
3	1	0	0	0	0
4	1	1	1	0	0
5	1	1	1	1	1

When the PIO produces the interrupt signal, its IEO signal goes low, inhibiting all other PIOs in the chain. A typical sequence is described in Table 12.18.

In the initial state, all interrupt enable signals are at a logic 1, because each PIO has IEI and IEO equal in the absence of an interrupt. In the second state, PIO3 has generated an interrupt forcing its IEO pin down to zero, inhibiting PIO4 and any others that may exist beyond point E.

In the third state, PIO1 has generated an interrupt, forcing its IEO pin to zero. PIO4's routine is put on hold and PIO1 interrupt is acknowledged and serviced by the CPU. Only when it is completed can PIO4 continue its interrupt service routine. In this way, A has priority over B, C, D and E, whilst B has priority over C, D and E, etc.

Summary

The whole subject of interrupts is quite complex as the reader will have observed. The use of $\overline{\text{NMI}}$ and $\overline{\text{INT1}}$ interrupts is fairly straightforward as we have seen, but the other two Z80 modes require more consideration and further analysis of the implementation of such systems is normally reserved for studies at a higher level. We conclude this chapter with a summary of the various Z80 interrupts.

Summary of Z80 interrupts

$\overline{\text{NMI}}$ Has a single mode response.

$\overline{\text{INT}}$ Has three modes; they are enabled through software and can be connected to multiple peripheral devices in a wired-OR configuration.

IM 0   (Interrupt mode 0); the interrupting device places an instruction on the data bus. This is usually a restart instruction which initiates a call to one of the eight available RST locations in zero-page memory.

IM 1   (Interrupt mode 1); this mode has only one restart location, namely, 0038H. In the 'MARC' this produces a JumP to 404FH; the user's interrupt vectors are placed in the two memory locations which follow as shown in Table 12.11.

IM 2   (Interrupt mode 2); in this mode, the interrupting peripheral selects the start address of the ISR by placing an 8-bit vector on the data bus during the 'interrupt acknowledge' cycle which forms the lsb and the CPU takes the contents of the interrupt register to form the msb. In this way, the peripheral device can select many different service routines which can be anywhere in memory.

Questions

1 Why are interfacing circuits necessary when using computer I/O ports?

2 (a) What is switch contact bounce and how can the effect be eliminated?

   (b) Describe, using suitable diagrams, two types of switch debounce circuits.

3 Why are control registers (Z80) or data direction registers (6502) necessary when using I/O ports?

4 6502 users will be aware that when a bit in the DDR is set to a 1, then it is set up for output. Z80 users will know that the control registers must be initialised to mode 3, followed by a 0 for each bit required for output, to obtain the same configuration. In either case, these instructions are followed by sending the actual DATA out to the data registers. State the bit patterns which would be obtained by sending the following hexadecimal values to the data registers:

   (a) 35H   (b) 11H   (c) AAH (d) FEH (e) 88H

   (Note, for example, that 62H would produce: 0110 0010)

The following questions will require a bank of eight LEDs, or some other indicating equipment, correctly interfaced, to be connected to one or other of the output ports:

5 Write a program which will produce a binary count on a bank of eight LEDs connected to port A.

6 Write a program which will produce the traffic light sequence. A simple display, consisting of red, yellow and green LEDs, may be devised to check the program. More advanced programmers may wish to consider four such sets, working in unison, simulating the action of the traffic lights at a crossroads.

7 (a) Write a program which will produce the 'Knightrider' effect. (This consists of a single LED in a row of eight being turned on at any one time. The first is illuminated and after a short delay is turned off; the second then comes on and so on down the line, returning in a similar sequence when it gets to one end. The effect was used on the front of a car in the *Knightrider* TV series, hence the name.)

     Try various delays until you obtain a suitable effect. Remember that the delay must be the same throughout, including that at each end.

   (b) Try rewriting the program so that *all* the LEDs are on but one, and a 'hole' seems to travel along the line of LEDs.

   (c) Combine both, so that the one or the other may be selected according to the setting of a switch on port B.

8 Modify your binary count program (Question 5) so that the setting of eight different switches on port B affects the speed of the sequence.

9 Write a program so that the setting of the switches on port B is reflected in the pattern of a set of LEDs connected to port A.

10 Write a program which will cause a stepper motor connected to port A to go through a series of predefined steps, backwards and forwards. (See Chapter 14 for details of construction of a suitable interface, if necessary.)

Interrupts

11 What is the main difference between:
    (a) (6502) $\overline{\text{NMI}}$ and $\overline{\text{IRQ}}$ interrupt signals?
    (b) (Z80) $\overline{\text{NMI}}$ and $\overline{\text{INT}}$ interrupt signals?
12 What is the purpose of interrupt vector memory locations?
13 (6502) In the $\overline{\text{IRQ}}$ demonstration program given in Table 12.6, state:
    (a) the effect of the CLI instruction;
    (b) the consequences of not including this instruction.
14 (6502/Z80) Modify the 'Knightrider' program to count the number
    of times b_4 flashes using an $\overline{\text{NMI}}$ interrupt. The count should be
    output to port B.
15 (6502/Z80) Modify the traffic lights program to count the number
    of times the lights are on green:
    (a) (6502) Use an $\overline{\text{IRQ}}$ interrupt and output the count to port B;
    (b) (Z80) Use an $\overline{\text{INT}}$ (mode 1) interrupt and output the count to
          port B.

Assignment 12.1

I/O data transfer: using Z80 mnemonics and identifying the addressing modes
used, produce suitable documentation and run a program to:

(a) Transfer the content of memory location 4500H to port A. The data
     transferred should be retained in the B register.
(b) Ttransfer the value 20H to memory location 4505H, retaining the value
     in the B register.
(c) Transfer the content of port B to memory location 4600H, retaining the
     data transferred in the B register.

Assignment 12.2

I/O data transfer: using 6502 mnemonics and identifying the addressing modes
used, produce suitable documentation and run a program to:

(a) Transfer the content of memory location &0250 to port A. The data trans-
     ferred should be retained in the X register.
(b) Transfer the value &20 to memory location &0255, retaining the value in
     the X register.
(c) Transfer the content of port B to memory location &0260, retaining the
     data transferred in the X register.

Chapter 13
Writing software – more advanced techniques

Z80 assembly language program documentation and layout

It is very useful if all programming documentation is of a standard type. This helps the programmer to get used to a particular style and form the habit of producing well-documented programs. An assembly language program should have a neat and logical layout so that a hard copy will be both easy to read and understand. The more information you include, the better. It is often the case that when you come back to a program you yourself have created, you may still find yourself asking: 'Now I wonder why I did that!' especially if you come back to it some months later. Program comments can prove to be extremely useful at such times!

A layout of assembly language programs has been developed by the US Army Electronics Command, and is the type of layout used by most assembly language practitioners. It is presented here as a guide; other layouts would also suffice for most programming requirements and you will note that we used something similar in the last chapter when presenting the program contained in Table 12.13.

The program in Table 13.1 illustrates the kind of layout which should be used when writing Z80 assembly language programs.

The program gives an accurate delay of one second for a processor clock rate of 2MHz. The operating time for each instruction is given in the comments column and the total for each section is shown on the full comment lines before that section.

The machine code for the instructions is also shown in the comments column.

Delay routine timing (Z80)

The Z80 instruction set usually gives execution times for a clock frequency of 4MHz. Since the MARC clock runs at 2MHz, these times must be multiplied by two as they take twice as long. A typical time delay program is shown below:

Label	Operator and operand	
DELAY:	LD	B,0FFH
LOOP2:	LD	C,0FFH
LOOP1:	DEC	C
	JR	NZ,LOOP1
	DEC	B
	JR	NZ,LOOP2
	RET	

Table 13.1 Preferred layout for Z80 assembly language programs

```
; ****************************************************************************************
; * DL1S              Delay for one second                                            *
; ****************************************************************************************
; JOB                Time critical routine to use one second's worth of clock cycles,
                     including the call to DL1S.
;
; ACTION             Set loop count = INT(Clock Hz/n)
;                    For count: [ use n cycles ].
;                    Fine tune for remainder of cycles.
```

; CPU	Z80 running at 2MHz
; HARDWARE	None.
; SOFTWARE	None.

; INPUT	None.
; OUTPUT	None.
; ERRORS	Inaccurate if interrupt occurs.
; REG USE	None.
; STACK USE	6.
; RAM USE	None.
; LENGTH	19.
; CYCLES	1999983 [as (42551 * 47) + 86]

```
;
LPCNT:    EQU    42551         ;loop counter value in decimal (0A637H)
;
; .... Save and initialise. Uses 32 cycles.
;
DL1S:     PUSH   AF            ;11.    Save registers and     F5
          PUSH   BC            ;11.    Flags used in DL1S.    C5
          LD     BC,LPCNT      ;10.    Set loop counter.      01 37 A6
;
; .... Main delay loop. Use up [ (47 * 42551) – 5 ] cycles.
;
DLOOP:    PUSH   HL            ;11.    Use 21 cycles.         E5
          POP    HL            ;10.                           E1
          DEC    BC            ;6.     Decrement counter      0B
          LD     A,C           ;4.     and test for zero.     79
          OR     B             ;4.     Repeat till count      B0
          JR     NZ,DLOOP      ;12.    done.                  20 F9
;
; Fine tune, restore and return. Uses 59 cycles.
;
          PUSH   HL            ;11.    Use 21 cycles.         E5
          POP    HL            ;10.                           E1
          NOP                  ;4.     Use 8 cycles.          00
          NOP                  ;4.                            00
          POP    BC            ;10.    Restore registers      C1
          POP    AF            ;10.    and flags used.        F1
          RET                  ;10.    Exit after 1 second.   C9
```

Timing analysis

LD B	(1.75×2)	=	3.5µs
LD C	(1.75×2)	=	3.5µs
DEC C + JR	$254(2 + 6)$	=	2032 µs
Final JR + DEC C	$(3.5 + 2)$	=	5.5 µs
so: for an LD C, DEC C, JR		=	2041 µs

EACH JR NZ needs 6µs, each DEC B, 2µs

so: each 'B DECREMENT' takes $(2041 + 6 + 2 + 3.5)$

$$= 2052.5 \text{ µs}$$

The 3.5µs is for the LD C

B is loaded with 0FFH (255D)

This is (2052.5×254) $\quad\quad\quad = \underline{521{,}335\text{µs}}$

The final time the length is 2050µs as the final jump does not succeed.

Total: $521{,}335 + 2050 + 3.5 + 5 = 523{,}393.5\text{µs}$

Alternative time delay program

The first delay routine has two loops, this is so that the C register can be reloaded with 0FFH for every time the B register is decremented. A slightly neater method is shown below, where the B register starts from zero. Decrementing a register containing zero results in its being 'reduced' to 0FFH. Subsequent decrements take the register contents back to zero so there's no need to reload it.

Label	Operator and operand		Comments
DELAY:	LD	C,64H	
	LD	B,00H	
LOOP:	DEC	B	;(255 × 8) + 5.5
	JR	NZ,LOOP	; = 2045.5µS
	DEC	C	
	JR	NZ,LOOP	
	RET		

Timing analysis

The inner loop of 2045.5µs + 2µs (for DEC C) + 6µs gives a total loop time of 2035.5µs. 64H = 100D so this loop will happen 99 times. The final time the length is 2051µs as the final jump does not succeed; this is $(2053.5 - 2.5)$µs as the fall through; JR NZ is $(6 - 3.5)$µs less.

This gives:

$$(99 \times 2053.5) + 2051\text{µs} = 205{,}347.5\text{µs}$$

Time for LD B, LD C and RET = $(3.5 + 3.5 + 5)$µs

$$= 12 \text{ µs}$$

Total time for subroutine = $\underline{205{,}359.5\text{µs}}$

Obtaining specific time delays

Suppose we want a time delay for half a second using the subroutine given above. We have seen that setting C to 64H (100D) produces a time delay just in excess of 205.4ms. Since we require 500ms, the value of 99 is changed to X as follows:

$$(X \times 2053.5) + 2051 + 12 = 500,000\mu s$$

From this we find that $X = 242.5$, but we can't have half a loop so we must use 242. This is 0F2H, but we'll need to load the C register with 0F3H as the final jump does not succeed and its shorter time is already accounted for in the equation (2051µs).

This gives:

$$(2053.3 \times 242) + 2051 + 12 = 499,010\mu s$$

This is not quite correct, so we 'pad' the loop with NOPs which take 2µs. Using two of these (before DEC C) gives:

$$(2057.5 \times 242) + 2055 + 12 = 499,982\mu s$$

We then need 9 NOPs in succession before the RET instruction to produce the final 18µs required, producing a time delay of exactly half a second. Note that this example is illustrative only; if the delay loop is called as a subroutine, you will also need to include the RET statement which takes 10 cycles and may need to use the stack to push and pop some registers.

6502 delay timing

The preceding comments on program documentation apply equally when writing programs in 6502 code. As an example, we'll look at some simple delay routines written in 6502 code. We'll assume that you're using a system such as the 'EMMA' which has a clock frequency of 1MHz.

Label	Operator and operand		Comments
.DELAY	LDY	#&64	; 2 cycles (2µs)
	LDX	#&00	; 2 cycles (2µs)
.LOOP	DEX		; (255 × 5) + 4
	BNE	LOOP	; = 1279µs
	DEY		; 2 cycles (2µs)
	BNE	LOOP	
	RTS		; 6 cycles (6µs)

Timing analysis

LDX, LDY, DEX and DEY all need 2 cycles each. BNE needs 3 cycles when it branches (which is every time but the last) and only needs 2 cycles if it doesn't branch, i.e. if it 'falls through'. The inner loop consists of DEX and BNE which normally take 5 cycles, there are 255 of these. The final one falls through so, since BNE only needs 2 cycles, we add on 4 cycles for it and the DEX. This loop therefore takes 1279 cycles as shown in the comments column.

When the inner loop fall through occurs, the processor encounters DEY and BNE LOOP again, so this loop needs (1279) + 5 cycles, 1284 cycles in all. We have set Y to &64 (64H) which is 100D, so the complete loop occurs 99 times. The final, fall through BNE, needs 1 cycle less, so this takes 1283 cycles.

So now we have:

(99 × 1284) + 1283	=	128,399 cycles
LDY, LDX and RTS	=	10 cycles
TOTAL delay time:		128,409 cycles
This is therefore:		128,409 µs

For a 1s delay, we'll need one million cycles. So we'll need:

1,000,000 / 1284 loops = 778.8 (approx)

This could be done, even though 778 won't fit into any register. We could have another loop which will take the delay routine around three times, or simply call the routine three times. This isn't really satisfactory, however, and a better way is to decrement a memory address. DEX and DEY require 2 cycles, but a DEC (absolute) needs 6, and a DEC (zero page) needs 5. Can you modify the program so that, using DEC (memory), a delay of exactly 1 second is produced?

Structured programming

We have already noted the wisdom of using a structured approach to programming, so that even for the simplest of programs we always use a standard methodology. After formulating the algorithm we draw up a flow chart, a 'story-board', or a structured design such as Jackson Structured Programming (JSP). It is to this latter methodology that we now turn our attention, and we shall be using only this system from now on.

The use of flow charts as an aid to programming has been discredited by many modern programmers. JSP is now gaining in popularity, and the reader will soon find it is undoubtedly the superior technique. We have used flow charts to illustrate the sequence of compiling and interpreting (in Chapter 5) and the same technique to illustrate the fetch–execute sequence (Chapter 7). In such cases, the flow chart method offers simplicity and clarity, but for more complex programs, JSP is to be preferred. Which system you intend to adopt for your own programming is up to you – but read on first!

In all systems, once the general structure has been produced, the program may be written in assembly language. Although hand assembly still has its uses (for 'patching' and correcting code), we shall be concentrating on using automatic assembly from now on. The assembly language program consists of what is called source code; after hand or automatic assembly, the object code is obtained. Any errors are then detected and corrected (a process known as debugging); a monitor program may be used to step through the program line by line to assist in the process. The completed program may then be RUN, stored on disc or blown into an EPROM or similar device. This process is summarised in Table 13.2.

Jackson Structured Programming (JSP)

The programmer starts from the premise that the complete program is too difficult to comprehend in its entirety. The problem is broken down into a sequence of more manageable components. Each such component is then broken down into smaller components, and so on until a level is reached which cannot be further simplified.

Without this technique, a programming problem can be too large and complex in its complete form. The above process produces a program structure resembling a tree diagram, as shown in Figure 13.1.

Table 13.2   Steps in producing a machine code program

1   Write down the algorithm in plain English
2   Draw up a structured design using JSP
3   Check operation of proposed design against software requirement
4   Write the program in assembly language mnemonics – this produces the 'source code'
5   Use an assembler to convert the source code into object code – the object code is the machine code in hex
6   Debug, i.e. detect and correct any errors
7   Run completed program; store on disc or in PROM

The elementary components

Each level lists the procedures (processes) that describe the complete program. Level 1 describes the program, but this level is still too high a level of abstraction. Each procedure (P) in level 1 is therefore further decomposed into procedures, giving level 2. Notice that the program consists of procedures P1, P2 and P3 in that order. Procedure P1 consists of a sequence, procedures P11 (pronounced P one,one) followed by P12. Procedures P2 and P3 also consist of further procedures.

In its simplest form, a program may be represented as in Figure 13.2. As shown, you always start at the left (input data) and then move to the right, dealing with each elementary component as you go along.

If the elementary component is too complex, this can be broken down further. For example, P_{data} (P standing for process) could consist of three elementary components. These might be:

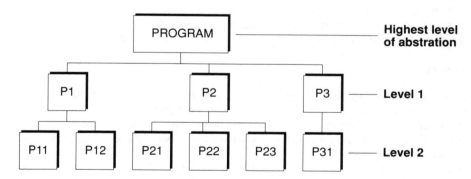

Figure 13.1   *Basic principle of Jackson Structured Programming.*

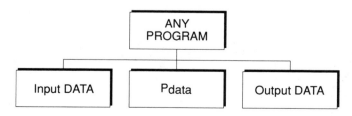

Figure 13.2   *Some of the basic elements of a JSP design.*

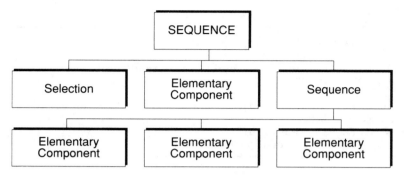

Figure 13.3 *An example of a sequence consisting of three parts.*

1   check for negative number and convert to absolute value;
2   calculate square root;
3   put into numerical order.

This is shown in a general way in Figure 13.3.

In this way, small sections of code can be built up as subroutines and each one called in turn.

Basic components of JSP diagrams

All computer programs can be described using only four basic components, they are:

1   *elementary components* which are not further subdivided into constituent parts;
2   a *sequence* of two or more parts, occurring in order;
3   a *selection* of one part from a number of alternatives;
4   an *iteration* in which a single part is repeated zero or more times.

Sequences

Each separate part of a sequence, a selection or iteration may be a sequence, a selection or iteration, so there is no limit to the complexity of structure which can be formed. For example, a sequence of three parts may be similar to that shown in Figure 13.4. The sequence itself may have two or more parts, occurring once each, *in order*.

A good way to describe a sequence is to consider the construction of an electronic unit (let's say a power supply unit); the main components will be the printed circuit board (PCB), transformer, resistors, capacitors, diodes and transistors.

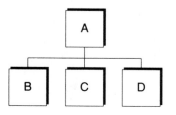

Figure 13.4 *A sequence of three parts.*

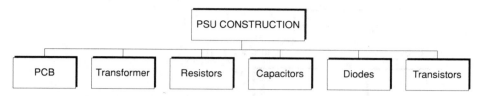

Figure 13.5 *First stage JSP design for a microelectronic system which will control the construction of a PSU. The automated system first selects the correct PCB, then goes on to select the correct transformer followed by all the other components, in sequence.*

Writing software to drive an automated system that will produce the PSU would be a daunting task, but with JSP it's relatively easy. The first stage is simply to define the overall process: 'PSU construction'. This is the process which occupies the main box at the top. The second stage is to define the sequence that must take place; this is shown in Figure 13.5.

The main box could, in theory, be labelled with any process. The procedures which follow it are software defined in the program; the hardware of the microelectronic system could be capable of producing any number of different electronic units.

Selection

The selection component indicates what alternatives are available when a choice is to be made of one single part. The selection process is shown in Figure 13.6. With this information it is now possible to put further detail on the PSU construction chart as shown in Figure 13.7. For clarity the number of alternatives has been limited.

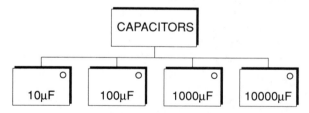

Figure 13.6 *For purposes of illustration we have chosen just one box from the sequence shown in Figure 13.5. Each box may have several selections and there may of course be more or fewer selections available. The small circles in the boxes indicate that ONLY ONE of the alternatives is to be selected. Where (in this case) more than one capacitor may be required, more boxes would need to be inserted in the sequence, e.g. C1, C2, C3, etc.*

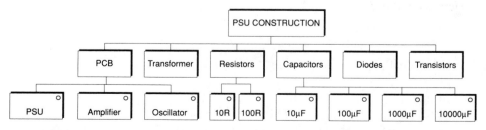

Figure 13.7 *Added detail on the PSU construction chart. In a real program there may be selections for the transformer, diode and transistor. These possibilities have been omitted simply for clarity.*

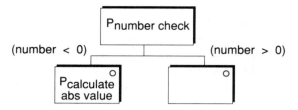

(number < 0)   (number > 0)

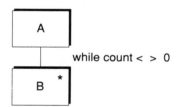

Figure 13.8 *The point is that you can't find the square root of a negative number. The process simply multiplies the number by −1 producing the absolute value (numerically the same but no sign); the next box contains nothing (there being nothing more to be done here), it's called a 'dummy box'. Program execution would therefore continue with the next box in the sequence.*

Sometimes a selection will consist of only one part which may be of interest, while any other possibility is of no interest. For example, it is always prudent to check a number to determine if it is negative, before the square root of the number is to be calculated. If it is negative, then the absolute value of the number can be determined. If the number is positive then no action needs to be taken. This can be done as shown in Figure 13.8.

Iteration

An iteration consists of one part which is repeated zero or more times. This is shown diagrammatically in Figure 13.9. The asterisk indicates that B is repeated zero or more times; A represents the complete process.

```
┌─────────────┐
│             │
│      A      │
│             │
└──┬──────────┘
   │           while count < > 0
┌──┴───────┐ *
│          │
│    B     │
│          │
└──────────┘
```

Figure 13.9 *Representation of an iterative process. The asterisk indicates that B is repeated zero or more times; A represents the complete process.*

Example 1

To see how JSP can be used, let's use the first example in Chapter 5, adding two numbers together, 8 and 6.

The machine code is usually written down by writing the code in up to three (sometimes four) columns to the left of the mnemonics, and after the address in memory where the program is placed. In the Marc Z80 system this is often at address 5C00H; in the EMMA 6502 system, 0200H.

Hence for the Z80:

```
5C00H   3E   08     LD   A,8
5C02H   C6   06     ADD  A,6
```

and for the 6502:

```
0200H   A9   08     LDA  #8
0202H   18          CLC
0203H   69   06     ADC  #6
```

Now it's possible to draw up complete, structured, design documentation for our simple addition program according to Table 13.2, including a JSP chart.

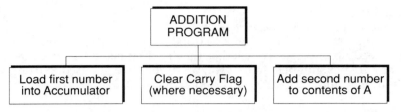

Figure 13.10   *Typical JSP design for ADDITION PROGRAM.*

1   *Algorithm*
    Program which adds two single integer numbers together.
2   *JSP*
    A JSP design is given in Figure 13.10.
3   *Check operation*
4   *Assembly mnemonics*
    (Z80):   LD A,8   ADD A,6
    (6502):   LDA #8   CLC ADC #6
5   *Machine code*
    (Z80):   5C00H   3E   08   C6   06
    (6502):   0200H   A9   08   18   69   06
6   *Debug*
    This is not really necessary with such a simple program.
7   *Run*
    The program operates according to the design specification.

Example 2
Let's add two numbers together again, but this time, we'll assume that we
want to retain each of the numbers as well as the result of the addition. The
documentation, in line with Table 13.2, may be as follows:

1   *Algorithm*
    Program which accepts two numbers, adds them together, and saves the
    two numbers and their sum in three consecutive memory locations.
2   *JSP chart*
    The charts are given in Figures 13.11 and 13.12.

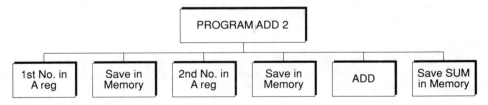

Figure 13.11   *JSP design for Example 2 (Z80 version).*

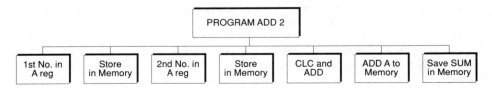

Figure 13.12   *JSP design for Example 2 (6502 version).*

3 *Check operation*
4 *Assembly mnemonics* and 5 *Machine code*
In this example, let's use the numbers 11 and 15. Since both are greater than 9, we need now to specify whether they're decimal or hexadecimal. Assume they are hex numbers so that we can continue to get used to using the notation: 11H and 15H (Z80), &11 and &15 (6502).

Z80

Machine code	Mnemonics	Comments
3E 11	LD A,11H	;first number into A
32 00 5D	LD (5D00H),A	;save in memory
47	LD B,A	;save in B reg
3E 15	LD A,15H	;second number into A
32 01 5D	LD (5D01H),A	;save in memory
80	ADD A,B	;add numbers together
32 02 5D	LD (5D02H),A	;save in next memory
76	HALT	;halt processor

6502

Machine code	Mnemonics	Comments
A9 11	LDA #&11	;first number into Acc
8D E0 02	STA &02E0	;store in memory
A9 15	LDA #&15	;second number into Acc
8D E1 02	STA &02E1	;store in memory
18	CLC	;clear carry flag
6D E0 02	ADC &02E0	;perform addition
8D E2 02	STA &02E2	;store result in memory
00	BRK	;breakpoint

6 *Debug* and 7 *Run*
The reader can proceed with 6 if required. After RUNning the program, examine the appropriate memory locations to see that the correct effect has been achieved.

Example 3
This is another example taken from Chapter 5, showing how numbers can be moved in and out of CPU registers, manipulated and stored in selected areas of memory. Try drawing up a JSP chart for this example (see Tables 13.3 and 13.4).

As a starting point, we've broken down the program as far as it can go. In practice, the 'addition program' is more likely to be a subroutine – just a small part of a much larger program. As a general rule, there should be no more than seven levels on a JSP chart. If more are needed, then they should be started on a separate sheet from one, more complicated part of the program. Hence, program 'ADD2' would reduce to a subroutine perhaps named 'SUM', containing three boxes, say 'LOAD DATA', 'ADD' and 'SAVE' as shown in Figure 13.13.

Table 13.3   Data for Example 3. We want to move the data from block 1 to block 2, adding up the numbers on the way and putting the sum at the end of block 2, i.e. at 5E0AH

Block 1		Block 2	
Memory location	Data	Memory location	Data
5D00H	00	5E00H	**
5D01H	05	5E01H	**
5D02H	0A	5E02H	**
5D03H	0F	5E03H	**
5D04H	14	5E04H	**
5D05H	19	5E05H	**
5D06H	1E	5E06H	**
5D07H	23	5E07H	**
5D08H	28	5E08H	**
5D09H	2D	5E09H	**
5D0AH	**	5E0AH	??

Table 13.4   Data for Example 3, 6502 version

Block 1		Block 2	
Memory location	Data	Memory location	Data
&02D0	00	&02E0	**
&02D1	05	&02E1	**
&02D2	0A	&02E2	**
&02D3	0F	&02E3	**
&02D4	14	&02E4	**
&02D5	19	&02E5	**
&02D6	1E	&02E6	**
&02D7	23	&02E7	**
&02D8	28	&02E8	**
&02D9	2D	&02E9	**
&02DA	**	&02EA	??

The great value of this is that the JSP chart can be directly translated into the code which may be written as follows:

```
;
        CALL  SUM
;
SUM:    CALL  LOAD
        CALL  ADD
        CALL  SAVE
;
NEXT:
```

In some cases it would be appropriate simply to have a box labelled 'SUM', since an experienced programmer would have no difficulty in writing the code for the routine, and would not need such a detailed chart in order to achieve it.

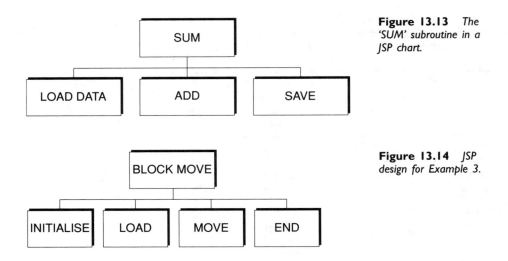

Figure 13.13   *The 'SUM' subroutine in a JSP chart.*

Figure 13.14   *JSP design for Example 3.*

Solution to Example 3

With previous comments in mind, the block move program could be represented as shown in Figure 13.14, resulting in the code given in Table 13.5. This is a truly structured programming style which you are encouraged to use in all future programming. It will also be a useful exercise to change the programs appearing in the next few chapters to this style before loading and running them on your machine.

We have incorporated some user routines and equates (EQU) into the program, and other devices used in automatic assembly and these are explained below.

Using an automatic assembler

The use of the assembler has been hinted at in Chapters 5 and 6 and some readers may already be familiar with the techniques involved. The assembler converts the source code (the assembler mnemonics) into machine code, works out relative jumps automatically and checks for errors. It also allows the use of labels, built in subroutines and 'pseudo ops'. The MARC system, for example, has subroutines to blank out the screen, remove the cursor, provide delays, scan the keyboard for input and set up titles, prompts and messages simply and easily.

Pseudo-ops

A pseudo-op is a mnemonic used within a program to be automatically assembled. We have already seen 'ORG' and 'ENT'. They don't actually generate machine code instructions as do LDs and JPs, etc. so they're called pseudo-ops. Many micro systems also have additional pseudo-ops, such as EQU, DEFB and DEFW. We discuss EQU below; DEFB generates one byte of data rather than an instruction, so one way of putting messages on the screen is to use the DEFB (DEFine Byte) instruction. The program shown in Table 13.6 shows one example of its use. There are quite a few other features in this program as follows:

1   The entry point (ENT) may be defined using a label; 'COMM' (commence) is shown in the example. A commonly used routine may be inserted before this so that, although program code may start at 5C00H, the entry

Table 13.5 Possible Z80 program solution to Example 3 using JSP

Address	Machine code	Label	Mnemonics	Comments
		INITS:	ORG 5C00H	;origin
			ENT COMM	;entry point
		;		
		CLS:	EQU 01B9H	;clear screen routine
		CRSOFF:	EQU 02C9H	;cursor off routine
		VIDEO:	EQU 0E000H	;first video RAM address
		BLOCK1:	EQU 5D00H	;first source address
		BLOCK2:	EQU 5E00H	;destination address
		;		
5C00	CD B9 01	COMM:	CALL CLS	;clear screen
5C03	CD 0C 5C		CALL LOAD	;
5C06	CD 17 5C		CALL MOVE	;
5C09	CD 26 5C		CALL END	;
		;		
		;BODY		
		;		
5C0C	21 00 5D	LOAD:	LD HL,BLOCK1	;start address
5C0F	01 00 5E		LD BC,BLOCK2	;destination address
5C12	16 0A		LD D,0AH	;number of items
5C14	1E 00		LD E,00H	;for running total
5C16	C9		RET	
		;		
5C17	7E	MOVE:	LD A,(HL)	;first data value
5C18	02		LD (BC),A	;move it
5C19	B3		ADD A,E	;create running total
5C1A	5F		LD E,A	;save running total
5C1B	23		INC HL	;point to next data byte
5C1C	03		INC BC	;point to next destination
5C1D	15		DEC D	;reduce loop counter by 1
5C1E	C2 17 5C		JP NZ,MOVE	;do again
5C21	7B		LD A,E	;running total to A
5C22	32 0A 5E		LD (5E0AH)	;store in memory
5C25	C9		RET	
		;		
5C26	C3 26 5C	END:	JP END	

point is at COMM. This is useful because you can tell the assembler where ENT is without actually knowing the address.

2 CLRSCR is a routine stored in memory starting at 01B9H. Once 'equated', using EQU as shown, the routine may be called at any time in the program using the label. The same applies to 2C9H which is a routine to turn the cursor off. 'VIDEO' is used to store the address of the first video RAM memory location.

3 To use DEFB, the label 'MESSGE' is used at the end of the program as shown. The message is inserted in single inverted commas. This is followed by a zero which is used in conjunction with the CP 0 instruction to indicate the end of the message.

The message is pointed to by the HL register pair; DE is used to store the video RAM address. The loop causes the first byte of the message to be stored in the accumulator; CP 0 checks for end of message; the byte in the accumulator is stored in video RAM using the instruction LD (DE),A. HL and DE are then incremented in order to point to the next byte of the message and next video RAM location respectively.

Table 13.6 Z80 program showing the use of DEFB and other routines

Address	Machine code	Label	Mnemonics	Comments
		INITS:	ORG 5C00H	;origin
			ENT COMM	;entry point
		CLS:	EQU 01B9H	;clear screen routine
		CRSOFF:	EQU 02C9H	;cursor off routine
		VIDEO:	EQU 0E000H	;first video RAM address
		;		
5C00	CD B9 01	COMM:	CALL CLS	;clear screen
5C03	CD C9 02		CALL CRSOFF	;cursor off
		;		
5C06	CD 09 5C		CALL MAIN	
		;		
5C09	CD 12 5C	MAIN:	CALL MESS	;message
5C0C	CD 24 5C		CALL CHARS	;characters
5C0F	CD 3F 5C		CALL END	
5C12	21 42 5C	MESS:	LD HL,MESSGE	;load message pointer
5C15	11 32 E0		LD DE,VIDEO+50	;load video RAM pointer
5C18	7E	LOOP:	LD A,(HL)	;first byte to accumulator
5C19	FE 00		CP 0	;check for end of message
5C1B	28 06		JR Z,DONE	;JumP if all done
5C1D	12		LD (DE),A	;first byte into Video RAM
5C1E	23		INC HL	;point to next message byte
5C1F	13		INC DE	;point to next video loc.
5C20	C3 18 5C		JP LOOP	;do again
5C23	C9	DONE:	RET	
		;		
5C24	DD 21 A5 E0	CHARS:	LD IX,VIDEO+165	;indexed addressing
5C28	06 20		LD B,20H	;immediate load
5C2A	DD 70 00	LOOP2:	LD (IX+0H),B	;
5C2D	78		LD A,B	;
5C2E	D6 FF		SUB 0FFH	;check for 256 chars
5C30	FE 00		CP 0	;
5C32	28 0A		JR Z,END	;
5C34	04		INC B	;next character code
5C35	DD 23		INC IX	;
5C37	DD 23		INC IX	;
5C39	DD 23		INC IX	;
5C3B	C3 2A 5C		JP LOOP2	;
5C3E	C9	FINISH:	RET	
		;		
5C3F	C3 3F 5C	END:	JP END	;
5C42	4D 49 43 52 4F 50 52 4F 43 45 53 53 4F 52 20 54 45 43 48 4E 4F 4C 4F 47 59 00			
		MESSGE:	DEFB 'MICROPROCESSOR TECHNOLOGY',0	
5C5C	76		HALT	

4 We've also shown that registers can be loaded by using labels as in LD HL,MESSGE and LD DE,VIDEO+50. The latter instruction shows how the label can be modified by adding a displacement.

5 The second half of the program gives an example of indexed addressing. In this case, it allows the ASCII codes to be displayed.

Table 13.7 Z80 program showing the use of DEFB, VDU and MOVCSR. We've omitted some of the message computer generated code for clarity. The actual spacing of the text on the screen may have to be changed

Address	Machine code	Label	Mnemonics	Comments
		INITS:	ORG 5C00H	;origin
			ENT COMM	;entry point
		CRSOFF:	EQU 02C9H	;cursor off routine
		CLS:	EQU 01B9H	;clear screen routine
		VIDEO:	EQU 0E02AH	;video RAM address
		VDU:	EQU 0525H	;enables DEFB to be used
		MOVCSR:	EQU 02FDH	;moves cursor
		;		
5C00	CD C9 02	COMM:	CALL CRSOFF	;call subroutine
5C03	CD B9 01		CALL CLS	;clear screen
		;		
5C06	CD 09 5C		CALL MAIN	
		;		
5C09	CD 0F 5C	MAIN:	CALL MESS	;message
5C0C	CD 3F 5C		CALL LOAD	
		;		
5C0F	21 07 01	MESS:	LD HL,0107H	;new cursor co-ordinates
5C12	CD FD 02		CALL MOVCSR	;move cursor
5C15	21 47 5C		LD HL,MESSGE	;first message
5C18	CD 25 05		CALL VDU	;get it printed on screen
5C1B	21 00 05		LD HL,0500H	;new cursor co-ordinates
5C1E	CD FD 02		CALL MOVCSR	;move to new position
5C21	21 61 5C		LD HL,MESS1	;point to next message –
5C24	CD 25 05		CALL VDU	;and print it on the screen
5C27	21 6C 5C		LD HL,MESS2	;do again
5C2A	CD 25 05		CALL VDU	;
5C2D	21 A8 5C		LD HL,MESS3	;
5C30	CD 25 05		CALL VDU	;
5C33	21 D4 5C		LD HL,MESS4	;
5C36	CD 25 05		CALL VDU	;
5C39	21 E1 5C		LD HL,MESS5	;
5C3C	CD 25 05		CALL VDU	;
		;		
5C3F	3E 20	LOAD:	LD A,20H	;immediate load
5C41	32 00 5E	START:	LD (5E00H),A	;store in memory
5C44	C3 41 5C		JP START	;repeat endlessly
5C47	4D 49 43 ..	MESSGE:	DEFB 'MICROPROCESSOR TECHNOLOGY',1EH	
5C61	43 50 55 ..	MESS1:	DEFB 'CPU TEST 2',1EH	
5C6C	20 20 20 ..	MESS2:	DEFB ' Repeatedly loads memory to check WR line.',1EH	
5CA8	20 20 43 ..	MESS3:	DEFB ' Compare with CLOCK waveform.',1EH	
5CD4	20 20 4D ..	MESS4:	DEFB ' Monitor M1',1EH	
5CE1	20 20 20 ..	MESS5:	DEFB ' line on pin 27.',1EH	

6   The code at 5C34H has been generated as a result of the DEFB statement. It contains all the ASCII codes for the message to be displayed.

You can learn much about user routines, labels, displacements, equates (EQU) and so on, by modifying this program and observing the results on the screen.

Table 13.8   Z80 program showing the use of direct keyboard input. Once again, the message machine code has been omitted for clarity

Address	Machine code	Label	Mnemonics	Comments
		INIT:	ORG 5C00H	;origin
			ENT COMM	;entry point
		CLS:	EQU 01B9H	;clear screen routine
		CRSOFF:	EQU 02C9H	;cursor off routine
		VIDEO:	EQU 0E0F0H	;video RAM address
		MOVCSR:	EQU 02FDH	;routine to move cursor
		WRSPCE:	EQU 037FH	;routine to write a space
		KYBD:	EQU 0455H	;keyboard input routine
		WRCHR:	EQU 0381H	;write to screen routine
		VDU:	EQU 0525H	;DEFB routine
		;		
5C00	CD B9 01	COMM:	CALL CLS	;clear screen
5C03	CD 06 5C		CALL MAIN	
5C06	CD 15 5C	MAIN:	CALL MESS	;message
5C09	CD 2D 5C		CALL INPUT	;keyboard input routine
5C0C	CD 48 5C		CALL BIZ	;select largest
5C0F	CD 5B 5C		CALL DISP	;display largest number
5C12	CD 5E 5C		CALL END	;finish
		;		
5C15	21 03 01	MESS:	LD HL,0103H	;cursor co-ordinates
5C18	CD FD 02		CALL MOVCSR	;
5C1B	21 61 5C		LD HL,MESS1	;load message 1
5C1E	CD 25 05		CALL VDU	;print it out on screen
5C21	11 00 5E		LD DE,5E00H	;start of memory for data
5C24	06 07		LD B,7	;number of data items
5C26	21 03 03		LD HL,0303H	;screen co-ordinates
5C29	CD FD 02		CALL MOVCSR	;move cursor
5C2C	C9		RET	;RETurn from subroutine
		;		
5C2D	CD 55 04	INPUT:	CALL KYBD	;scan keyboard for input
5C30	CD 81 03		CALL WRCHR	;display input on screen
5C33	12		LD (DE),A	;store first number in mem
5C34	13		INC DE	;point to next mem loc
5C35	CD 7F 03		CALL WRSPCE	;put in a space
5C38	05		DEC B	;decrement towards zero
5C39	20 F2		JR NZ,INPUT	;repeat if not at zero
5C3B	21 03 05		LD HL,0503H	;new screen co-ordinates
5C3E	CD FD 02		CALL MOVCSR	;move cursor
5C41	21 76 5C		LD HL,MESS2	;next message
5C44	CD 25 05		CALL VDU	;put on screen
5C47	C9		RET	
		;		
5C48	21 00 5E	BIZ:	LD HL,5E00H	;go back to first data
5C4B	0E 07		LD C,7	;number of data items
5C4D	7E	MAX:	LD A,(HL)	;store largest number in A
5C4E	0D	NEXT:	DEC C	;examine next number
5C4F	CA 5B 5C		JP Z,DISP	;display largest number
5C52	2C		INC L	;look at next number
5C53	BE		CP (HL)	;compare with present max
5C54	DA 4D 5C		JP C,MAX	;store if larger –
5C57	C3 4E 5C		JP NEXT	;otherwise continue
5C5A	C9		RET	
		;		
5C5B	32 F0 E0	DISP:	LD (VIDEO),A	;display largest number
5C5E	C3 5E 5C	END:	JP END	
5C49	49 4E 50 ..	MESS1:	DEFB 'INPUT 7 NUMBERS < 10',1EH
5C5E	54 48 45 ..	MESS2:	DEFB 'THE LARGEST NUMBER IS: ',1EH

Moving around the screen

In the next program, TEST2, shown in Table 13.7, we introduce two more routines, VDU and MOVCSR. VDU is an alternative (and probably easier) method of using the DEFB instruction. The message is given a label at the end of the program; the instruction is DEFB as before and the message itself is placed inside single inverted commas. The end of message terminator is the instruction 1EH.

MOVCSR allows the cursor to be moved to any spot by loading HL with the co-ordinates. On the MARC, the video screen has 24 rows and 42 columns which means there are 24 lines with a possible 42 characters on each line. To move the cursor, load the H register with the row number, and the L register with the column number. The top left-hand corner is therefore 0000H whilst the third row, six columns in, would be 0306H. The technique becomes obvious once you try a few examples.

Input via the keyboard

In the next program (see Table 13.8) we show how it's possible to input numbers (and other data) via the keyboard and have those numbers processed by the computer. In this simple example, the user is asked to input seven numbers less than 10. The computer then checks for the highest number and prints it out on the screen.

Input is accepted via the keyboard using the routine 'KYBD'. This causes the keyboard to be scanned continuously until a key is pressed; the appropriate ASCII code is then stored in the accumulator. 'WRCHR' causes the character in the accumulator to be displayed on the screen.

The small routine 'BIZ' compares each number the user has input with a provisional maximum (originally the first number it comes across). This is compared using 'CP (HL)'; if the number being compared is larger, the carry flag is set so that JP C,MAX causes this new maximum to be stored at the current memory position. This is compared with the next number and so on through all the numbers input. 'JP Z,DISP' causes the maximum to be displayed consequent upon the counter (Reg C) reaching zero.

Questions

Use the JSP methodology when producing programs asked for in the questions that follow. Answers to these questions are not given but all the basic information needed to write the programs are contained in this chapter or elsewhere in the book. A system such as the MARC with user routines available is assumed.

1 Construct a JSP chart for the timing delay shown below:

Z80

Label	Operator and operand	
DELAY:	LD	B,0FFH
LOOP2:	LD	C,0FFH
LOOP1:	DEC	C
	JR	NZ,LOOP1
	DEC	B
	JR	NZ,LOOP2
	RET	

6502

Label	Operator and operand	
.DELAY	LDY	#&64
	LDX	#&00
.LOOP	DEX	
	BNE	LOOP
	DEY	
	BNE	LOOP
	RTS	

2　Write a program, with full documentation, which will produce a 'checker-board' pattern on the screen. The precise design is unimportant, but the completed display should look something like a chess board. On the MARC, the ASCII code for a square is 0A1H. Arranging 12 of these (four across and three down) produces a suitable individual square for the checker-board pattern.

3　Write a program which produces a display on the screen inviting the user to input ten numbers. The program should then sort the numbers into numerical order and print them out on the screen.

4　Do the same as in Question 2, but input letters of the alphabet and sort them into alphabetical order.

5　Extend the program in Question 3 to sort ten *words* into alphabetical order.

6　Write a program which can hold ten names and ten telephone numbers. A prompt should ask for either name *or* number and then output the name associated with a particular number, or the number associated with the name.

7　Write a program which will accept a binary number (via a set of eight switches, for example) on ports A and B and display that number on the screen in binary, decimal, and hexadecimal as shown below:

	BINARY	DECIMAL	HEX
Port A:	0000 0000	00	00H
Port B:	0000 0000	00	00H

8　Write a program which produces a frame on the screen in which a graphics byte is made to move from one spot to another diagonally and bounce off the frame when it meets it.

9　Write a program which displays ten numbers, input via the keyboard, as a bar chart.

10　Write a program which produces a real-time clock with an appropriate display on the screen.

Chapter 14
Computer interfacing

Introduction

A microelectronic control system is of little use if it can't communicate with the outside world. To be useful, the system must be able to sense external events and respond to them. The response may be in a variety of forms; varying illumination, changing temperature, increasing or decreasing sound level or speed of motors, for example.

Chapter 12 showed how the output ports may be accessed by software, that is to say, writing the appropriate programming instructions, and there was some discussion about the need for and construction of the requisite hardware. This section expands on what has already been given.

Transistor drivers

The transistor can perform two major functions:

1   it can amplify an electrical signal;
2   it can act as a switch.

We are presently interested in the transistor's behaviour as a switch and in this context it can be compared with an ordinary light switch. This has a lever or 'toggle' which joins a pair of contacts to turn the switch on, or separates them to turn the switch off. In a similar way, the base of the transistor is equivalent to the lever. The base can be used to 'join together' the other two terminals of the transistor (the collector and the emitter) or to separate them. When the collector and emitter are joined, the transistor is turned on; this is achieved by putting a 'high' input on the base. 'High' in this context means +5V (a TTL logic 1) but the connection must not be made direct. If it were, the base would draw excessive current and the transistor would be irreparably damaged. The solution is to fit a current limiting resistor and we see later that values ranging from about 1K to 33K are frequently found suitable for the purpose.

Basic transistor theory

Figure 14.1 illustrates a transistor showing the n- and p-type areas. Electrons are *emitted* from the emitter and they travel towards the base region, a very narrow and lightly doped section of p-type material, which is made slightly positive with respect to the emitter; the base/emitter voltage is nearly always somewhere between 0.6V and 0.7V. However, because of the nature of the base region, and the very high positive voltage present on the collector, most (around 99 per cent) of the electrons travel to the collector.

Figure 14.2 shows the biasing arrangements for a single transistor in what is known as common emitter mode. The three electrodes are labelled. The

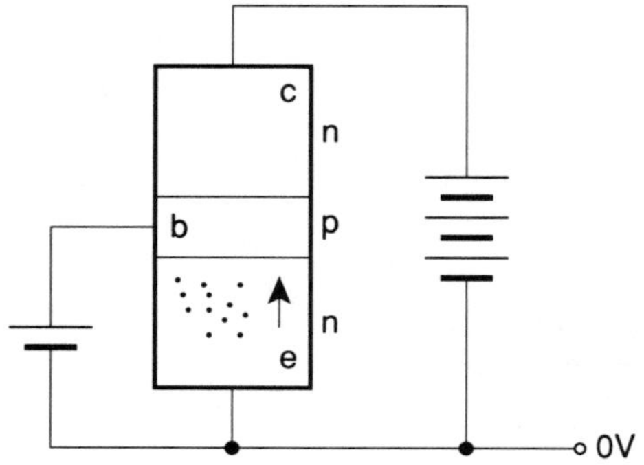

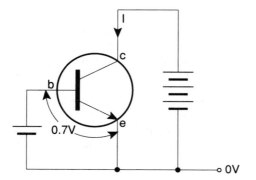

Figure 14.1 *Simple representation of a transistor.*

Figure 14.2 *Biasing arrangements for an npn transistor.*

arrow indicates the emitter and points in the direction of conventional current flow. The electrons which constitute the current actually flow in the opposite direction, and this is clearly shown in Figure 14.1. The collector/emitter current is some hundred times larger than the base/emitter current.

A simple transistor switch was described and illustrated in Chapter 1. As a general rule, npn transistors tend to be much more commonly used than pnp types as they use electrons as majority carriers – which are much more mobile than 'holes'. However, this is of little consequence if the transistor is being used as a simple interfacing switch and both npn and pnp transistors can be used in this application. Figure 14.3 shows a typical circuit using a pnp device; the npn version is discussed below and typical circuits are shown in Figure 14.4.

Transistor gain

The gain of a transistor is given by the manufacturers and is typically in the region of 100 to 250. You may find the current gain (I_c/I_b) denoted by the symbol β (beta), which we'll use in this book. However, you may also come across the hybrid parameters, h_{FE} and h_{fe}. The 'h' is for hybrid – a mixture of units – the 'f' indicates forward gain and the 'e' indicates common emitter mode. It is quite inappropriate to say more about this here; it won't be needed and there are lots of other excellent texts which cover the subject in detail.

Refer now to the simple transistor switch shown in Figure 14.4.

The component values are worked out as follows.

Starting points: the supply is the standard TTL value of 5V. Assume the LED to have a forward current (I_f) of 20mA and a forward voltage drop (V_f) of 2V. Assume the gain of the transistor to be 100 and there to be no voltage drop across the collector/emitter junction when the transistor is fully conducting.

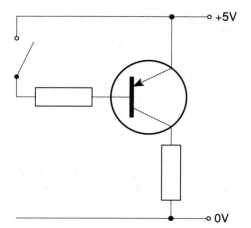

Figure 14.3   *Biasing arrangements for a pnp transistor.*

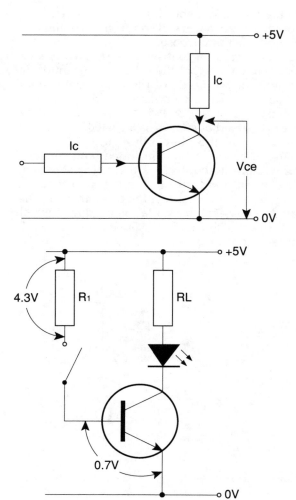

Figure 14.4   *Transistor switch driving an LED indicator.*

Under these conditions, R_L must drop 3V across it at 20mA, since the voltage must be the supply (5V) minus the voltage drop across the LED. The value of R_L can then be easily obtained from Ohm's Law ($R = V/I$):

Hence: $R = 3/0.02 = 150R$

Current gain, β, is I_c/I_b so:

$$
\begin{aligned}
I_b &= Ic/gain \\
&= 0.02/100 \\
&= 0.2mA
\end{aligned}
$$

Assuming 0.7V across the b/e junction, R1 must drop $(5 - 0.7)V = 4.3V$. Given that we have already calculated base current as 0.2mA, the value of R1 is:

$$
\begin{aligned}
R1 &= V/I_b \\
&= 4.3/0.2 \times 10^{-3} \\
&= 21,500\Omega \\
&= 22K \text{ (preferred value)}
\end{aligned}
$$

In practice, the value is by no means critical, and in this simple circuit any value between 1K0 and 33K would probably work all right. It is common to see values of either 10K or 15K used for the purpose.

A similar configuration may be used in order to drive a relay. Two precautions must be taken here: the first is to ensure that the collector current of the transistor you are using is sufficient for the needs of the relay. Many small relays require of the order of 50mA for correct operation and this is well within the capability of most transistors of the BC108, BC182 (etc.) variety. The second precaution to observe is to fit a diode (such as a 1N4002) in inverse parallel with the relay. Such an arrangement is shown in Figure 14.5.

The old relay symbol is still often used so we have included both versions. In all honesty, the older symbol is more descriptive, as the coil is shown as an inductor would be. Since it is a coil (and you can call it a solenoid or electromagnet as well, if you like; they all mean much the same thing in this context), when the current through it ceases, a back EMF is produced which may be large enough to destroy the transistor. The inclusion of the diode protects the transistor under these circumstances.

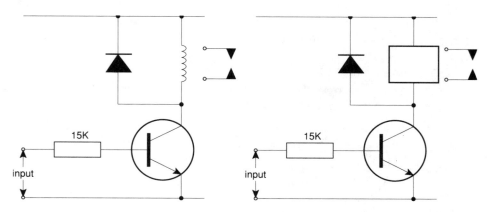

Figure 14.5 *The circuit on the left uses the old relay symbol. The one on the right shows the correct British Standard symbol.*

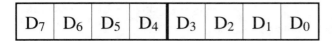

D_7	D_6	D_5	D_4	D_3	D_2	D_1	D_0

Figure 14.6 *The CPU output port data register.*

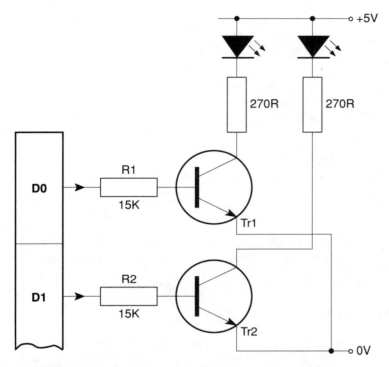

Figure 14.7 *Connecting the transistor switches to an output port.*

Microprocessor output ports

We'll continue to assume the use of 8-bit ports (which are still very common). In order to interface the port in a simple way will require the use of eight transistors. Figure 14.6 represents the data register of one of the CPU's output ports. The bits are designated D_0 to D_7 with D_0 being the lsb and D_7 being the msb. Each bit can be set to either a zero or a one. The output levels cannot drive even a small LED which is why the interface circuit is needed, one for each data line.

Figure 14.7 shows two transistor switches connected to the two least significant bits of a CPU output port data register. The remaining parts of the circuit and its connections are similar.

Darlington drivers

Where a more powerful and sensitive interfacing device is needed, a two-transistor arrangement known as a 'super-alpha pair' or Darlington driver is frequently used. Connected in this way the transistors provide much increased gain and effectively a very high input impedance. A typical circuit is shown in Figure 14.8.

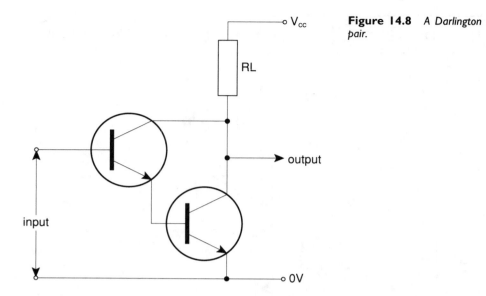

Figure 14.8 *A Darlington pair.*

The input impedance of a transistor is the b/e resistance which the input 'sees', plus any series input element (R and/or C) that may be connected. This value is doubled when using the Darlington pair. The gain is also much increased, being (gain of Tr1 × gain of Tr2), giving high sensitivity.

Darlington driver arrays

For microprocessor applications, an integrated circuit array of eight Darlington drivers is available, complete with protection diodes (for use if required), all in an 18 pin DIL package. The ULN2801A and ULN2803A octal

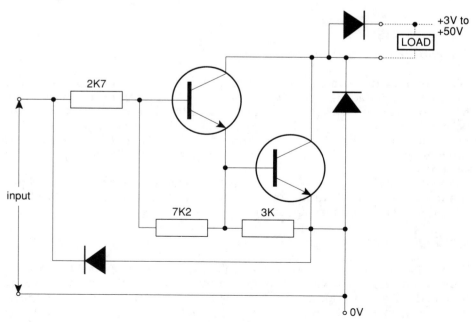

Figure 14.9 *Circuit diagram of a ULN2801A Darlington pair.*

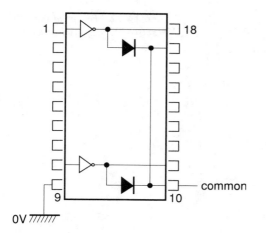

Figure 14.10 *The ULN2803A Octal Darlington driver array. Only two are shown for clarity.*

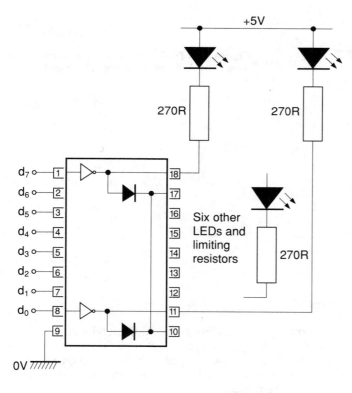

Figure 14.11
Wiring details for the ULN2803A octal driver. The computer output port is connected to pins 1–8 (d_7–d_0), the LEDs are connected common anode as shown.

Darlington driver arrays are examples of such devices. Figure 14.9 shows the circuit diagram of one of the drivers, each of which can supply 500mA at up to 50V. They can easily drive a 24V motor, for example. Figure 14.10 shows details of the IC package.

In order to use one of these devices in a simple way, consider the circuit diagram of Figure 14.11. This is an interfacing unit which can be connected to a computer port and which will light up an array of LEDs. The reader is encouraged to construct such a device and hence become familiar with its design and operation.

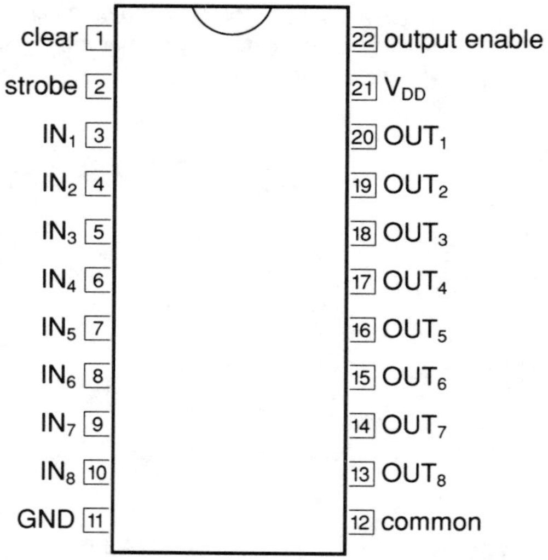

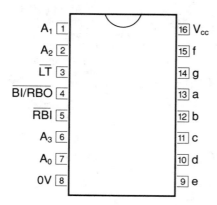

Figure 14.12 *Pin-out of the UCN5801A latching octal driver.*

Figure 14.13 *A 7448 seven-segment display driver. Pins 3, 4 and 5 will normally be tied to V_{cc}.*

Latching Darlington arrays

Figure 14.12 shows the pin-out of a UCN5801A latching octal driver. This device will latch (hold) data appearing at its inputs when the strobe pin is taken high, whilst pins 1 and 22 (clear and output enable respectively) are kept low. The device allows a chosen output to be displayed, even when the signal that produced it has been removed.

Seven-segment display drivers

We introduced seven-segment displays in Chapter 2. For those who had a go at the assignment, you may have discovered that an IC type 7448 (or the CMOS 4033) can be used for interfacing. The pin-out is shown in Figure 14.13, where A_0 to A_3 are the BCD inputs and the outputs are marked with the appropriate segment letter of the display. These must be connected via current limiting resistors, a typical value being 330R.

The device requires a BCD input and produces the necessary output to drive the seven-segment display. Hence, if a BCD output is made available at

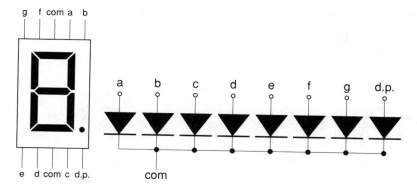

Figure 14.14 *A typical common cathode, seven-segment display. Sectors a–g denote the main segments, the decimal point is marked d.p. and the pins marked 'com' connect to all the cathodes.*

port A, two of these devices would enable that output to be converted into a decimal display which could count up to a maximum of 99.

It only takes a little thought to see how, if port B is also utilised, four decoders could produce a display up to 9999. Software could be evolved which would make the displays show a count up, or a count down, from any chosen values and with real-time accuracy. All the information needed to do this has already been given either here or elsewhere in the book.

The seven-segment displays

These displays have seven segments (as the name implies) each of which is a discrete light emitting diode (LED). There's nearly always an eighth element in the form of a right-hand decimal point, and some displays have two (a left-hand one as well). Apart from size and colour, there are two distinct types: common anode and common cathode. In our experience, common cathode types tend to have pins top and bottom whereas common anode types have pins on each side. The 7448 device is designed to source the display and so must be used with common cathode types. The details are shown in Figure 14.14. Note, however, that we have seen slightly different pin-outs on later versions, so it's important to check.

Direct connection of displays

The use of BCD to seven-segment display decoders is of limited value and it is more common to need direct interfacing. One way of doing this is to use an octal buffer similar to the ULN2803A previously described (Figure 14.9) driving a common anode display. Because the ULN2803A is an inverting buffer, it sinks the current so we couldn't use a common cathode display unless we added another inverting buffer. One suitable arrangement is shown in Figure 14.15.

Constructional details

We put one of these together on stripboard with great success. The fact that the pins are down the sides makes construction a little more difficult than

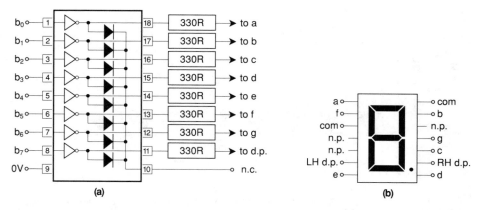

Figure 14.15 (a) The ULN2803A octal Darlington driver. The internal diodes aren't being used, so they are not connected to anything. (b) A typical common anode display. If pin 3 is connected to V_{cc}, pin 14 can be ignored; 'np' means 'no pin', i.e. it doesn't exist. Connect RH d.p. (right-hand decimal point) to pin 11 of the buffer. Ignore the LH d.p. The type we used fitted into a standard 14-pin DIL holder.

would be the case if we used common cathode types, but it can be done. More skilful readers may care to attempt a pcb.

We advise using a colour code for the interconnecting wires between port and buffer. It always makes sense to document your choice of code for future

Table 14.1 Z80 program to make the outer segments of a seven-segment display rotate in a circle, clockwise, using RLA. Note the use of ADD A,0 to clear the carry flag

Address	Machine code	Label	Mnemonics	Comments
			ORG 5C00H	;origin
			ENT 5C00H	;entry point
5C00	3E 0F	START:	LD A 0FH	;port A mode 0
5C02	D3 02		OUT (2),A	;i.e. to OUTPUT
5C04	3E 01	RPT:	LD A,01H	;1st segment ON
5C06	D3 00	GO:	OUT (0),A	;
5C08	CD 16 5C		CALL DELAY	
5C0B	C6 00		ADD A,0	;clears carry flag
5C0D	17		RLA	;
5C0E	FE 40		CP 40H	;does A = 40H?
5C10	28 F2		JR Z,RPT	;if so, goto RPT
5C12	C3 06 5C		JP GO	;if not, GO!
5C15	76		HALT	
				;DELAY
5C16	06 33	DELAY:	LD B,33H	;immediate load
5C18	0E FF	LOOP2:	LD C,0FFH	;immediate load
5C1A	0D	LOOP1:	DEC C	;decrement
5C1B	20 FD		JR NZ,LOOP1	;relative jump
5C1D	05		DEC B	;decrement
5C1E	20 F8		JR NZ,LOOP2	;relative jump
5C20	C9		RET	;return

Table 14.2 6502 program to make the outer segments of a seven-segment display rotate in a circle, clockwise, using ROL A

Address	Machine code	Label	Mnemonics	Comments
			ORG &0200	;origin
			ENT &0200	;entry point
0200	A9 FF	.START	LDA #&FF	
0202	8D 03 09		STA &0903	;port A to output
0205	A9 01	.RPT	LDA #&01	;
0207	8D 01 09	.GO	STA &0901	
020A	48		PHA	;push A onto stack
020B	20 20 02		JSR DELAY	;delay subroutine
020E	68		PLA	;pull A off stack
020F	18		CLC	;clear carry flag
0210	2A		ROL A	;rotate left acc.
0211	C9 40		CMP #&40	;got to bit 7?
0213	F0 F0		BEQ RPT	;branch back if so
0215	4C 07 02		JMP GO	;jump back to GO
				;DELAY
0220	A9 33	.DELAY	LDA #&33	;immediate load
0222	85 30		STA &30	;zero-page memory
0224	A9 FF	.LOOP1	LDA #&FF	;fill accumulator
0226	85 31		STA &31	;store in memory
0228	C6 31	.LOOP2	DEC &31	;decrement in
022A	D0 FC		BNE LOOP2	;inner loop
022C	C6 30		DEC &30	;decrement in
022E	D0 F4		BNE LOOP1	;outer loop
0230	60		RTS	;return from ;subroutine

reference and we chose to make segment 'g' a green wire and segment 'b' blue. The others were a compromise.

Once completed, the unit may be tested by applying a logic 1 (the 5V supply via a 1K resistor, for example) to each input. The appropriate segment should light up. If all is well, connect the unit to port A of your system and write a program which will light up the outer segments individually, one after another, in a clockwise direction. This involves lighting up segment a for, say, half a second, turning it off, then lighting up segment b and so on round to segment f, and ignoring segment g in the middle. We have included a possible solution in Z80 and 6502 codes in Tables 14.1 and 14.2 respectively. You should attempt to obtain your own solution before turning to these.

Programming a count

You may notice that we have not used the convention for storing the seven-segment display codes mentioned in Chapter 2. It is of no importance unless you are writing software for an established protocol. If you are producing your own hardware then you are free to use whatever system you wish. To keep things simple we are connecting b_0 to segment a, b_1 to segment b, etc. as shown in Figure 14.15. You will need to work out the resulting codes for yourself, though some are listed in Table 14.3.

Table 14.3 Some seven-segment display codes

Number	dp	g	f	Segments on: e	d	c	b	a	Hex code
0	0	0	1	1	1	1	1	1	3FH
1	0	0	0	0	0	1	1	0	06H
2	0	1	0	1	1	0	1	1	5BH
.	.	.	.	.	.	.	.	.	. .
9	0	1	1	0	1	1	1	1	6FH

Table 14.4 6502 program to count up continuously from 0 to 9. We have used the same delay routine at address &0220, but changed the first LoaD to #&EE for a slightly longer delay. The main program starts at address &0250. The code for the numbers 4 to 8 have been omitted. They follow a logical pattern and you should be able to work them out easily for yourself

Address	Machine code	Label	Mnemonics	Comments
			ORG &0200	;origin
			ENT &0200	;entry point
0250	A9 3F	.DATA	LDA #&3F	;code for '0'
0252	85 40		STA &40	;store in zero page
0254	A9 06		LDA #&06	;code for '1'
0256	85 41		STA &41	
0258	A9 5B		LDA #&5B	; '2'
025A	85 42		STA 42	
025C	A9 4F		LDA #&4F	; '3'
. .	. .	.	. .	
0274	A9 6F		LDA #&6F	; '9'
0276	85 49		STA &49	
0278	A9 FF	.START	LDA #&FF	;Set up port A
027A	8D 03 09		STA &0903	;for output
027D	A2 00	.RPT	LDX #&00	;index offset
027F	B5 40	.GO	LDA &40,X	
0281	8D 01 09		STA &0901	;OUT to port A
0284	48		PHA	;save A contents
0285	20 20 02		JSR &0220	;DELAY
0288	68		PLA	
0289	E8		INX	;increment index
028A	C9 6F		CMP #&6F	;got to '9'?
028C	F0 EF		BEQ RPT	;if so branch -17
028E	4C 7F 02		JMP GO	;if not, do again ;by JMP to 027F
				;DELAY
0220	A9 EE		LDA #&EE	;longer delay
0222	85 30		STA &30	;zero-page memory
0224	A9 FF	.LOOP1	LDA #&FF	;fill accumulator
0226	85 31		STA &31	;store in memory
0228	C6 31	.LOOP2	DEC &31	;decrement in
022A	D0 FC		BNE LOOP2	;inner loop
022C	C6 30		DEC &30	;decrement in
022E	D0 F4		BNE LOOP1	;outer loop
0230	60		RTS	;return from ;subroutine

Table 14.5 Z80 program to count up continuously from 0 to 9. We have used the same delay routine but changed the first LoaD to 0DDH for a slightly longer delay. The main program starts at address 5C00H. The code for the numbers 3 to 8 have been omitted. They follow a logical pattern and you should be able to work them out easily for yourself

Address	Machine code	Label	Mnemonics	Comments
				;COUNTUP
			ORG 5C00H	;origin
			ENT 5C00H	;entry point
5C00	21 00 5E	DATA:	LD HL,5E00H	
5C03	36 3F		LD (HL),3FH	; '0'
5C05	23		INC HL	
5C06	36 06		LD (HL),06H	; '1'
5C08	23		INC HL	
5C09	36 5B		LD (HL),5BH	; '2'
5C0B	23		INC HL	
..	...	..	. . .	. .
5C1E	36 6F		LD (HL),6FH	; '9'
5C20	3E 0F	START:	LD A,0FH	;port A mode 0
5C22	D3 02		OUT (2),A	;i.e. to OUTPUT
5C24	21 00 5E	RPT:	LD HL,5E00H	;first number
5C27	7E	GO:	LD A,(HL)	;number into A
5C28	D3 00		OUT (0),A	;output number
5C2A	23		INC HL	;point to next
5C2B	CD 35 5C		CALL DELAY	
5C2E	FE 6F		CP 6FH	;is it 9?
5C30	28 F2		JR Z,RPT	;Yes? JR
5C32	C3 27 5C		JP GO	;if not, GO!
				;DELAY
5C35	06 DD	DELAY:	LD B,0DDH	
5C37	0E FF	LOOP2:	LD C,0FFH	
5C39	0D	LOOP1:	DEC C	
5C3A	20 FD		JR NZ,LOOP1	
5C3C	05		DEC B	
5C3D	20 F8		JR NZ,LOOP2	
5C3F	C9		RET	

When devising a system that doesn't use decoders, the count can be achieved by storing the seven-segment display codes (some of which are given in Table 14.3) in memory and then using an iterative process to step through them, sending them out to the appropriate I/O port(s) on the way.

Simple programs in 6502 and Z80 code are shown in Tables 14.4 and 14.5 respectively.

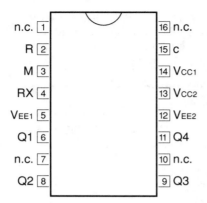

Figure 14.16   *Pin-out of SAA1027 stepper motor driver IC.*

Pin		Pin	
n.c.	1	16	n.c.
R	2	15	c
M	3	14	Vcc1
RX	4	13	Vcc2
Vee1	5	12	Vee2
Q1	6	11	Q4
n.c.	7	10	n.c.
Q2	8	9	Q3

Other applications

Careful study of the foregoing should enable you to design and build the kind of interfacing equipment necessary for a particular application together with the appropriate software programming requirements. Competence in this area comes best from careful study, thought and experimentation; but remember, document everything meticulously. You will certainly be glad you did.

Stepper motor drivers

We've used two different methods to drive seven-segment displays, a BCD to display decoder (a 7448, for example) and an octal Darlington driver (the ULN

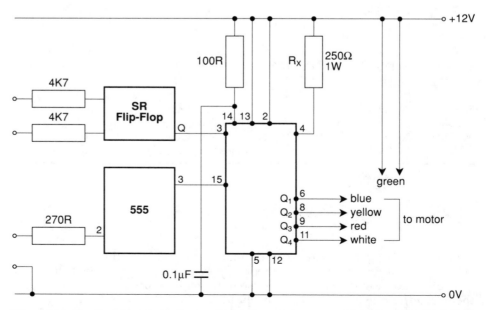

Figure 14.17   *Complete circuit diagram for small stepper motor interface. Note that R_x (250Ω) must be rated at 1W or more.*

Table 14.6 Suggested program for driving a stepper motor using a dedicated IC (MARC Z80)

Address	Machine code	Label	Mnemonics	Comments
			ORG 5C00H	;origin
			ENT 5C00H	;entry point
		LONDEL:	EQU 018E3H	;long delay
		CRA:	EQU 02H	;control register
		DRA:	EQU 00H	;data register
5C00	3E 0F		LD A,0FH	;port A output
5C02	D3 02		OUT (CRA),A	
5C04	06 30	START:	LD B,30H	;number of steps
5C06	CD 23 5C	LOOP1:	CALL FORWD	;forward
5C09	05		DEC B	
5C0A	C2 06 5C		JP NZ,LOOP1	
5C0D	CD E3 18		CALL LONDEL	;long delay
5C10	06 30		LD B,30H	;number of steps
5C12	CD 2F 5C	LOOP2:	CALL BACK	;backward
5C15	05		DEC B	
5C16	C2 12 5C		JP NZ,LOOP2	
5C19	CD E3 18		CALL LONDEL	;long delay
5C1C	CD E3 18		CALL LONDEL	
5C1F	C3 04 5C		JP START	;repeat sequence
5C22	76		HALT	
5C23	3E 11	FORWD:	LD A,11H	;set forward bit
5C25	D3 00		OUT (DRA),A	;and pulse bit 0
5C27	CD 3B 5C		CALL DELAY	
5C2A	3E 10		LD A,10H	
5C2C	D3 00		OUT (DRA),A	
5C2E	C9		RET	;return
5C2F	3E 21	BACK:	LD A,21H	;set reverse bit
5C31	D3 00		OUT (DRA),A	;and pulse bit 0
5C33	CD 3B 5C		CALL DELAY	
5C36	3E 20		LD A,20H	
5C38	D3 00		OUT (DRA),A	
5C3A	C9		RET	;return
				;DELAY
5C3B	16 A0	DELAY:	LD D,0A0H	
5C3D	0E A0	LOOP3:	LD C,0A0H	
5C3F	0D	LOOP4:	DEC C	
5C40	C2 3F 5C		JP NZ,LOOP4	
5C43	15		DEC D	
5C44	C2 3D 5C		JP NZ,LOOP3	
5C47	C9		RET	

2803A) to drive the segments independently. The latter requires more software but is more flexible. We can identify two analogous methods for driving a stepper motor.

Using a dedicated chip

Figure 14.16 shows the pin-out of an SAA1027 stepper motor driver IC. This can be used with smaller stepper motors; larger ones require more current

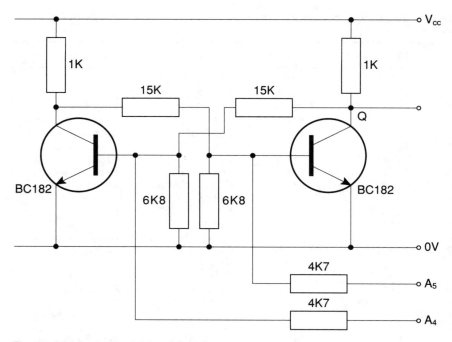

Figure 14.18   *Simple S-R bistable circuit.*

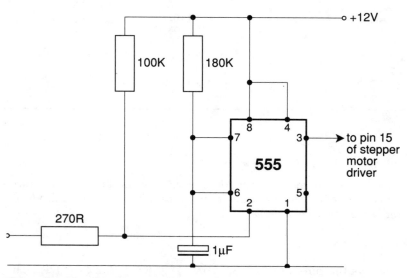

Figure 14.19   *A 555 timer IC wired as a monostable.*

Table 14.7 Suggested program for a driving a small stepper motor using a dedicated IC (EMMA 6502)

Address	Machine code	Label	Mnemonics	Comments
			ORG &0200	;origin
			ENT &0200	;entry point
0200	A9 FF		LDA #&FF	;set up port A
0202	8D 03 09		STA &0903	
0205	A2 0A	.START	LDX #10	;number of steps
0207	20 1E 02	.LOOP1	JSR FORWD	;forward
020A	CA		DEX	
020B	D0 FA		BNE LOOP1	
020D	20 60 02		JSR LONDEL	;long delay
0210	A2 05		LDX #5	;number of steps
0212	20 2C 02	.LOOP2	JSR BACK	;backward
0215	CA		DEX	
0216	D0 FA		BNE LOOP2	
0218	20 60 02		JSR LONDEL	;long delay
021B	4C 05 02		JMP START	
021E	A9 21	.FORWD	LDA #&21	;set direction and
0220	8D 01 09		STA &0901	;pulse bit 0
0223	20 40 02		JSR DELAY	
0226	A9 20		LDA #&20	
0228	8D 01 09		STA &0901	;OUT to port A
022B	60		RTS	;return
022C	A9 11	.BACK	LDA #&11	;set direction and
022E	8D 01 09		STA &0901	;pulse bit 0
0231	20 40 02		JSR DELAY	
0234	A9 10		LDA #&10	
0236	8D 01 09		STA &0901	
0239	60		RTS	;return
				;DELAY
0240	A9 88	.DELAY	LDA #&88	;short delay
0242	85 30		STA &30	;zero-page memory
0244	A9 FF	.LOOP1	LDA #&FF	;fill accumulator
0246	85 31		STA &31	;store in memory
0248	C6 31	.LOOP2	DEC &31	;decrement in
024A	D0 FC		BNE LOOP2	;inner loop
024C	C6 30		DEC &30	;decrement in
024E	D0 F4		BNE LOOP1	;outer loop
0250	60		RTS	;return from
				;subroutine
0260	20 40 02	.LONDEL	JSR &0240	;long delay
0263	20 40 02		JSR &0240	
0266	60		RTS	

than this IC can supply on its own. It's been designed so that it can be used with hard-wired electronics (for example, by clocking with a 555) but with minor modifications can be adapted for use with a microprocessor. Figure 14.17 shows the complete circuit.

The outputs to the motor, Q_1 to Q_4, appear at pins 6, 8, 9 and 11. The colour codes refer to the wires going to the motor and it will be necessary to check that both the codes and the motor itself are compatible with this IC. Pins 5

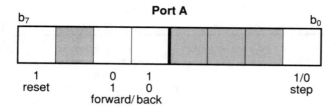

Figure 14.20 *Stepper motor bit allocation of Port A data register. Shaded portions are unused; we allocated bit 7 to the reset pin for future development, but it has not been used in the hardware or software described.*

and 12 (V_{EE1} and V_{EE2}) are the 0V connections, V_{CC1} and V_{CC2} are connected to the 12V line, V_{CC2} directly and V_{CC1} via a 100Ω resistor. The resistor R_x connected to pin 4 needs to carry all the motor current and we used four 1K resistors in parallel to obtain the appropriate power rating. Pin 2 is for a resetting function if required; we just connected it to the 12V line.

Now, with pin 3 taken 'high', the motor goes one way, when pulsed on pin 15. With pin 3 'low' the motor goes in the other direction. A 12V supply is adequate and the pulses to pin 15 should be square waves of approximately 12V amplitude. To obtain the high/low direction signal, we used the simple S-R bistable circuit shown in Figure 14.18. The bases are taken to b_4 and b_5 of port A, as will become obvious when we describe the software.

The pulses are supplied by a 555 wired as a monostable, triggered by the changing logic state appearing at b_0 of port A, under software control. This is a neat way of obtaining clean pulses of approximately 12V amplitude. The microelectronic system controls the frequency of the pulses and the 555 shapes them and increases the voltage. The circuit is shown in Figure 14.19.

The software

A suitable program to drive this circuit is shown in Table 14.6 (Z80) and Table 14.7 (6502). Each version has two delays: 'LONDEL' (long delay) utilises the

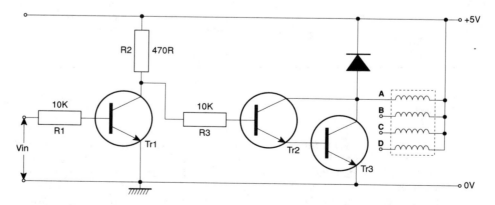

Figure 14.21 *Typical arrangement for driving a stepper motor directly. Only winding A is shown connected; duplicate circuits are required for windings B, C and D.*

Table 14.8 'Roger bleep' for the MARC. The program repeatedly gives a 'dah-di-dah' – Morse code for the letter 'K'

Address	Machine code	Label	Mnemonics	Comments
			ORG 5C00H	;origin
			ENT 5C00H	;entry point
5C00	CD 1F 5C		CALL PORT	;set up output port
5C03	16 45	START:	LD D,45H	;dash
5C05	CD 26 5C		CALL MAIN	
5C08	CD 3A 5C		CALL DELAY	
5C0B	16 15		LD D,15H	;dot
5C0D	CD 26 5C		CALL MAIN	
5C10	CD 3A 5C		CALL DELAY	
5C13	16 45		LD D,45H	;dash
5C15	CD 26 5C		CALL MAIN	
5C18	CD 3A 5C		CALL DELAY	
5C1B	CD 3A 5C		CALL DELAY	
5C1E	C3 03 5C		JP START	;repeat sequence
				;
5C21	3E 0F	PORT:	LD A,0FH	;control, port A
5C23	D3 02		OUT (02),A	;mode 0, byte output
5C25	C9		RET	
				;
5C26	3E 80	MAIN:	LD A,80H	;bit 7 set to 1
5C28	26 02		LD H,02H	;counter for cycle
5C2A	2E 40	LOOP2:	LD L,40H	;frequency
5C2C	D3 00	LOOP1:	OUT (00),A	;O/P port address
5C2E	2D		DEC L	;
5C2F	20 FB		JR NZ,LOOP1	;
5C31	3E 00		LD A,0	;
5C33	25		DEC H	;
5C34	20 F4		JR NZ,LOOP2	;
5C36	15		DEC D	;duration
5C37	20 ED		JR NZ,MAIN	;
5C39	C9		RET	;
				;
				;DELAY
5C3A	01 00 28	DELAY:	LD BC,2800H	;delay for space
5C3D	0B	LOOP3:	DEC BC	
5C3E	78		LD A,B	;B into A
5C3F	B1		OR C	;sets flag when A=0
5C40	20 FB		JR NZ,LOOP3	;repeat
5C42	C9		RET	

1 second delay routine at memory location 18E3H (MARC Z80), whilst 'DELAY' is a fairly short delay, implemented in software in the manner we've seen many times before. The 6502 'DELAY' is similar and 'LONDEL' simply calls 'DELAY' twice. The short delay is used between each step of the motor, the long delay between changes of direction. By alternating between 10H and 11H, the motor goes in one direction, 20H and 21H drive it in the opposite direction. This is shown diagrammatically in Figure 14.20.

The values given are for example only and the reader is, of course, free to change the number of steps in each direction and the length of the delays.

Table 14.9 'Roger bleep' for the EMMA

Address	Machine code	Label	Mnemonics	Comments
			ORG &0200	;origin
			ENT &0200	;entry point
0200	20 E0 02		JSR PORT	;I/O port set up
0203	A9 70	.START	LDA #&70	;dash
0205	20 E6 02		JSR MAIN	
0208	20 00 03		JSR DELAY	;delay
020B	A9 30		LDA #&30	;dot
020D	20 E6 02		JSR MAIN	
0210	20 00 03		JSR DELAY	;delay
0213	A9 70		LDA #&70	;dash
0215	20 E6 02		JSR MAIN	
0218	20 00 03		JSR DELAY	;delay
021B	20 00 03		JSR DELAY	;delay
021E	20 00 03		JSR DELAY	;delay
0221	4C 03 02		JMP START	
02E0	A9 FF	.PORT	LDA #&FF	;set up port
02E2	8D 03 09		STA &0903	
02E5	60		RTS	
02E6	85 30	.MAIN	STA &30	
02E8	A9 80	.RPT	LDA #&80	;
02EA	A2 02		LDX #&02	;
02EC	A0 80	.LOOP2	LDY #&80	
02EE	8D 01 09	.LOOP1	STA &0901	;output
02F1	88		DEY	;PA7 is ON...
02F2	D0 FA		BNE LOOP1	;..for loops in &30
02F4	A9 00		LDA #00	;turns PA$_7$ OFF
02F6	CA		DEX	
02F7	D0 F3		BNE LOOP2	
02F9	C6 30		DEC &30	
02FB	D0 EB		BNE LOOP3	
02FD	60		RTS	
				;DELAY
0300	A9 44	.DELAY	LDA #&44	;short delay
0302	85 20		STA &20	;zero-page memory
0304	A9 FF	.LOOP3	LDA #&FF	;fill accumulator
0306	85 21		STA &21	;store in memory
0308	C6 21	.LOOP4	DEC &21	;decrement in
030A	D0 FC		BNE LOOP4	;inner loop
030C	C6 20		DEC &20	;decrement in
030E	D0 F4		BNE LOOP3	;outer loop
0310	60		RTS	;return from
				;subroutine

Driving the motor directly

Figure 14.21 shows a typical arrangement for driving a stepper motor. We leave it up to the reader to write the appropriate software.

Audio interfaces

By switching a single bit of the output port on and off rapidly, a square wave in the audible range can be produced. Even a small audio amplifier would suffice in order to allow such a signal to drive a loudspeaker. As an example we are using a program written to produce Morse code dots and dashes and called by the author, 'Roger bleep'. As it stands it produces a 'dah-di-dah' – Morse code for the letter 'K' – but readers with some musical knowledge can easily adapt the program to play a tune. The sound is quite harsh because it consists of square waves and you'll need a lot of patience; a lengthy amount of code will be required to produce even a modest tune.

A suitable program for the MARC is shown in Table 14.8 and a similar one for the EMMA is given in Table 14.9.

Interfacing to the mains

Safety notes

This is one of those occasions where it's tempting to say: 'It goes without saying . . .' but it's so important, we can't resist saying it anyway: the MAINS can be **LETHAL** and utmost care must be exercised when using it. Although adequately rated relays can be used to drive mains lamps and motors, etc. (and methods by which this can be achieved have already been described), we shall now be concentrating on the use of optically coupled devices and triacs in order to ensure complete isolation from the mains. Whatever you choose to do, the following basic guidelines should be adhered to when building computer interfaces to the mains.

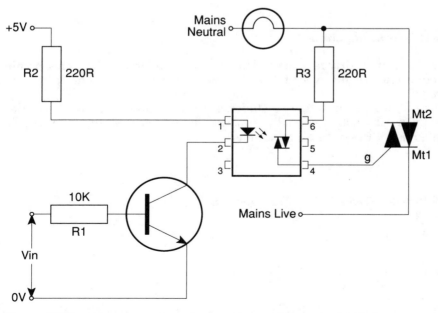

Figure 14.22 *Using an optotriac and power triac to drive a mains light bulb.*

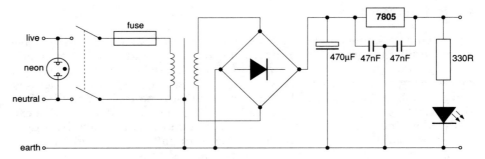

Figure 14.23   *A typical 5V power supply. The 47nF capacitors should be physically as close to the 7805 regulator as possible. The bridge rectifier may be a monolithic device or be implemented from four diodes such as the 1N4002.*

1   The whole unit should be built into an earthed, metal box.
2   As well as bolting on an earth connection, the mains earth wire should be directly soldered to the metal box.
3   Double wound transformers must be used.
4   A physical metal barrier should be placed between mains and low voltage sections.
5   The mains lead should be secured by an approved cable cleat.
6   A neon lamp should be connected to the mains input, such that it glows as soon as the mains is connected, i.e. before any switch the unit may contain.
7   'Star' washers should be used with any fixing screws, especially for any lid or cover which provides access to the circuitry inside.

Readers may also wish to include double-pole switches, mains and low tension fuses and LED indicators, etc. in addition.

A mains interfacing circuit

A typical circuit for driving a mains lamp is shown in Figure 14.22. This circuit must be duplicated for every bit in the port's data register; eight of them if they're needed.

Generally, school and college students are *not* allowed to work on the mains. In the absence of special permission being obtained, the following is presented as a theoretical exercise only.

Typical Triac   **Figure 14.24**   *Typical power triac.*

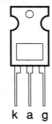

k a g

The opto-isolator usually comes in a standard 6-pin DIL package (rather like a 741 or 555 but with six pins instead of eight). A typical device number is MOC 3020, but many suppliers simply list them as 'Triac Isolators'. Each device offers something like 5000V isolation between input and output. The LED has forward voltage and current values (V_F and I_F) of around 1.3V and 30mA respectively, and 15mA is required to latch the output. The resistor values given in the circuit have proved to be entirely satisfactory.

Circuit operation

The signal from one bit of a port's data register is applied as V_{in}. It is connected *only* as shown, and comes nowhere near any mains input. The opto-isolator's LED is connected into the collector circuit of the transistor as shown.

For a single unit, the computer's own power supply could conveniently be used. It is a matter of personal choice and the dictates of design and application whether or not a separate power supply is required. It does not have to be a 5V supply, but if a value other than this is chosen, the value of the collector load resistor, R2 (220R), must be adjusted accordingly. A 12V supply might typically require a value of 680R for R2. A suitable power supply circuit is shown in Figure 14.23.

When V_{in} assumes a logic 1, the transistor is turned ON, collector current flows and the opto LED glows. This is sensed by the opto triac which conducts and connects the power triac gate, via R3 (coincidentally, also 220R in this circuit) to the 'cold' end of the mains lamp neutral supply. The power triac is sent into conduction and the mains lamp comes on. Details of the triac are given in Figure 14.24, the current rating being chosen according to the size of the load.

Using this type of circuit, you can control almost anything. Resistive loads can simply replace the lamp shown in Figure 14.22. Inductive loads (such as motors) require a 'snubber network', consisting of a 100nF (1000V working) capacitor in series with a 100R resistor across Mt1 and Mt2 of the triac. Further discussion on these circuits is not appropriate here.

In order to test the Figure 14.22 circuit, connect V_{in} to bit 0 of port A and turn back to Chapter 12 and run the program given in Table 12.1 (6502) or Table 12.9 (Z80).

Programmable devices

The interfacing circuits described here are relatively simple to construct and use. There is also a whole family of programmable peripheral interfaces (PPIs) often also called peripheral input/output (PIO), peripheral interface adapter (PIA) or versatile interface adapter (VIA) devices and these are discussed in some detail in Chapter 17.

Questions

1 Draw a circuit diagram of a simple transistor switch suitable for interfacing a standard LED to a typical computer output port:
   (a) using an npn transistor;
   (b) using a pnp transistor;
   (c) using a Darlington driver.

2 Draw a diagram of a typical octal buffer which contains Darlington transistors and diodes.
   (a) state the type number of the device;
   (b) state the purpose of the diodes;
   (c) show how one of the buffer elements would be connected in order to drive a relay.

3 State two methods of driving seven-segment displays.
   Draw circuit diagrams to show how a computer may be interfaced to drive:
   (a) a common cathode seven-segment display;
   (b) a common anode seven-segment display.

4 A greengrocer wants a flashing sign saying 'PEAS'. A technician realises he can do this by using standard seven-segment displays, and a microprocessor with the necessary interfacing hardware, to light up the appropriate segments. Outline what the technician needs to do as follows:
   (a) what codes would be used for each letter of 'PEAS';
   (b) how the hardware could be designed and built;
   (c) how the software would be developed;
   (d) how the complete system would be connected.
   Ideally, build and run such a system and indicate how it could be modified to say 'CHIPS' for the take-away next door.
   Could you build a system that reads 'OPEN' and 'CLOSED'?

5 Describe how a stepper motor may be driven by a microelectronic system using:
   (a) a dedicated IC;
   (b) Darlington transistors.

6 Outline at least eight safety precautions which should be considered when building a computer interface for mains driven devices.

7 What is an opto-isolator and why is it particularly important to incorporate such a device in computer/mains interfacing?

8 Describe how a 240V electric motor may be controlled by a computer using opto and power triacs.

9 Design (and build if possible) a microelectronic system to the following specifications:
   (a) a heater is turned on when an $\overline{\text{NMI}}$ interrupt is received;
   (b) a flashing LED indicates when the heater is on;
   (c) an $\overline{\text{INT}}$ interrupt turns the heater off.

10 Design a microelectronic system which will turn a 240V mains fan on, and a 12V heater off, when a sensor detects a critical temperature. The system should then turn the fan off and the heater back on when the temperature falls below the critical temperature.

Assignment 14.1

Compare simple transistor switches (such as the one shown in Figure 14.4) using transistors in:

(a) common emitter mode;
(b) common collector (emitter follower) mode.

Describe how the transistor operates in each mode and hence state any advantages of using a particular mode.

Assignment 14.2

Investigate the design of a microelectronic interfacing system which can operate four, seven-segment LEDs, but which uses only one decoder to feed all of them.

Assignment 14.3

Investigate the nature and design of LED displays such as the Maplin Electronics 5 by 7 red dot matrix array (order code FT61R). These devices can display the complete ASCII character set, and have slots and tongues in the sides so that rows of displays may be assembled for moving messages, etc. Show how a bank of about six such devices can be used to display a moving message, writing the appropriate software and building the appropriate hardware.

Assignment 14.4

Investigate the design considerations of an LM380 (or similar) amplifier and construct a typical circuit to enable the use of the 'Roger bleep' program (Tables 14.8 and 14.9). Study the program carefully and describe how it works.

N.B. Ensure that the amplifier presents a high impedance to the port's data register to prevent any possibility of damage to it (use a 100n coupling capacitor, for example).

Chapter 15
Integrated circuits and logic families

Basic logic gates

Familiarity with the basic logic gates, NAND, AND, OR, NOR, EX-OR, EX-NOR and NOT, is assumed. For further information the reader is referred to Appendix IV.

TTL was the first type of logic circuit specifically designed to be manufactured in IC form; the most common form of TTL is shown in Figure 15.1 in which the output transistors form what is known as a 'totem pole' arrangement. This has a low output resistance for both '1' and '0' logic outputs.

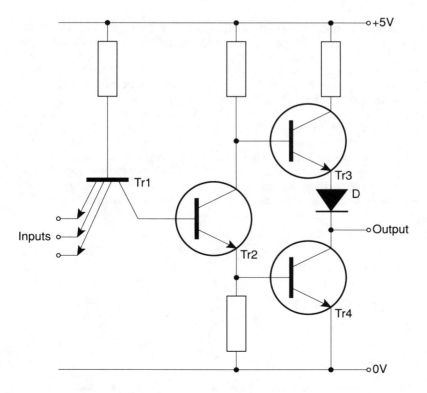

Figure 15.1   *Detail of a TTL NAND gate.*

For fast switching and high fan-out, the output resistance of a logic gate should be very low. Since all capacitances associated with the inputs to the next logic level must be charged or discharged as the voltage level changes, the output circuit must be capable of supplying large currents in an upward transition or sinking large currents in a downward transition.

The output resistance of a saturated transistor (Tr3 in Figure 15.1) is inherently low (about 10 ohms typically), providing the desired low output resistance when the output goes LOW. When the output is HIGH, Tr4 is active and provides the low output resistance characteristic. In IC form, TTL gates are small, reliable and cheap; because of their excellent characteristics they are used in a great variety of IC devices.

Wired logic

Generally; the outputs of TTL NAND gates should not be connected together. If, for example, two TTL NAND gates have their outputs connected together and only one gate has a zero output, then Tr4 in Figure 15.1 must sink, not only the current from the load, but also the current from Tr3 of the other gate. This is likely to overload Tr4. However, if this were not a problem, there are some advantages to using wired logic where possible.

When the outputs of two TTL NAND gates are connected together and one of them produces a zero output, the combined output must also be zero. This is because the output transistor (Tr4) will maintain a zero output even if the other output, independently, would have produced a 1. This is the same as saying that the output is the AND function of the two outputs $\overline{AB}$ and $\overline{CD}$, i.e.

OUTPUT = $\overline{AB} \cdot \overline{CD}$

This is why the arrangement is called WIRED-AND; Figure 15.2 illustrates the effect and shows how an AND gate may be saved by using such an arrangement. To avoid overloading Tr4, 'open collector' versions of TTL are available, the circuit of which is shown in Figure 15.3. When a number of these are wired together, a single external pull-up resistor must be provided, whose value depends on the number of gates wired together.

Since wiring ordinary TTL gates together is undesirable, where this is considered necessary, gate outputs should be connected via OR gates. Open collector devices can be connected, however, and such gates have another advantage; they can be used to switch loads that require a higher supply voltage than the logic circuits themselves. Figure 15.4 shows one application; note, however, that this does *not* mean that open collector TTL gates can be powered from supplies other than 5V; TTL must always be supplied with 5V,

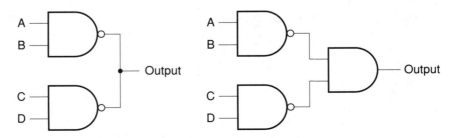

Figure 15.2 *On the left, a WIRED-AND arrangement. On the right, the alternative arrangement using an extra AND gate.*

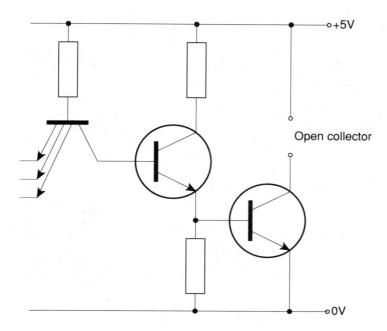

Figure 15.3 *An open collector gate.*

Open collector

but it does mean that a TTL gate can be used to switch a load which requires a greater voltage than the TTL 5V, in this case 24V. Table 15.1 lists some practical open collector gates.

Tri-state devices

Tri-state logic devices are used in microelectronic systems to allow parts of the system to be temporarily isolated from the buses. Two of the logic states are of course '1' and '0'; the third state occurs when the output is effectively disconnected, 'high impedance', or 'open circuit'. This provides the isolation required. The application of a suitable voltage to the enable pin of the device allows it to take up the appropriate output logic state.

The buffers may be inverting or non-inverting, and the device may be enabled by either a logic 1 or a logic 0. Some of the possibilities are shown in Figure 15.5. As an example, Figure 15.6 illustrates the effect of tri-stating a NAND gate, although any type of gate may be tri-stated.

Table 15.1 Some practical open collector gates. There are, of course, many other examples. The reader is encouraged to look at manufacturers' details to gain familiarity with the design and pin-out of such devices

IC	Description
7401	Quad 2-input NAND
7409	Quad 2-input AND
7406	Hex inverting buffer
7407	Hex non-inverting buffer (30V)
7417	Hex non-inverting buffer (15V)

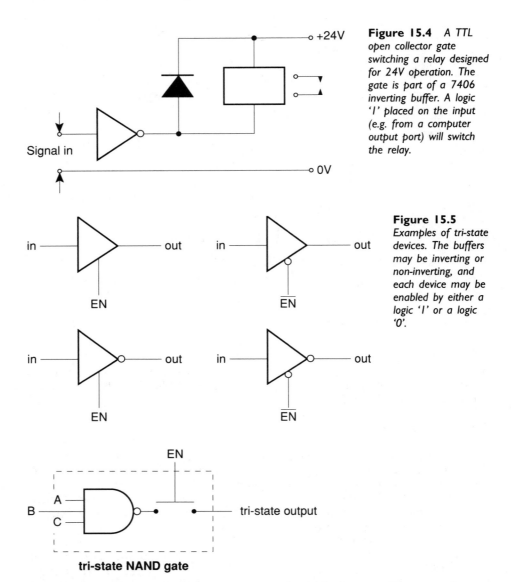

Figure 15.4 *A TTL open collector gate switching a relay designed for 24V operation. The gate is part of a 7406 inverting buffer. A logic '1' placed on the input (e.g. from a computer output port) will switch the relay.*

Figure 15.5 *Examples of tri-state devices. The buffers may be inverting or non-inverting, and each device may be enabled by either a logic '1' or a logic '0'.*

tri-state NAND gate

Figure 15.6 *A tri-stated NAND gate. When the appropriate enable (EN) signal is applied, the switch is closed, allowing the logic state of the NAND gate to reach the output. As soon as the enable signal is removed, the output is totally disconnected or 'high impedance (Z)'. The output could then float to any logic value to which it happened to be connected.*

Three-state operation is used widely in many other logic elements associated with microelectronic systems, including most common logic gates, latches (flip-flops) and I/O devices. Most importantly, a random access memory unit has all its data connections tri-stated. Every single bit in every memory location must be connected in this way, as illustrated in Figure 15.7.

Analysis of logic employed in data flow

If the $\overline{\text{CS}}$ pin is high, all tri-state devices are disabled. With $\overline{\text{CS}}$ low, an enabling signal is available depending on the state of the OR gate inputs. If

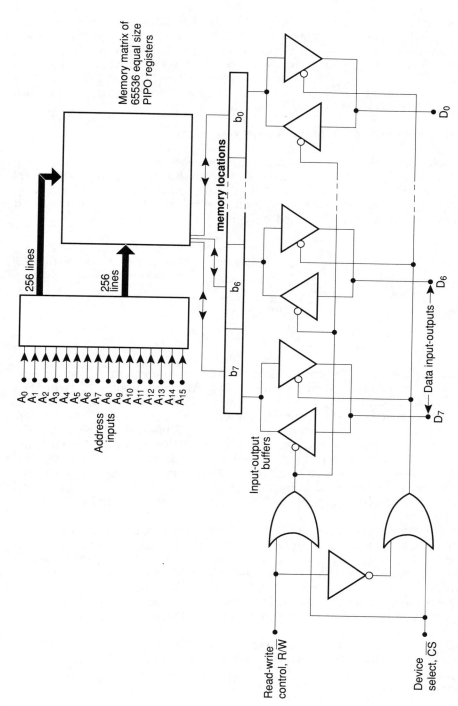

Figure 15.7 Tri-state devices in use to control data flow in and out of a 64K RAM store.

the R/$\overline{W}$ line is high, data can flow out of the RAM store but not in, since the tri-state buffers producing data output are enabled, but the input ones are not. (The READ signal is inverted before being passed to the lower OR gate.)

If the R/$\overline{W}$ line goes low, the reverse occurs, the lower OR gate now has a logic 1 on it disabling all data output buffers; meanwhile, data input buffers are enabled by the low R/$\overline{W}$ signal. Hence, the outputs from the OR gates enable either the input buffers or the output buffers.

Multiplexers, demultiplexers and decoders

Multiplexing is used in many signal transmission systems, not only for transmission of logic signals. It is a method by which different signals from several sources may be transmitted down a single line, synchronised switching being used to ensure that each signal reaches its destination. A simple, switch-based multiplexing system is illustrated in Figure 15.8.

Figure 15.8   *A simple multiplexed communications system.*

Imagine that three telephone users wish to share a common line. As long as sender A is connected to receiver A' by correct positioning of switches 1 and 2, normal communication can take place. If the position of the switches is changed, sender B may communicate with receiver B'; in the final position, sender C communicates with C'. If the switch positions are changed rapidly, all three pairs of communicators can talk to each other without even being aware of the existence of the others. An electronic means of switching and synchronising is available which could operate such a system. In fact the system could be expanded to include dozens of different subscribers. Digital information may be communicated in a microelectronic system by exactly the same means.

In a microprocessor, data and address lines may be multiplexed. This has the advantage that fewer pins are required on the CPU package and only one set of bus wires is necessary to convey both data and addresses. The disadvantage is the reduction in speed as it takes time to switch between the two signals.

In the case of a microcomputer data bus, it will be necessary to provide a bidirectional path, i.e. the data bus not only accepts data from an input device via a tri-state buffer, but must also be able to transmit data in the opposite direction.

Such a system is described as a bidirectional bus driver. The logic diagram of a 4-bit data bidirectional bus driver is shown in Figure 15.9.

Decoders

Within logic circuits the term 'multiplexer' is usually used for the circuit which selects and connects one source from several to the transmission system. The circuit which connects the signal to the required destination is known as a demultiplexer. In cases of single direction connections the multiplexer is also known as a 'data selector'. A data selector is a simple combinational logic circuit.

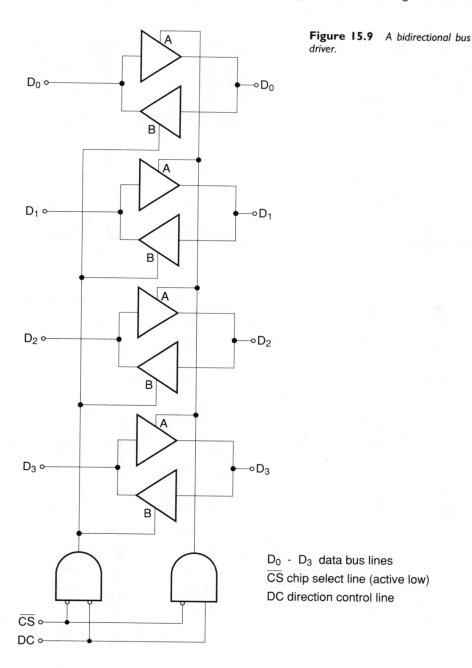

Figure 15.9 *A bidirectional bus driver.*

D_0 - D_3 data bus lines
$\overline{CS}$ chip select line (active low)
DC direction control line

When signals may travel in either direction along some connection in a multiplexed system, the routing circuits have to pass signals in both directions; the bidirectional switching unit is another type of multiplexer. Essentially it consists of a 'data selector' and a 'demultiplexer' combined with tri-state devices to control the direction of the signals.

Within processor systems a binary number is often used as a code to define which particular member of a group of similar items is to be used. That is, each item in the group is given a unique number, an 'address', by which it

Table 15.2   Truth table for a 2-line to 4-line decoder

Inputs			Decoded outputs			
Enable	Address					
EN	B	A	S_3	S_2	S_1	S_0
0	0	0	1	1	1	0
0	0	1	1	1	0	1
0	1	0	1	0	1	1
0	1	1	0	1	1	1
1	0	0	1	1	1	1
1	0	1	1	1	1	1
1	1	0	1	1	1	1
1	1	1	1	1	1	1

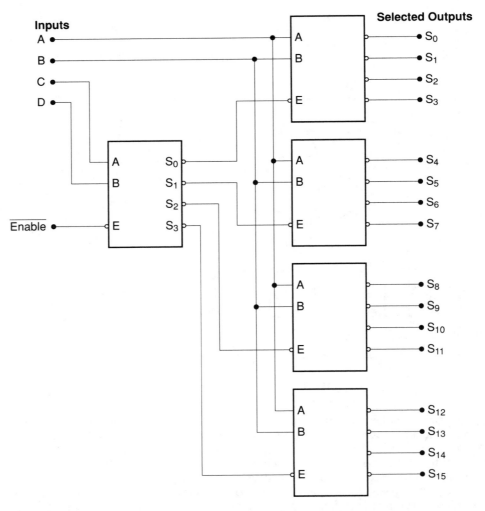

Figure 15.10   *4-line to 16-line decoder utilising five 2-line to 4-line decoders.*

can be identified. Table 15.2 shows the truth table for a two-line to four-line decoder.

The truth table shown in Table 15.2 indicates that the decoder has an active low enable input. Unless the enable pin is low (logic 0) the device gives no output. When enable is low, each of the outputs goes to a zero when the appropriate 2-bit binary address is supplied; e.g. with enable at zero, and address 10 applied, the output is 1011, i.e. bit S_2 is at 0. The effect of applying other address values is shown. If the enable is at a logic 1, all outputs are at 1 regardless of the address used.

A 4-line to 16-line decoder may be implemented by connecting five, simple 2-line to 4-line decoders together as in Figure 15.10.

Address decoders are required in a number of places in a microcomputer system, e.g. to derive the 'chip select' (CS) signal for memory devices, interfacing circuits, counters and timers, etc. A typical decoder chip takes in a number of binary-coded input lines and produces separate output 'select' lines to drive the CS inputs of memory and I/O circuits. In many systems the CS lines of memory and I/O circuits require *active low* signals so most decoder outputs go low when active.

The great benefit to be gained in a computer system by the use of decoding is that it allows the number of interconnections between units to be reduced. Special decoder integrated circuits, having the necessary logic to decode three inputs to eight outputs, for example on a single chip, are available as support chips for microcomputer systems.

Address decoding using basic logic gates

To see how the decoder chips operate, consider the circuit shown in Figure 15.11, which uses four AND gates with two inverters. The resulting truth table is given in Table 15.3. Although it's possible to implement this kind of logic, it's much easier to use a single chip designed for the purpose. The 74LS139 IC is a dual 2-line to 4-line decoder. It contains two, quite independent decoders, where a specific code on the 'select' inputs will drive one of the four outputs low, providing the enable pin is low. The enable input can be used as a data input for demultiplexing.

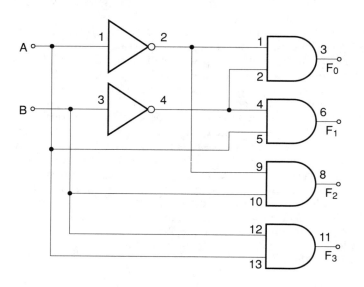

Figure 15.11 *Pin numbers on a 7404 inverter chip and a 7408 quad AND gate are shown.*

Table 15.3 Truth table for a 2-line to 4-line decoder. As an exercise, replace the AND gates in Figure 15.11 with NAND gates and then work out the new truth table. Why would a decoder generally be constructed in this manner?

B	A	F_3	F_2	F_1	F_0
0	0	0	0	0	1
0	1	0	0	1	0
1	0	0	1	0	0
1	1	1	0	0	0

The 74HCT138 is one of many 3-line to 8-line decoders, whilst the 74LS145 IC is a 4-line to 10-line decoder. In the latter, a specific code between binary 0 to 9 on the 4 input lines will switch on one of the ten outputs. Binary codes 10 to 15 switch all outputs off. This particular chip can supply 80mA sink current for directly driving lamps or relays. The reader should consult manufacturers' literature on these and other decoders to gain familiarity with their design and implementation in microelectronic systems.

In the next chapter, we see how decoders can be used to implement computer memory systems.

Logic gate fabrication

We conclude this chapter by looking at the fabrication of logic gates. Two main logic families exist: the 7400 series which uses TTL technology, based on the use of bipolar transistors and operating on 5V power supplies, and the 4000 CMOS family which are unipolar devices operating at up to 15V.

Field effect transistors (FETs)

Transistors of the type referred to in Chapter 1 (npn or pnp, 'collector–base–emitter' devices) and discussed in some detail in the last chapter, are known as bipolar transistors. This is because their operation relies upon the simultaneous motion of electrons and 'holes'.

When correctly biased, a tiny current is caused to flow between base and emitter producing a much larger (often at least 100 times larger) current to flow between collector and emitter. This is how the transistor amplifies; currents (and voltages) are not miraculously made larger, a tiny signal current flowing into the base stimulates a much larger current collector to emitter. Another family of transistors – field effect types – are not current devices at all. That is to say, no current need flow into the device in order to stimulate a larger current elsewhere. FETs rely upon the application of a voltage – an electric field – hence the term 'field effect'.

This is not meant to be a comprehensive review of transistor fabrication, merely an introductory note to explain some of the terms involved and cover the requirements of some microelectronics courses in this area.

Field effect transistor symbols are shown in Figure 15.12. There are two basic types, n-channel and p-channel, distinguished by the direction of the arrow on the gate. In this context, the 'gate' is the control terminal of the device, roughly equivalent to the base of a bipolar transistor. The term should not be confused with a logic gate which may be composed of several transistors (and other components) as we have seen.

The drain is equivalent to the collector of a bipolar transistor whilst the source may be compared to the emitter.

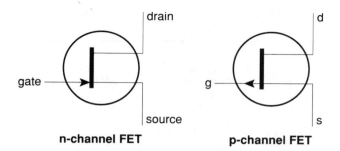

n-channel FET

p-channel FET

Figure 15.12 *n-channel and p-channel field effect transistors.*

NMOS, PMOS and CMOS devices

Transistor-transistor logic (TTL) is still very widely used in industry; it can operate at very high speeds and is now firmly established. The other family of integrated circuits which are in widespread use utilise metal oxide semi-conductor field effect transistor (MOSFET) technology. Although NMOS (N enhancement MOSFET) devices have superior characteristics to PMOS, there were initial difficulties in manufacturing them, so the first MOS digital ICs were PMOS.

However, research into NMOS fabrication was continued and eventually proved to be successful; NMOS devices have a higher packing density (more gates in a given volume), work at a higher speed and are compatible with TTL, having positive gate and drain voltages. In consequence, NMOS devices have now superseded PMOS.

CMOS devices contain both NMOS and PMOS elements; the combination results in fewer gates per unit volume (a lower packing density) and increased cost. However, the increase in speed over NMOS devices and the reduction in power dissipation makes them an attractive alternative. Nowadays most digital ICs which are not TTL are CMOS.

In summary, NMOS superseded PMOS because:

(a) its packing density is higher (electrons are more mobile than holes);
(b) speed is increased because of reduced size;
(c) gate and drain voltages are positive so are compatible with TTL.

CMOS – Complementary MOS:

(a) are faster than NMOS
(b) have lower power dissipation, BUT
(c) have lower packing density and
(d) higher cost

Packing density

The number of gates that can be fabricated on a single IC has risen dramatically over the last decade, the most spectacular development occurring in MOS technology which has led to the production of systems such as the microprocessors we are presently interested in. Packing density is often described in terms of 'scales of integration'. A useful basis for definition is 'a logic gate or circuit of similar complexity'; this is summarised in Table 15.4.

Table 15.4 IC packing density or scale of integration. VLSI is only feasible with MOS technology; the others can be bipolar or MOS

Scale of integration	Abbrev	Packing density
Small-scale integration	(SSI)	a single microcircuit containing less than 12 gates
Medium-scale integration	(MSI)	12 to 100 gates
Large-scale integration	(LSI)	100 to 1000 gates
Very large-scale integration	(VLSI)	Over 1000 gates

Questions

1 With reference to logic gates, what is a 'totem pole' arrangement and why is it used?
2 State one advantage and one disadvantage of using a 'WIRED-AND' configuration.
3 Tri-state devices:
   (a) what are they?
   (b) what are they used for?
   (c) draw diagrams of the devices, both inverting and non-inverting, with low and high enable.
4 (a) What does NMOS stand for?
   (b) Why did NMOS supersede PMOS devices?
   (c) State two advantages and two disadvantages of CMOS compared to NMOS devices.
5 State one advantage and one disadvantage of using multiplexed data and address lines in microelectronic systems.

Multiple choice questions

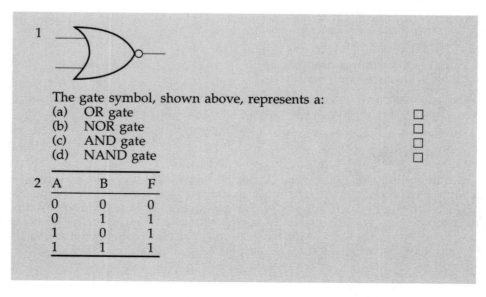

1

The gate symbol, shown above, represents a:
(a) OR gate   ☐
(b) NOR gate   ☐
(c) AND gate   ☐
(d) NAND gate   ☐

2

A	B	F
0	0	0
0	1	1
1	0	1
1	1	1

The truth table shown above is for a:
- (a) OR gate ☐
- (b) EX-OR gate ☐
- (c) NOR gate ☐
- (d) EX-NOR gate ☐

3 In a logic gate, a 'totem pole' arrangement is used:
- (a) to allow voltages greater than 5V to be used in TTL systems ☐
- (b) to allow open collector devices to be used ☐
- (c) to enable a WIRED-OR configuration to be used ☐
- (d) to provide low output resistances for both '1' and '0' logic outputs ☐

4 Open collector devices are useful to enable:
- (a) TTL gates to operate higher voltage devices ☐
- (b) TTL to operate at a higher voltage ☐
- (c) TTL to source more current ☐
- (d) TTL to sink more current ☐

5 Tri-state devices have:
- (a) three distinct logic levels ☐
- (b) positive and negative logic levels ☐
- (c) logic 1, logic 0 and 'high impedance' states ☐
- (d) low impedance outputs regardless of level ☐

6 Tri-state buffers may be:
- (a) inverting only ☐
- (b) non-inverting only ☐
- (c) inverting or non-inverting but enabled by a logic 0 only ☐
- (d) inverting or non-inverting but enabled by a logic 1 or a logic 0 ☐

7 A RAM unit must have tri-stated connections:
- (a) on all data lines ☐
- (b) only on multiplexed data lines ☐
- (c) only if PROM devices are also present in the system ☐
- (d) only if dynamic memory devices are employed ☐

8 PMOS was superseded by NMOS because:
- (a) its packing density is lower ☐
- (b) it has negative gate and drain voltages ☐
- (c) it is cheaper ☐
- (d) it is faster ☐

9 CMOS is preferred to NMOS because:
- (a) it is faster ☐
- (b) it has a higher packing density ☐
- (c) it is cheaper ☐
- (d) it is compatible with TTL ☐

10 A CPU with multiplexed data and address lines will have:
- (a) simplified circuitry ☐
- (b) increased speed ☐
- (c) larger addressing capability ☐
- (d) fewer pins ☐

11 The number of simple 2-line to 4-line decoders that would be needed to implement a 4-line to 16-line decoder is:
- (a) 2 ☐
- (b) 3 ☐
- (c) 4 ☐
- (d) 5 ☐

12 A single microcircuit which contains between about 12 and 100
    gates would have a scale of integration which is described as:
    (a)   small   ☐
    (b)   medium   ☐
    (c)   large   ☐
    (d)   very large   ☐

Chapter 16
Implementation of memory systems

Memory maps

In Chapter 4, we discussed the use of memory maps, noting how they can be used to show the address ranges of the various ROM, PROM, RAM and other memory chips contained within the system. We also saw that incomplete memory maps can exist, where some of the available addresses are simply not used.

In this chapter, we want to look at how these memory systems can be implemented using the actual memory chips. To take a simple example, Figure 16.1

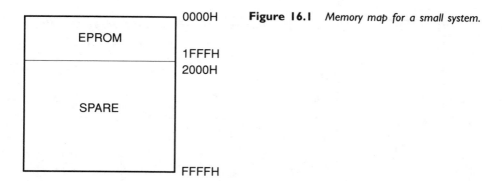

	0000H
EPROM	
	1FFFH
	2000H
SPARE	
	FFFFH

Figure 16.1 *Memory map for a small system.*

Address Lines																Address Range
A_{15}	A_{14}	A_{13}	A_{12}	A_{11}	A_{10}	A_9	A_8	A_7	A_6	A_5	A_4	A_3	A_2	A_1	A_0	
not used			0	0	0	0	0	0	0	0	0	0	0	0	0	0000H
			1	1	1	1	1	1	1	1	1	1	1	1	1	1FFFH

Figure 16.2 *The address range used by an 8K EPROM.*

Table 16.1 Commonly available EPROM devices

Type	Memory capacity	Arranged as:
2708	8K	1K × 8-bit words
2716	16K	2K × 8-bit words
2732A	32K	4K × 8-bit words
2764	64K	8K × 8-bit words
27128	128K	16K × 8-bit words
27256	256K	32K × 8-bit words
27512	512K	64K × 8-bit words

shows the memory map of a small microelectronic system. This consists of a single 8K EPROM. 8K is equal to $2^3 \times 2^{10}$ which is 2^{13} and we showed at the beginning of Chapter 4 that this indicates that 8K of memory will need 13 separate lines to address it. This is shown diagrammatically in Figure 16.2.

An 8K memory device will have an address range of zero to 1FFFH (&1FFFH). All memory chips have many things in common; they'll all have address and data lines, of course, but also $\overline{CS}$ (chip select) or $\overline{CE}$ (chip enable). RAM chips have read and write lines, the one or the other usually being taken low when that particular operation is required. ROM chips can't be written to, so they can be recognised by the absence of a $\overline{WE}$ (write enable) pin. We'll concentrate on EPROMS for the moment and Table 16.1 lists some of the more common devices presently available.

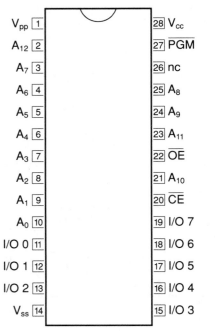

Figure 16.3 *Pin-out of a 2764, 8K EPROM.*

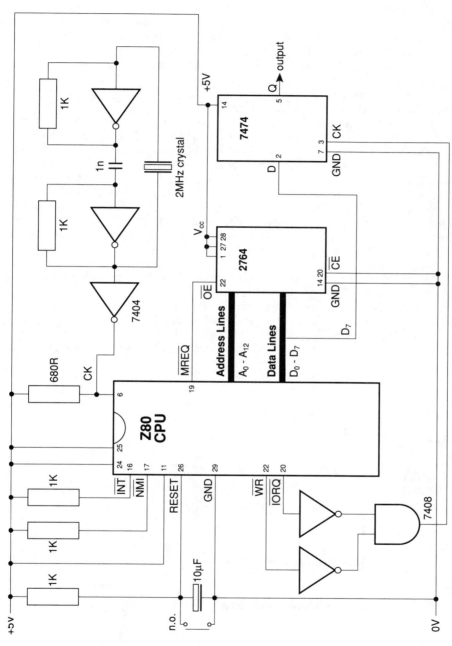

Figure 16.4 A microelectronic system using a Z80 CPU, a 2764 EPROM, a 7474 data latch and two logic chips (7408 AND gate and 7404 hex inverter).

EPROM connections

The address and data lines are taken directly to the corresponding pins of the CPU; the $\overline{\text{MREQ}}$ line goes to the $\overline{\text{OE}}$ (output enable) pin of the EPROM and (in this simple example) also to the $\overline{\text{CE}}$ (chip enable) pin via inverters and a NAND gate. A very simple microelectronic system can easily be built up using just these two chips and a clock oscillator. All that are needed besides are a few logic gates and the necessary I/O devices.

Figure 16.3 shows the pin-out of a 2764, 8K EPROM. The address lines are marked A_0–A_{12} and the data lines are shown as I/O lines. The power supply pins are marked V_{ss} (0V) and V_{cc} (+5V). V_{pp} and $\overline{\text{PGM}}$ are used to program the device. $\overline{\text{OE}}$ and $\overline{\text{CE}}$ are the output enable and chip enable pins respectively.

To show how the connections are made, study the circuit of Figure 16.4 which is probably the most basic Z80 microelectronic system it's possible to have.

The CPU is driven by the simple clock oscillator first described in Chapter 1; the frequency is not critical and could just as easily be 1MHz or 4MHz, for example. Many of the CPU connections are not used in this simple example and are permanently tied as shown. The reader will be able to identify the address and data line pins from the diagrams already given and so they're omitted from this circuit for clarity. The $\overline{\text{OE}}$ (output enable) pin is taken to the $\overline{\text{MREQ}}$ (memory request) pin of the Z80. Chip enable ($\overline{\text{CE}}$) is tied permanently to ground. There are no other memory chips in this design so there's no choice in the matter.

The system obviously has limited capability in its present form, but it is ideal to show how such a system can be implemented. To run the program in the EPROM, you simply press the reset button and the appropriate output appears on the Q pin of the 7474 D-type flip-flop. To do this, the program

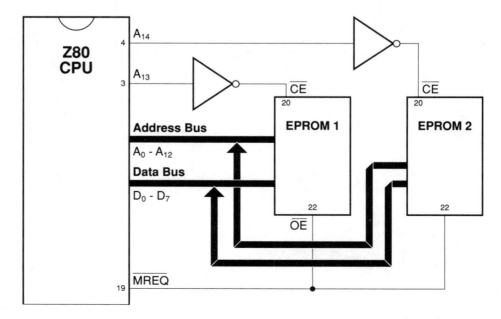

Figure 16.5   *Simple representation of a microelectronic system containing two 8K EPROMs.*

Table 16.2 The possible address ranges for EPROMs in a typical system. This results from linear address decoding

EPROM1		EPROM2	
0010 0000 0000 0000	2000H	0100 0000 0000 0000	4000H
0011 1111 1111 1111	3FFFH	0101 1111 1111 1111	5FFFH
1010 0000 0000 0000	A000H	1100 0000 0000 0000	C000H
1011 1111 1111 1111	BFFFH	1101 1111 1111 1111	DFFFH

puts the relevant logic level on D_7 data line, which in turn is connected to the D input of the flip-flop. When the CPU encounters the OUT instruction, $\overline{\text{WR}}$ and $\overline{\text{IORQ}}$ go low, providing the clock pulse for the 7474 and the data is latched into the Q output.

Adding another EPROM

This is OK as it stands, but how can we add another EPROM? Let's say we want another 8K of memory, how would we address it? One simple answer

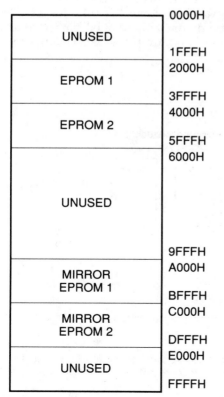

Figure 16.6 *Memory map showing how it remains incomplete and duplicate (mirror) addresses can arise.*

is to connect the 13 address lines (A_0–A_{12}) to each EPROM in parallel and use two of the unused address lines (say A_{13} and A_{14}) for the chip enable ($\overline{CE}$) pins of each memory chip. This is shown in Figure 16.5.

We've simplified the diagram because it's the same as Figure 16.4 except for the additional EPROM and the $\overline{CE}$ connections. The two inverters produce a chip enable signal whenever A_{13} or A_{14} assume a logic 1. So, to select EPROM1, we make A_{13} a one and A_{14} a zero, and the other addresses are as before. Hence, the address range for EPROM1 will be 10 0000 0000 0000 to 11 1111 1111 1111 or 2000H to 3FFFH (&2000 to &3FFFH). The address range for EPROM2 is therefore: 100 0000 0000 0000 to 101 1111 1111 1111 which is 4000H to 5FFFH. In this example, therefore, the address range 0H to 1FFFH is invalid. Furthermore, other values could address each chip. Since A_{15} is not used, setting it to a one or a zero will have no effect, but will give what are called 'mirror' addresses. Table 16.2 shows that EPROM1 can be addressed not only by addresses in the range 2000H to 3FFFH, but also by addresses in the range A000H to BFFFH.

This is called linear address decoding and it gives rise to a memory map like the one shown in Figure 16.6.

This state of affairs may be quite acceptable in a small system; it may even be advantageous as it obviates the need for complicated and expensive address decoding logic. However, in more complex systems, proper address decoding is required to allow for a greater number of memory chips and to ensure that each memory location has a unique address.

Address decoding

We saw in the last chapter how a 2-line to 4-line decoder could be implemented with AND or NAND gates and examined some ICs (like the 74139 and the 4555BE) which provide the function on one chip. A 2-bit binary code on the input drives each of four outputs low (high on the 4555). This means

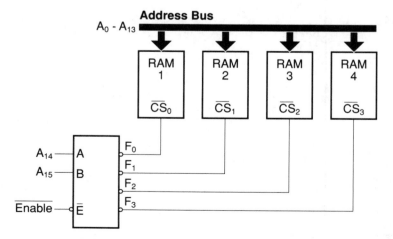

Figure 16.7 *A 2-line to 4-line decoder used to select each of four 16K RAM chips in order to implement a 64K RAM memory system. A type 74139 TTL device contains two independent 2- to 4-line decoders.*

Table 16.3 Truth table for a 2- to 4-line decoder

Inputs		Outputs			
B	A	F_0	F_1	F_2	F_3
0	0	0	1	1	1
0	1	1	0	1	1
1	0	1	1	0	1
1	1	1	1	1	0

Table 16.4 Columns A and B show the way in which the addresses are extended to provide the decoder addressing

	B	A															
	A_{15}	A_{14}	A_{13}	A_{12}	A_{11}	A_{10}	A_9	A_8	A_7	A_6	A_5	A_4	A_3	A_2	A_1
RAM 1	0	0	0	0	0	0	0	0	0	0	0	0	0	0	0
	0	0	1	1	1	1	1	1	1	1	1	1	1	1	1
RAM 2	0	1	0	0	0	0	0	0	0	0	0	0	0	0	0
	0	1	1	1	1	1	1	1	1	1	1	1	1	1	1
RAM 3	1	0	0	0	0	0	0	0	0	0	0	0	0	0	0
	1	0	1	1	1	1	1	1	1	1	1	1	1	1	1
RAM 4	1	1	0	0	0	0	0	0	0	0	0	0	0	0	0
	1	1	1	1	1	1	1	1	1	1	1	1	1	1	1

Table 16.5 Individual memory chip address ranges in hexadecimal, derived from the binary values shown in Table 16.4

Device	Hex address range
RAM 1	0000H to 3FFFH
RAM 2	4000H to 7FFFH
RAM 3	8000H to BFFFH
RAM 4	C000H to FFFFH

Table 16.6 Decoder connections

Input on		Decoder output	Connected to
B	A		
0	0	F_0	EPROM
0	1	F_1	RAM 1
1	0	F_2	RAM 2
1	1	F_3	nothing

Table 16.7 Address ranges for the system shown in Figure 16.8

	B	A													HEX
	A_{12}	A_{11}	A_{10}	A_9	A_8	A_7	A_6	A_5	A_4	A_3	A_2	A_1	A_0	
EPROM	0	0	0	0	0	0	0	0	0	0	0	0	0	0000H
	0	0	1	1	1	1	1	1	1	1	1	1	1	07FFH
RAM 1	0	1	0	0	0	0	0	0	0	0	0	0	0	0800H
	0	1	1	1	1	1	1	1	1	1	1	1	1	0FFFH
RAM 2	1	0	0	0	0	0	0	0	0	0	0	0	0	1000H
	1	0	1	1	1	1	1	1	1	1	1	1	1	17FFH
Spare	1	1	0	0	0	0	0	0	0	0	0	0	0	1800H
	1	1	1	1	1	1	1	1	1	1	1	1	1	1FFFH

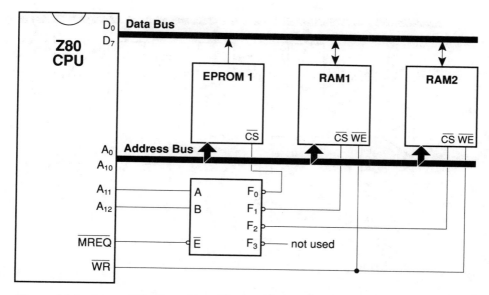

Figure 16.8 *The use of a 2- to 4-line decoder to select any one of three memory elements in a simple system.*

that we can set up the codes on any spare address lines and use the consequent low outputs for the chip selects of each memory device.

Obtaining a 64K byte memory system

A 64K memory system may be implemented in various ways as indicated in Chapter 4. Let's see how a decoder may be used to obtain a 64K memory from four, 16K RAM chips. A 16K memory needs 14 address pins, since:

$$16K = 2^4 \times 2^{10} = 2^{14}$$

This gives addresses 0 to 16383 decimal, 0 to 11 1111 1111 1111 binary, and 0 to 3FFF hexadecimal. (Decimal $16383 = 2^{14} - 1$.)

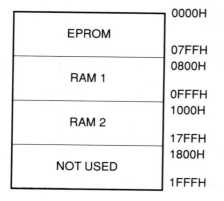

Figure 16.9 *Memory map for the simple system shown in Figure 16.8.*

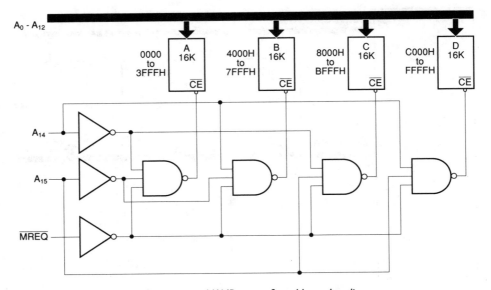

Figure 16.10   *The use of three-input NAND gates for address decoding.*

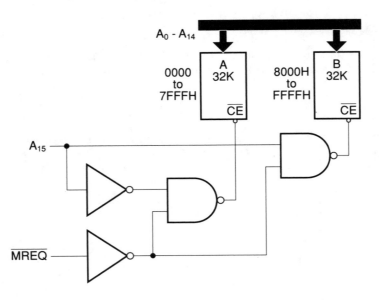

Figure 16.11   *Decoding A_{15} to produce two 32K pages.*

Table 16.8 Truth table for a 3- to 8-line decoder

Inputs			Outputs							
C	B	A	F_0	F_1	F_2	F_3	F_4	F_5	F_6	F_7
0	0	0	0	1	1	1	1	1	1	1
0	0	1	1	0	1	1	1	1	1	1
0	1	0	1	1	0	1	1	1	1	1
0	1	1	1	1	1	0	1	1	1	1
1	0	0	1	1	1	1	0	1	1	1
1	0	1	1	1	1	1	1	0	1	1
1	1	0	1	1	1	1	1	1	0	1
1	1	1	1	1	1	1	1	1	1	0

The first 14 address lines (A_0 to A_{13}) can be connected directly to each 16K RAM chip, via tri-state buffers. Assuming a 16-bit address bus, the final two lines (A_{14} and A_{15}) can then be used to address a 2- to 4-line decoder whose four outputs operate the chip selects ($\overline{CS}$) of each RAM chip, according to the truth table shown in Table 16.3.

The logic which results when implementing such a system is shown in Figure 16.7 and Table 16.4 shows the addresses of each portion of RAM. Table 16.5 gives the address ranges in hex which can be used to construct a memory map for the system.

Example

A simplified block diagram of the memory address decoding logic of a simple microcomputer system is shown in Figure 16.8. Let's work out the hexadecimal ranges of the EPROM and RAMS 1 and 2, and hence draw up a memory map for the system.

The address lines are marked: A^0 to A^{10}, which is 11 lines. This gives 2^{11} addresses (2048D or 2K).

The important information from Figure 16.8 may be summarised as shown in Table 16.6, whilst the address ranges in binary and hex are detailed in Table 16.7.

It is now a simple matter to draw up a memory map for the system since all the information has been provided, and this is shown in Figure 16.9.

Other address decoding ideas

In general, either $\overline{IORQ}$ or $\overline{MREQ}$ will be ANDed in with the decoded address (although generally NAND gates are used for practical reasons associated with active low $\overline{CE}$ inputs). We have seen that address decoding is often applied to the high end of the address bus to give a range of addresses that correspond to a single, valid logic state. This enables a mix of RAM and PROM memory and also allows a range of small chips to be accommodated.

The range for which the enable is valid is called a page. If A_{14} and A_{15} are decoded as shown in Figure 16.10, four 16K pages are obtained. (This is an alternative method using three-input NAND gates as opposed to the use of a 2 to 4-line decoder seen previously.)

The decoding of a range of addresses as pages is usually applied to the memory map, i.e. ANDed with $\overline{MREQ}$, and the decoding of single locations

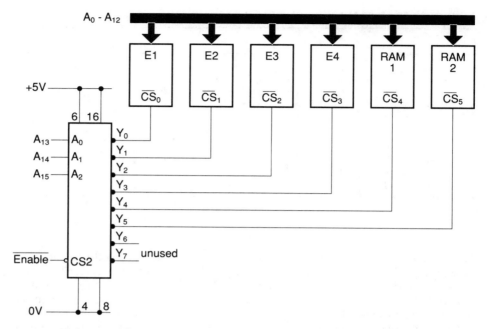

Figure 16.12 *A microelectronic system containing four 8K EPROMs and 16K of RAM.*

is more usually in the I/O map, i.e. ANDed with $\overline{\text{IORQ}}$, although there are always exceptions.

It is not unusual to find holes in the memory map where an 8Kb or 4Kb chip is in a space decoded to 16 or 32Kb. This is because the smaller the hole, the more complex is the decoding logic, and we all try to keep circuits as simple as possible. In practice, large chips are now available at realistic prices, so the days of putting 6116 (2Kb static ram) into 32Kb holes to save cost and keep the decoding simple now seems in the distant past. It is more likely to see a system like that shown in Figure 16.11. Here, A_{15} is decoded and ANDed with $\overline{\text{MREQ}}$. This means that chip A will be enabled for addresses 0000 to 7FFFH and chip B will be enabled for addresses 8000H to FFFFH. The memory map is broken up into just two, 32Kb pages, so a 32K EPROM and a 32K RAM chip could easily be accommodated.

3- to 8-line and 4- to 16-line decoders

To enable more complex memory systems to be built up, ICs with greater decoding ability may be used. The 74137 chip is a typical 3- to 8-line decoder, whilst the CMOS 4515BE chip is one example of a 4- to 16-line decoder. The truth table of the former appears below in Table 16.8. Readers are invited to draw up their own truth table for a 4- to 16-line decoder.

Figure 16.12 shows the use of a 74137 3- to 8-line decoder in a practical microelectronic system.

The next chapter is mainly devoted to programmable peripheral interface devices, but as an example, a complete microelectronic system – SAMSON – based upon the circuit of Figure 16.4, is described. The system is expanded to include a programmable I/O device and RAM memory; the memory map that results is given. This is just one example of how the theory described in this chapter may be applied in practice.

Questions

1   State and describe how many lines are required to address the following memories:
   (a)   a 2816A, 2Kb EEPROM;
   (b)   4Kb made up from two 2716, 16K EPROMS;
   (c)   10Kb made up from a 6116, static RAM and 2 × 2716 EPROMS;
   (d)   two 2764 EPROMS;
   (e)   a 27256, 32Kb EPROM.

2   (a)   What is linear address decoding?
   (b)   What are the advantages and disadvantages of using linear address decoding?

3   A simple microelectronic system is made up from four 8K EPROMS and 16K of RAM. Sketch a diagram showing how the system may be implemented using readily available chips and draw up the resulting memory map.

4   Show, using appropriate diagrams, two different methods of addressing three 8K memory devices in a simple Z80 or 6502 microelectronic system.

5   A special purpose microelectronic system is built up from 15 × 4Kb EPROMS and 1 × 4Kb static RAM. Sketch a diagram to show how such a system may be implemented and draw the resulting memory map.

Chapter 17
Programmable I/O devices

Parallel I/O devices

We introduced the PPI (programmable peripheral interface) in Chapter 12 by describing the setting up and operation of the I/O ports in the 'EMMA' and 'MARC' microcomputer systems. That text shows how the programmer can use various codes in software to drive LEDs (and other transducers) on the output side, and make use of debounced switches, for example, on the input side. Details about Z80 interrupts concluded with a brief discussion on the Z80 PIO (peripheral input/output) device. We now look at the actual chips so that we can program and use them in a system of our own design.

The 6522 VIA

The 'EMMA' uses a 6522 VIA (versatile interface adapter) to implement its two ports, designated A and B. The device has two 8-bit buffer registers to access the I/O ports. In addition, the 6522 also has two interval timers, a serial to parallel–parallel to serial shift register and input data latching on the peripheral ports. The pin-out of the 6522 is shown in Figure 17.1.

Pins 35 to 38 are connected to the register select lines. Which register is accessed depends upon the sort of address decoding used in a particular system; generally, however, the lower 4 bits of the address bus are connected to the register select lines, the actual address being formed from this and the decoding used for the chip selects which are on pins 23 and 24. The 6522 is thus selected when CS1 is high and CS2 is low. We saw in Chapter 12 that the 'EMMA' address decoding system puts I/O registers and data direction registers A and B at addresses 0900 to 0903 in the main memory map.

RES, R/W, IRQ, Φ_2 clock and all data lines are normally connected in parallel with the similar CPU lines. The port connections themselves represent one standard TTL load in the input mode and will drive one standard TTL load in the output mode. More precise details of the 6522 are given in the 'EMMA' technical manual and in data sheets available from the manufacturers.

The 8255 PIA device

The Z80 PIO was covered in Chapter 12 so we'll look at another useful, programmable I/O chip, the 8255. This device has 24 I/O lines which can be set up as inputs or outputs. It has three modes of operation which are summarised in Table 17.1, although we will only be considering mode 0 here.

The 24 I/O lines are in three groups called ports A, B and C. The pin-out diagram is shown in Figure 17.2.

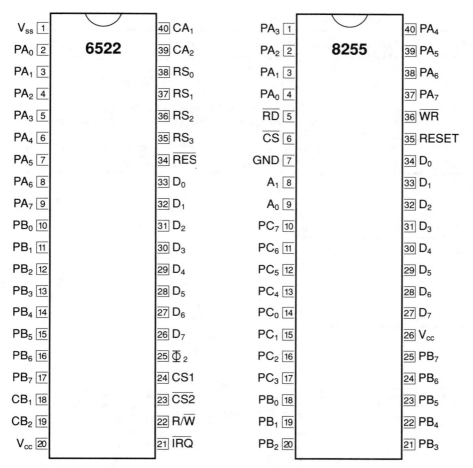

Figure 17.1 *Pin-out of the powerful 6522 versatile interface adaptor (VIA).*

Figure 17.2 *Pin-out of the 8255 PIA.*

Port A can be set to all inputs or all outputs. Port B is the same, all in or all out. Port C is more flexible; it can be all inputs, all outputs, or four inputs and four outputs. The directions of the ports are set by sending a mode word to the control register as described in Chapter 12 for the EMMA and MARC micros. The word is calculated in binary as shown in Figure 17.3 and the port addresses are given in Table 17.2.

Before looking at some sample programs, we'll describe 'SAMSON', a complete microelectronic system based upon the Z80 CPU with an 8255 PIA and a 2764 EPROM.

SAMSON – a Z80/8255 micro system

The name SAMSON came from 'Simple All-purpose Microelectronic System Outputting Numbers'; a suggested circuit is shown in Figure 17.4. Naturally, some measure of practical skill, together with the appropriate facilities, will be required to construct such a system. The reader will have to decide to what extent construction of the project is valid from that point of view. The circuit is, in fact, a modified version of Figure 16.4.

Table 17.1 Summary of the 8255 PIA modes of operation

8255	PIA – programmable interface adapter.
Modes	May be individually programmed in two groups of 12. There are three modes:
Mode 0	Each group of 12 I/O pins may be programmed in sets of four to be input or output.
Mode 1	Each group may be programmed to have eight lines of I/O, and of the remaining four, three are used for handshaking and interrupt control signals.
Mode 2	A bidirectional bus mode which uses eight lines for the bus and five lines (one borrowed from the other group of 12) for handshaking.

Table 17.2 8255 port addresses

Port	Address
A	00
B	01
C	02
CONTROL	03

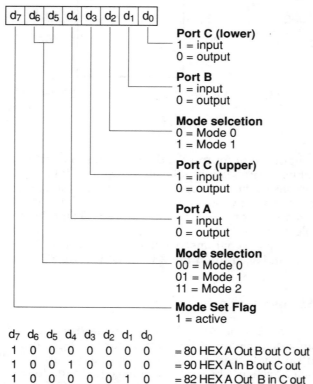

Figure 17.3 *Planning the I/O lines and setting the ports accordingly.*

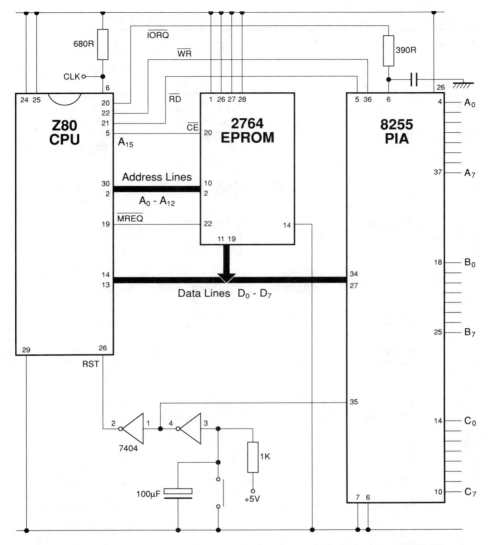

Figure 17.4 *A complete microelectronic system based on a Z80 CPU and an 8255 PIA.*

This simple system, whilst basic and minimal in many respects, is sufficient to control a whole variety of domestic and industrial processes. The imaginative use of the 8255's output ports will enable the user to satisfy the needs of literally hundreds of possible microelectronic control systems. All that is left is to produce the necessary software and we look at an example now.

Software

The first thing to note is that this simple system is initiated by pressing the reset button which is connected to both CPU and 8255 reset pins. A high to low transition is required for the CPU and a low to high transition for the 8255. However, the latter takes much longer (in microelectronic terms!) to recover from a reset; any attempt to send control codes to the 8255

Table 17.3 'Count up': a sample program for the Z80/8255 system

Address	Machine code	Label	Mnemonics	Comments
				;SAMSON 'Count up'
			ORG 0000H	;origin
			ENT 0000H	;entry point
0000	01 00 50		LD BC,5000H	;load BC register
0003	0B	LOOP1:	DEC BC	;reduce BC by 1
0004	78		LD A,B	;B into A
0005	B1		OR C	;sets flag when A=0
0006	C2 03 00		JP NZ,LOOP1	;repeat
				;
0009	ED 56		IM 1	;interrupt mode 1
000B	F3		DI	;disable interrupts
				;
000C	01 03 00		LD BC,0003H	;control port
000F	3E 90		LD A,90H	;B & C ports o/p
0011	ED 79		OUT (C),A	;send data to
				;control register
				;
0013	21 00 01	RPT:	LD HL,0100	;look-up table
0016	7E	GO:	LD A,(HL)	;
0017	01 02 00		LD BC,0002	;
001A	ED 79		OUT (C),A	;
				;
001C	06 DD		LD B,0DDH	;delay
001E	0E FF	LOOP2:	LD C,0FFH	;
0020	0D	LOOP3:	DEC C	;
0021	C2 20 00		JP NZ,LOOP3	;
0024	05		DEC B	;
0025	C2 1E 00		JP NZ,LOOP2	;
				;
0028	23		INC HL	;
0029	FE 6F		CP 6FH	;is it 9?
002B	CA 13 00		JP Z,RPT	;Yes? then repeat
002E	C3 16 00		JP GO	;No? then GO!
				;
0100	3F 06 5B	TABLE:		;look up table with
0103	4F 66 6D			;hex codes for
0106	7D 07 7F			;7-segment display
0109	6F			;

immediately after a reset may therefore fail, so a delay loop is included at the beginning of the program which is shown in Table 17.3. This may not be necessary if the clock frequency is 2MHz or less but it's still a useful point to bear in mind.

The program is almost the same as that given in Table 14.5, which produced an up count on a seven segment display. It has been modified so that the 8255 ports are set up and accessed (rather than the Z80 PIO). The control word, 90H, is sent to the accumulator via the C register. This makes port A all inputs whilst ports B and C both become outputs.

In more sophisticated programs, a switch unit connected to port A may be used to select different programs in memory. More advanced systems may be created by using a calculator style key-pad. This is continuously scanned by a software routine which detects a specific key press and acts upon it.

The possibility of interrupts occurring is negligible as we have tied the $\overline{\text{NMI}}$ and $\overline{\text{INT}}$ CPU pins to +V in hardware. However, it's good advice to select an interrupt mode (we chose IM 1) and then disable interrupts, just to be sure. We have also used direct JumP instructions (rather than relative jumps) as they are much easier to debug.

Because the only memory in the system is EPROM, which cannot, of course, be written to, all memory LoaDs in the original 'Count up' program have been eliminated. In order to send the seven-segment display character codes to the accumulator, the HL register pair is used as a pointer and a 'look-up table' has been created at address 0100.

It should also be noted that no stack can exist in the system, as this would be implemented in RAM (we show how to do this later). For this reason the CALL instruction cannot be used. CALLs cause automatic stack action pushing current PC contents onto the top of the stack. The delay routine therefore is embedded inside the main program, rather than being CALLed when required.

Comments on 'Count up'

A look-up table containing the codes for the display digits is included in memory starting at location 0100. It is not possible to load them from memory in an EPROM! The delay is integrated into the program; it can't be CALLed since the CALL instruction uses a stack to save the return addresses and there is no stack. We have also used direct jumps (rather than relative jumps), because they are much easier to debug where this becomes necessary.

This simple example and the foregoing provides all the necessary information to enable a programmer to use any of the 8255 ports, so that configurations such as those described in Chapter 12 for the MARC and EMMA may be implemented in a system based upon the circuit of Figure 17.4. All that's necessary is to write the relevant software.

Before considering any other possibilities, however, be careful to use the appropriate port interfacing hardware; if the 8255 is loaded with a low impedance, that bit of the PIO will be destroyed.

Constructional details

Our prototype panel is pictured in Figure 17.5. The data lines from CPU to EPROM and CPU to PIA were implemented using ordinary wires and 'molex' connectors. This is not ideal, but constructed carefully works adequately and greatly simplifies the pcb design. We used separate 7404 chips to implement the clock oscillator and reset logic, simply as a matter of convenience. The clock oscillator (not shown in Figure 17.4) is the same as the one given in Chapter 1 (Figure 1.13) running at 2MHz. The reset circuit provides a negative-going transition on pin 26 to reset the CPU and a positive-going transition on pin 35 to reset the PIA. The switch is a simple pcb mounting 'push to make' type.

'A suitable pcb design is shown in Figure 17.6. In our version, a molex connector was used for data connections to the PIO; the connections to the EPROM were made by drilling another set of holes down the side of the molex socket and then soldering the wires in directly. Once built and tested, simple though it is, SAMSON can be used to control an infinite variety of powerful microelectronic systems.

Adding RAM to SAMSON

SAMSON can become even more versatile if a RAM chip is included and we now describe the addition of a 6264 8K static memory. This device may be

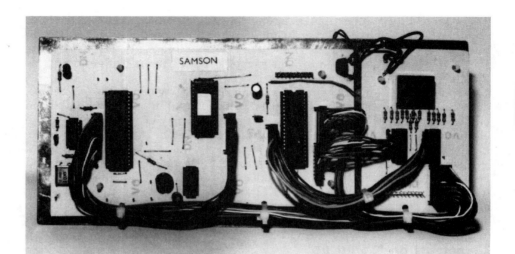

Figure 17.5 *The prototype SAMSON.*

regarded as the RAM version of the 2764 EPROM used previously. In fact, the 6264 is almost pin-compatible with the 2764 EPROM, so all that's required is to extend the data and address buses to the RAM. The same applies to the output enable, only the chip enable is different. This is so that the RAM can be placed in a different part of the memory map from the EPROM. This has been achieved by using A_{15} to enable the EPROM when it is low, and (via an inverter) to enable the RAM when it is high.

The memory map that results is EPROM memory from 0000 to 1FFF and RAM memory from 8000 to 9FFF. It would be possible to use a 27256 EPROM and 32K RAM chip, and, by connecting A_{13} and A_{14} from the CPU to the two chips, we would have a full memory map, i.e. EPROM from 0000 to 7FFF and RAM from 8000 to FFFF. Use of the smaller chips, however, simplifies the hardware and keeps the cost down. It also means that we can illustrate in software how you can work with an incomplete memory map.

Address decoding

Figure 17.7 shows the various inter-chip connections which perform enabling and chip select functions. Address line A_{15} provides the chip enable signal. When it's high, RAM is selected; when A_{15} is low, the EPROM is selected.

Table 17.4 Control signal states for the various READ/WRITE operations

R/W	EPROM (2764)	RAM (6264)	PIA (8255)
READ	A_{15}: low $\overline{\text{MREQ}}$: low	A_{15}: high $\overline{\text{WR}}$: high $\overline{\text{MREQ}}$: low	$\overline{\text{RD}}$: low $\overline{\text{IORQ}}$: low
WRITE		A_{15}: high $\overline{\text{WR}}$: low $\overline{\text{MREQ}}$: low	$\overline{\text{WR}}$: low $\overline{\text{IORQ}}$: low

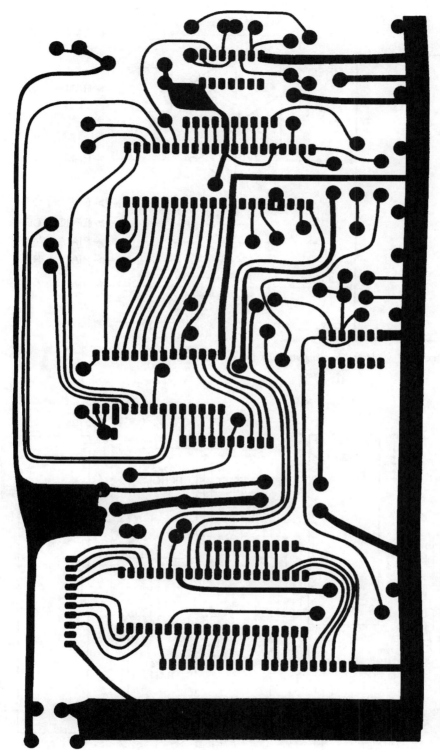

Figure 17.6 Suggested PCB design for SAMSON.

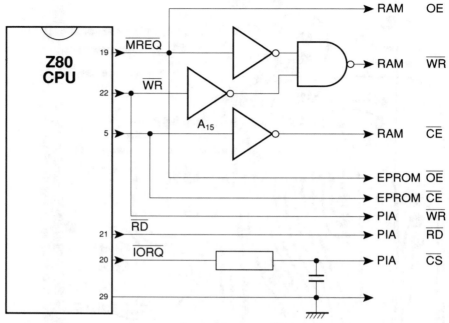

Figure 17.7 *Address decoding structure for SAMSON with added RAM.*

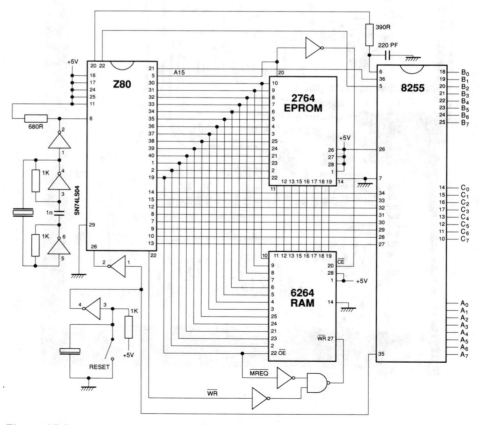

Figure 17.8 *Circuit diagram of SAMSON with added RAM.*

$\overline{\text{WR}}$ is active low and indicates to the RAM chip that it should store the word on the data bus at the location in the address bus A_0 to A_{12}. The address decoding and control signal states are summarised in Table 17.4.

The circuit diagram of SAMSON with added RAM is shown in Figure 17.8.

Further study of programmable parallel I/O devices may be continued by looking at the end of chapter questions and assignments; we have, however, already supplied sufficient information for the assembly and programming of quite complex microelectronic systems, based upon the Z80.

You will learn more about the capabilities and use of such systems by building and using them. In the next chapter we discuss some simple fault-finding techniques which will help in getting your system working – if it doesn't burst into life as soon as it's plugged in!

Serial I/O devices

When data is made available to an output port where the distances between control system and peripherals is quite small, parallel transmission is quite advantageous. It is straightforward and simple to implement and very much faster than comparable serial systems. However, where distances between control and peripheral exceed around 2 metres, difficulties associated with degradation of signal begin to arise, and the sheer bulk of so many parallel connections often renders the system impractical.

By comparison, serial communications can take place with as little as two wires and satisfactory transmission and reception may take place over quite long distances. Naturally, there must be parallel to serial conversion at the computer end and serial to parallel conversion at the peripheral end, but this is a relatively simple matter since there are a whole variety of devices which exist to perform the function. Most important amongst these are the UART and the ACIA. The UART (universal asynchronous receiver/transmitter) is a typical, industry standard device. The ACIA (asynchronous communications interface adapter) is more or less exactly the same thing. However, before looking at typical devices it will be useful to investigate the nature and format of standard serial data transmission.

Serial data transmission

As the name implies, serial transmission allows data to be transmitted one bit at a time only. It is therefore not as high speed as parallel transmission where data might typically be transferred 8 bits at a time. In serial transmission, where this is the word length, the 8 bits are sent in one group, one bit following the next, asynchronously. This means that successive data transmissions are unrelated in time and may be sent whenever they are required, provided that any previous transmissions have been completed.

If there is no transmission, a serial link is maintained at a logic 1 level. When this level falls to a zero, it indicates a break or some kind of fault in the system. The format for one character of asynchronously transmitted data is shown in Figure 17.9.

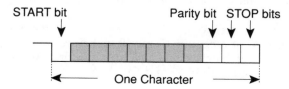

Figure 17.9 *Format for one character of asynchronously transmitted data.*

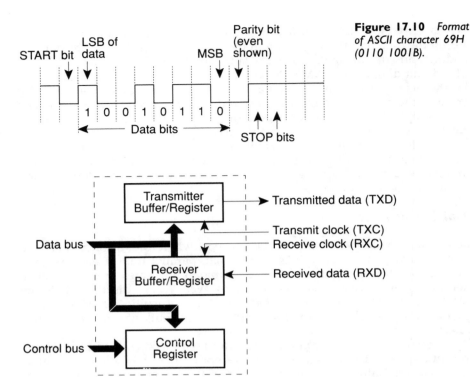

START bit | LSB of data | MSB | Parity bit (even shown)

1 0 0 1 0 1 1 0

◄——— Data bits ———► STOP bits

Figure 17.10 *Format of ASCII character 69H (0110 1001B).*

Figure 17.11 *Simple representation of a typical, programmable, serial I/O device. The transmit (TX) register is PISO and the receive (RX) register is SIPO.*

The character consists of a 'start' bit, followed by seven data bits, a 'parity' bit (used for error checking) and one or two 'stop' bits. The start and stop bits are used in asynchronous serial transmission to detect the start and finish of the data characters being transmitted. The start bit consists of a logic 0, preceding the data bits which are transmitted least significant bit (lsb) first.

Some simple links do not use a parity bit but where it is included (to check on transmission accuracy), 0 = even parity and 1 = odd parity. This is derived from the number of logic ones in the character. If the number is even, the parity bit is zero; if the number is odd, it will be a one. Figure 17.10 shows the format of a character assuming ASCII code 69H (&69) is being transmitted.

The transmitted character is completed by 1, 1.5 or 2 STOP bits consisting of a logic 1. All the bits occupy the same bit time and the data rates (bits per second) are often given in terms of 'BAUD' rate, named after the telegraphy engineer, Baudot. 600, 1200 and 2400 baud are common values in present use.

Transmitter/receiver devices

Figure 17.11 shows a simple representation of a typical, programmable, serial I/O device.

The device will need to provide parallel to serial conversion for transmission, and serial to parallel conversion for reception. It might be helpful to look at two possible configurations of shift registers which can perform these operations and these are shown in Figure 17.12(a) and (b).

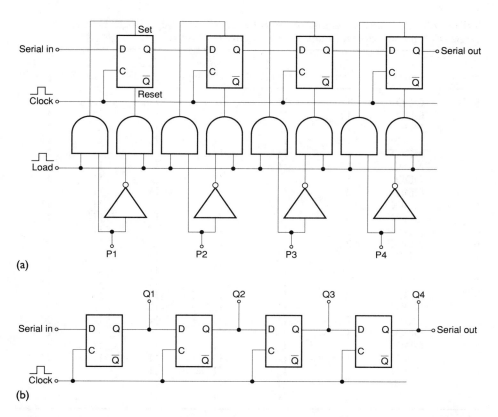

Figure 17.12   (a) Parallel-in/serial-out (PISO) shift register. (b) Serial-in/parallel-out (SIPO) shift register.

A typical UART

The 6402 IC is an industry standard universal asynchronous receiver transmitter (UART). The device can convert parallel data into a serial bit stream and serial data into parallel form. It can operate at any of the standard baud rates, including 600, 1200 and 2400 baud. The oscillator which provides the clock signal must be 16 times the baud rate, as there is a divide-by-16 circuit within the device.

The precise configuration of the UART will depend upon what it's being used for. Table 17.5 shows how the control pins must be connected in order to produce the required format. Use this in association with Figure 17.13 which gives the 6402 pin-out.

The most popular word format uses 8 data bits, one stop bit and no parity; alternative formats are given in the table. Note that if, for example, you select a format that uses 6 data bits, then only 6 bits will be available at the output. In this case, the most significant bits will be unused.

A typical serial/parallel application

Most computer printers are connected in parallel to the host computer, usually via the standard 'Centronics' interface. However, some are connected serially

Table 17.5 Formatting details for the 6402 UART. H is high (logic 1), L is low (logic 0), X means 'doesn't matter' (we usually connect to logic 1)

Data bits	Parity	Stop bits	PI pin 35	SBS pin 36	CLS2 pin 37	CLS1 pin 38	EPE pin 39
5	ODD	1	L	L	L	L	L
5	ODD	1.5	L	H	L	L	L
5	EVEN	1	L	L	L	L	H
5	EVEN	1.5	L	H	L	L	H
5	NONE	1	H	L	L	L	X
5	NONE	1.5	H	H	L	L	X
6	ODD	1	L	L	L	H	L
6	ODD	2	L	H	L	H	L
6	EVEN	1	L	L	L	H	H
6	EVEN	2	L	H	L	H	H
6	NONE	1	H	L	L	H	X
6	NONE	2	H	H	L	H	X
7	ODD	1	L	L	H	L	L
7	ODD	2	L	H	H	L	L
7	EVEN	1	L	L	H	L	H
7	EVEN	2	L	H	H	L	H
7	NONE	1	H	L	H	L	X
7	NONE	2	H	H	H	L	X
8	ODD	1	L	L	H	H	L
8	ODD	2	L	H	H	H	L
8	EVEN	1	L	L	H	H	H
8	EVEN	2	L	H	H	H	H
8	NONE	1	H	L	H	H	X
8	NONE	2	H	H	H	H	X

and most VDUs tend to be connected via the world standard RS232 interface. The RS232 standard uses −12V for a logic 1 and +12V for a logic 0, so when a UART is employed to interface to RS232 devices, the logic must be inverted and changed to TTL levels. This can be achieved quite simply using a single transistor. As an example, consider Figure 17.14 which shows the circuit diagram of an RS232 compatible serial to parallel converter.

We've used a 4047BE CMOS multivibrator chip to provide the clock. The values shown will give a baud rate of 1200. An RS232 interface for transmission is shown in Figure 17.15. This can be used in conjunction with the 6402 UART and we leave it up to the reader to evolve the final design.

The ACIA

The asynchronous communications interface adaptor (ACIA) is simply another manufacturer's version of a UART. A typical ACIA block diagram is shown in Figure 17.16 and the pin-out (with some explanation) of a typical device (the 6850) is shown in Figure 17.17.

This device is often used in 6502 systems; the BBC computer, for example, used a 6850 ACIA to interface between the microprocessor and its serial port.

The information given in this section is necessarily limited, although all the basic ideas are covered. The reader is encouraged to obtain manufacturers' data sheets on the various devices available and construct a working system.

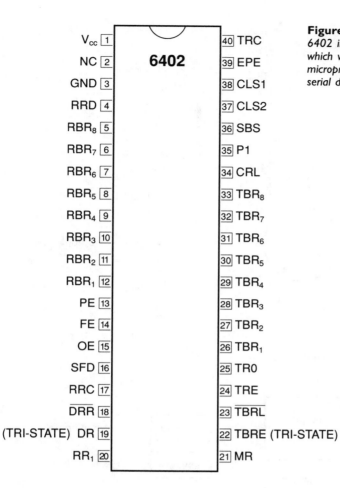

V_{cc}	1	40 TRC
NC	2	39 EPE
GND	3	38 CLS1
RRD	4	37 CLS2
RBR_8	5	36 SBS
RBR_7	6	35 P1
RBR_6	7	34 CRL
RBR_5	8	33 TBR_8
RBR_4	9	32 TBR_7
RBR_3	10	31 TBR_6
RBR_2	11	30 TBR_5
RBR_1	12	29 TBR_4
PE	13	28 TBR_3
FE	14	27 TBR_2
OE	15	26 TBR_1
SFD	16	25 TR0
RRC	17	24 TRE
$\overline{DRR}$	18	23 $\overline{TBRL}$
(TRI-STATE) DR	19	22 TBRE (TRI-STATE)
RR_1	20	21 MR

Figure 17.13 *Pin-out of the 6402 industry standard UART which will interface microprocessors to asynchronous serial data channels.*

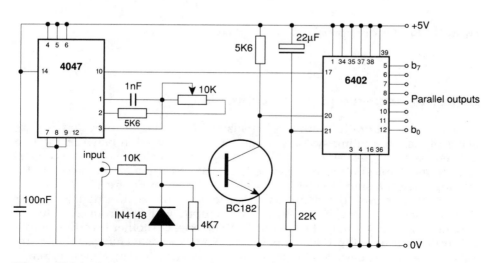

Figure 17.14 *Suggested circuit diagram of an RS232 compatible, serial to parallel converter.*

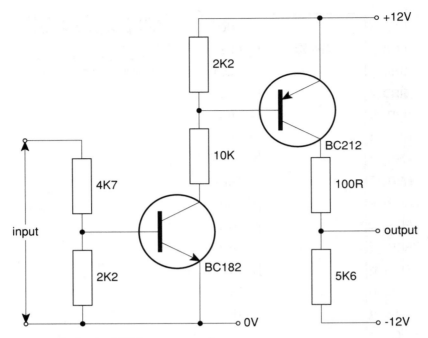

Figure 17.15 *An RS232 transmit interface.*

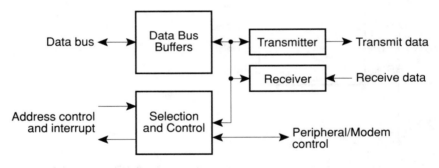

Figure 17.16 *Block diagram of a typical ACIA device.*

Digital to analogue (D to A) conversion

Because most naturally occurring signals are of analogue form and we are using digital computing systems, some form of conversion between the two is often necessary. We start by looking at digital to analogue (D to A) conversion. In its simplest form, this can be implemented using a weighted resistor op-amp circuit, a simple version of which is shown in Figure 17.18.

You will recognise the circuit as an analogue summer. If a digital counter were connected to the input (b_0–b_3), the output will fall from zero to the maximum negative value (the op-amp is an inverter). Another inverting amplifier at the output will produce a positive going ramp waveform as the binary input value increments. The ramp becomes periodic if the system is automatically reset when the counter reaches 1111. This is shown in Figure 17.19.

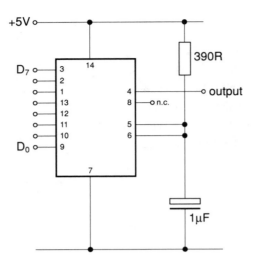

Figure 17.17 *Pin-out and details of the 6850 ACIA. When CS0 and CS1 are high and $\overline{CS2}$ is low, the ACIA is selected.*

Figure 17.18 *A simple, weighted resistor op-amp D to A converter.*

Figure 17.19 *The ramp output from a D to A converter whose input is an incrementing, 4-bit binary value.*

Figure 17.20 *Simple D to A ramp generator using a ZN426E D to A converter IC.*

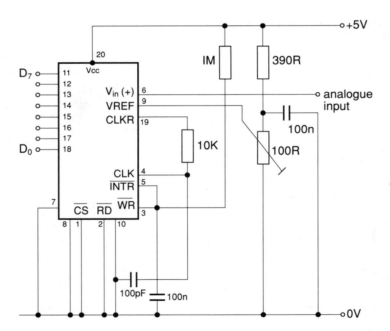

Figure 17.21 *Simple A to D circuit employing an ADC0804 A to D converter IC.*

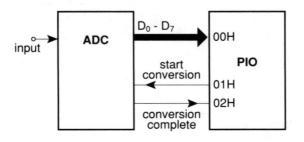

Figure 17.22 *Simple block diagram of an ADC in a microelectronic system. The ADC is an ADC0804, the PIO may be an 8255.*

This simple converter is not very good in practice as it is difficult to obtain resistors of adequate tolerance. Obviously, for a Z80/6502 based system, 8-bit definition would be needed; this could be achieved by the addition of more input resistors, following the sequence, i.e. 40R, 80R, etc. The lowest value (2R5 in this case) becoming the msb. This means that the lsb resistor would have a value of 320Ω – 128 times larger than the msb resistor value. Obtaining this precise ratio is overcome in commercially available chips such as the ZN426E. Figure 17.20 shows a simple circuit using the chip to produce a ramp waveform. The input data lines may be connected to the output port of the system you are using. The frequency of the ramp depends upon the length of delay between each increment of the binary count.

There are several other commonly available D to A converters including the DAC 0801, ZN425E and ZN428E, the latter having input latches to facilitate updating from a data bus. The reader is encouraged to investigate these devices, obtaining data sheets where possible, in order to determine the device's suitability in specific applications.

Analogue to digital (A to D) conversion

As an example, we have chosen the ADC0804 analogue to digital converter IC. The $\overline{CS}$ and $\overline{RD}$ pins are tied to ground in this simple application, as are the analogue and digital ground connections. The analogue input voltage range is 0V to 5V with a single 5V supply and 2.5V applied to pin 9. However, the voltage reference on 9 can be any voltage under 2.5V and we have made this variable by using a 100R preset as shown in Figure 17.21.

The device has an internal clock which requires the 10K and 150pF components connected to pins 4 and 19. This is a standard arrangement suitable in this circuit. The $\overline{WR}$ pin is connected to pin 5 and to the 100n/1M arrangement which ensures that the circuit starts up as soon as it's switched on and goes into continuous ('free run') operation. Immediately after a conversion has been completed, pin 5 tells pin 3 to output new data. Sometimes, these pins are under microprocessor control, as are $\overline{CS}$ and $\overline{WR}$, and a simple block diagram is shown in Figure 17.22.

The ports of the PIO have addresses as shown and the 'start' and 'complete' lines are connected to bit 0 of their respective ports. A software routine can cause the system to request a conversion and then poll the ADC and read the 8-bit signal. The ADC has output latches which can directly drive a microprocessor data bus. The IC looks like a memory location or I/O port to the microprocessor so no interfacing logic is required.

Questions

1 Why are I/O devices needed in microelectronic systems?
2 State the meaning of the following terms:
   (a) PPI (b) PIO (c) PIA (d) VIA
   What is the difference between them?
3 What is the difference between parallel and serial I/O devices?
4 In a simple Z80/8255 microelectronic system, what start/resetting signals are required and how may they be implemented?
5 State a typical clock frequency for a simple Z80/8255 system. Is the clock frequency critical?
6 State and detail a suitable EPROM device for use in a simple Z80/8255 system.

7 Why can't CALL instructions be used in a small system whose memory consists of a single EPROM?

8 What is a 'look-up table' and why may it be necessary to include one in the program contained in a single EPROM memory, microelectronic system?

9 What is the main difference between the pin-out detail of an EPROM and a RAM chip with the same memory capacity?

10 (a) Construct a memory map for the microelectronic system we have called 'SAMSON' in this chapter, which contains an 8K EPROM and 8K of RAM.
   (b) Why is the memory map incomplete?
   (c) Why don't the RAM addresses follow the EPROM addresses?
   (d) Show how, with the addition of extra EPROMs, the memory map could be made complete.

11 State two advantages and two disadvantages of using parallel data transmission in microelectronic systems.

12 What is the difference between synchronous and asynchronous data transmission?

13 What logic level is a serial link maintained at if there is no transmission down the line?

14 Draw a diagram to illustrate the format of one character of asynchronously transmitted data.

15 Why are start and stop bits used in asynchronously transmitted data?

16 Draw a simple representation of a typical, programmable, serial I/O device.

17 (a) State the meaning of the following terms:
       (i) UART   (ii) ACIA
   (b) What is the difference between them?

18 State three typical serial transmission baud rates.

19 What voltages are used for logic 1 and logic 0 in a standard RS232 serial communications link?

20 What is a standard 'Centronics' interface?

Assignment 17.1

(a) Obtain manufacturers' data sheets for the Z80 PIO, the 6522 VIA and the 8255 PPI and compare the relative merits of each.

(b) Either:
   (i) design and build a microelectronic system based upon a CPU (Z80 or 6502), EPROM and PPI device, to perform a simple function, or
   (ii) build a microelectronic system based upon 'SAMSON' described in the text.

Assignment 17.2

(a) Obtain manufacturers' data sheets for the 6402 UART and the 6850 ACIA and compare the two devices.

(b) Design and build a simple serial data link using a pair of the devices of your choice.

Chapter 18

Fault-finding in microprocessor systems

General fault-finding techniques

Fault-finding is one of the most difficult techniques to acquire in electronics. Fortunately, there are many aids both theoretical and practical, and we end this book by looking at some basics, with a particular emphasis on micro-electronics. First of all, there are several rules to bear in mind when learning fault-finding techniques, and these should be remembered at all times:

1 Always use service manuals and other relevant documentation; make notes.
2 Always adopt a logical approach.
3 Check – and then recheck.
4 The practice doesn't always fit the theory.
5 Things aren't always what they seem.
6 You can put a fault on a piece of equipment whilst attempting to repair it.
7 A second fault can occur on a piece of equipment, whilst looking for the first.
   And finally:
8 NEVER TAKE ANYTHING FOR GRANTED!

These are important rules you should think carefully about; unfortunately, only after some practice and experience will you take them seriously. We encourage you to do so right from the start. It's no good buying a brand-new computer and trying to make sense of it by trial and error; then, after hours of frustration, opening the instruction book. Have everything on your side right from the start.

I hate books (text books in particular) that ramble on about all sorts of things before getting down to the nitty gritty. I find myself skipping whole paragraphs and thinking, 'Why doesn't he get on with it?' So, having introduced some basic rules, let's look at some simple, general, introductory fault-finding techniques.

A Schmitt trigger

Figure 18.1 shows the circuit of a Schmitt trigger which contains only three components: a 741 op-amp and a brace of resistors. What could be easier?

The Schmitt trigger is an electronic circuit which switches between its output maximum and minimum at a critical level of input. The effect is that, regardless of the input, the output is a square wave (assuming that the input varies above and below the critical levels).

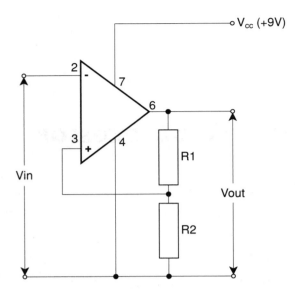

Figure 18.1 *Simple Schmitt trigger circuit using a 741 op-amp.*

The easiest way to check the circuit, therefore, is to apply a sine wave and use a CRO to monitor both input and output. Assuming there to be a fault present, we need to develop a strategy for locating it in the shortest possible time. In the case of this simple circuit, it wouldn't take long to replace all the components and hence cure the fault, but this is not good practice. It may work for this circuit, but supposing it had fifty ICs and hundreds of associated components, the method would clearly be untenable. Here is one suggestion for tackling the problem:

Step 1

Check that the correct voltages are present. A good starting point is to measure across pins 4 (0V) and 7 (V_{cc}). If all is well, move on to step 2. If not, check other points for V_{cc}, gradually moving back to the power source. Note that it's important to measure across the pins of the IC and then work backwards; obtaining the correct voltage at source tells you only that it's present there, it doesn't mean it's on the IC pins themselves. There could be a dry joint at one of several places, the IC socket (where fitted) could be faulty, there may be a break in the print. Working backwards as suggested will reveal any such possibilities; further checks on the 'ohms' range (continuity) will reveal its location.

Step 2

If 9V are present across pins 4 and 7, check for a signal on pin 2 of the IC. Comments given in the last paragraph apply here; a signal on the input to the panel does not prove the presence of a signal on the chip itself. If it's absent, work backwards to locate it.

Step 3

If a signal is present on pin 2 and there is still no output at pin 6, it's almost certain that the IC is faulty. This should be checked by substitution.

Step 4

If there is still no output, remember that a higher voltage on pin 2 than that present on pin 3 should send the output down to zero. If pin 3 is higher than pin 2, a high voltage should appear at the output.

Step 5

The possibility that either R1 or R2 is short circuit is about a million to one against (I've never known a carbon resistor go S/C). Resistors tend to go high resistance or O/C and neither is likely in this case. Nevertheless, if all else fails, the resistor values should be checked.

Step 6

The procedure is given in some detail here as it is a logical way of going about things. Generally, however, you'll need only to check for appropriate supply voltages and input signal and if they're OK then change the IC.

Component faults

Having jumped in at the deep end with an example, let's take stock.

Resistors

Carbon types may increase in value or go open circuit. Examples of both possibilities usually occur only when the component passes consistently higher current than it's designed for (unlikely – the manufacturers would use a higher power type) or momentarily higher current owing to a fault elsewhere. In signal circuits, where little current flows, resistors tend to outlive the equipment they form part of. Wirewound resistors frequently fail because of the heat they dissipate. It is almost unknown for any type of resistor to go short circuit.

Capacitors

All types can go S/C, O/C or low in value. Electrolytic types can leak (literally) and the physical manifestation of this is easy to see.

Diodes and transistors

Signal types seem to go on forever; power types frequently go open circuit or short circuit.

Testing components

The easiest and quickest method of testing discrete (separate) components is to use a multimeter. There are two main types of meter, analogue and digital. In order to describe the procedure, we'll use an 'AVO' meter (analogue) and a 'Beckman' meter (digital). A simple representation of an AVO is shown in Figure 18.2.

The AVO has a 15V internal battery as shown. Rather confusingly, the positive terminal of the meter is connected to the negative terminal of the battery. If a short circuit (S/C) is placed across the meter terminals, the needle of the meter should be deflected across the scale and read ZERO ohms on the

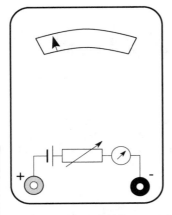

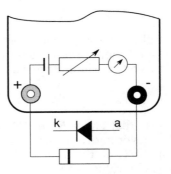

Figure 18.3 *Testing diodes.*

Figure 18.2 *The AVO terminals showing how the battery is connected internally.*

right-hand side. A good fuse is a virtual S/C, so this is one way of testing it. No movement of the meter needle indicates an open circuit (O/C) or a reading of infinity (∞) ohms.

Testing diodes

Set the AVO to the ohms × 1 range and connect the diode as shown in Figure 18.3.

With the cathode on the positive terminal of the meter, a reading of about 20–30Ω, up to a few hundred ohms should be obtained, depending on the type of diode. Connected the opposite way, the resistance should be several million ohms (MΩ) and it is usual to observe no reading at all.

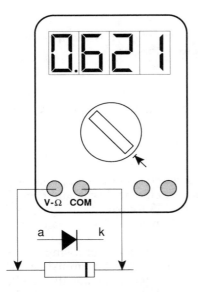

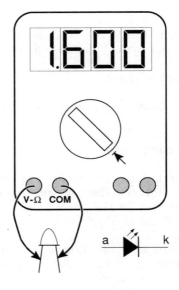

Figure 18.4 *Using a DMM to test diodes.*   **Figure 18.5** *Testing LEDs.*

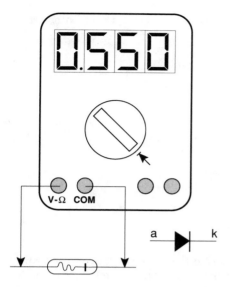

Figure 18.6   *Testing germanium diodes.*

Use of a digital multimeter (DMM)

The procedure is similar to that described above, except that on *most* digital meters, the polarity of the internal battery is the same as the terminals on the front of it. Some DMMs (e.g. the Beckman) have a special diode testing circuit. Connected as shown in Figure 18.4 on the diode range (marked with the diode symbol) a reading of about 0.62V should be obtained. In the reverse direction, the meter displays OL (overload).

With a standard (red) LED connected as shown in Figure 18.5, the meter should indicate a value of about 1.6V. This may be slightly different for diodes of different types or different colours. In the reverse direction, the meter gives the OL signal. The diode itself should glow slightly during this operation.

Germanium point contact diodes such as the OA91 will give a reading of around 0.55V. This may vary slightly from one diode to another. The value also varies as a function of temperature. A simple representation is given in Figure 18.6. Bridge rectifiers and many other types of diode may also be measured in this way.

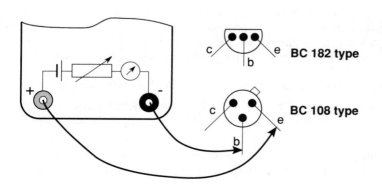

Figure 18.7
Testing npn transistors.

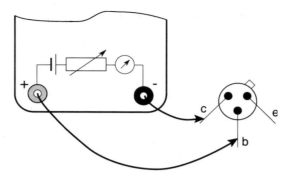

Figure 18.8   *Testing pnp transistors.*

Transistor testing

Testing transistors is more difficult to describe than the procedure for testing diodes, because there are so many different types. Here, we'll just consider bipolar, collector–base–emitter type transistors. Common npn transistors include the BC107, BC108, BC109 and the familiar BC182, BC183 and BC237A plastic encapsulated types. The pnp equivalents are BC186, BC187, BC478 and BC212. To test npn transistors, connect as shown in Figure 18.7 using an AVO type meter.

The base of the transistor is connected to the negative of the meter. When the positive lead is connected to either emitter or collector, a reading of between 20 and 30Ω indicates a good transistor. No other combination of connections should give a reading. For pnp transistors, connect the base to the positive lead of the meter (see Figure 18.8). A reading should be obtained when the negative lead is connected to either emitter or collector. Any other readings would indicate a faulty transistor.

When using a digital multimeter, the testing of transistors is in the opposite sense to that described for the AVO. For an npn transistor, when the base is connected to the positive terminal and the emitter to the negative, a reading of about 0.62V on the diode range should be obtained, as in the case of the analogous diode tests.

Meters should always be returned to the OFF position when not in use or being transported. In the case of the AVO, switching off shorts out the meter movement protecting it from vibration, and in the case of the DMM, it saves the battery.

Testing a simple digital circuit

We now return to the astable circuit first introduced in Chapter 1 and shown slightly modified in Figure 18.9. The procedure for finding a fault in this circuit depends on what the symptom is. If neither LED lights, suspect the power supply connections; if one or other (or both) lights, clearly a power check is unnecessary.

The fault-finding approach is rather different in the following cases:

1   Where a circuit has been built from scratch and has never worked.
2   Where a circuit has been known to work and has subsequently developed a fault.

In the first case, components could have been inserted in the wrong place or the wrong way round. The number of possibilities reduces where a manufac-

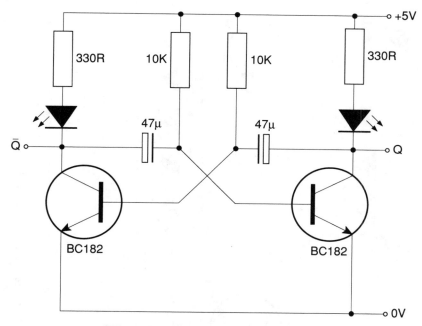

Figure 18.9 *Simple two-transistor astable circuit.*

tured pcb has been used. Home-made pcbs and stripboard layouts are obviously more susceptible to faulty connections. In any case, the following preliminary checks should be made:

1   Are all the components of the correct value?
2   Are (electrolytic and tantalum) capacitors, transistors and diodes connected the right way round?
3   Have all necessary wire links ('jumpers') been included?
4   On stripboard designs, have all necessary breaks in the copper strip been made (e.g. between capacitor terminals) and are all tracks electrically insulated?

     If all visual checks appear satisfactory, measure across the power leads. On an AVO meter, readings above a few kΩ indicate that it's safe to connect up the circuit.
5   Check that 5V exists from each transistor emitter to the V_{cc} line. (The emitter may seem to be connected to 0V but this is the only way to be sure.)
6   Short each transistor collector down to 0V in turn. In each case, the LED should glow. If it doesn't, recheck LED polarity, resistor value and then all associated connections.
7   If all is well, one LED should be 'ON' and the other 'OFF'. The one that is off should come on when its base is connected to V_{cc} *via a 1k resistor*. If either transistor fails this test, check it with a meter as described earlier.
8   If each transistor switches as it should, check the capacitor polarity, and finally check these components by substitution.

In most cases, the above checks will reveal the fault.

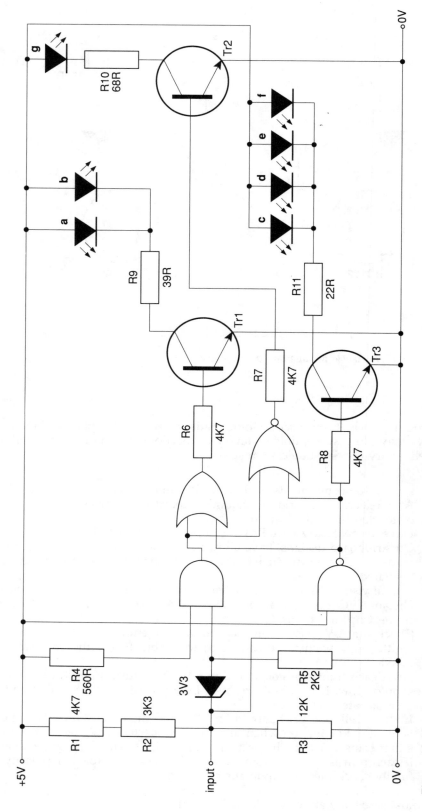

Figure 18.10 Circuit diagram for a logic probe. The input senses not only a logic 0 and logic 1, but also a 'floating' input which is neither.

Digital circuits

The astable is a simple digital circuit. When the LED is on, its associated collector will be 'LOW' (i.e. less than 0.8V and usually around 0.2V or less in a circuit such as this). When the LED is off, the associated collector rises towards V_{cc} (a logic 1), and about 5V in this case. It is usual and convenient to measure logic levels using a logic probe.

Many commercial designs are available and we have produced two novel designs ourselves. The logic probe usually has a three-LED display, often having a red LED to indicate a logic 1, a green LED to indicate a logic 0 and a yellow LED to show a train of pulses. However, there are lots of variations, including those that give an audible bleep on encountering a logic 1. One of our designs appears in Figure 18.10 and the reader is encouraged to study its design and operation.

This is an important exercise in terms of understanding the operation of logic circuits generally, but actually building the unit will provide a greater insight and enable the reader to produce a piece of test equipment that can actually be used.

The logic probe

Our design uses a common cathode seven-segment display for the output. It produces a 1 or 0 according to the input, but also indicates a 'floating' or 'high impedance' input by lighting up the central (g) element on its own, when neither a 1 nor 0 is detected. The operation is best described using Figure 18.11 and the associated truth table given in Table 18.1.

When the input is a logic 1, the AND gate produces an output at D. This gives a 1 at J via the OR gate and lights up segments a and b to produce a 1 on the display. The central segment is inhibited by the 1 at D which is connected at K to the NOR gate. When the input is a 0, the AND gate

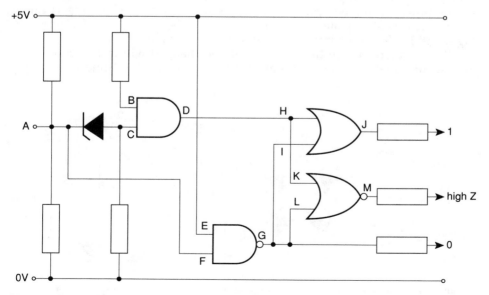

Figure 18.11   *The logic probe circuit diagram. The letters enable a truth table to be constructed as shown in Table 18.1.*

Table 18.1 Truth table for the logic probe

Input A	B	C	D	E	F	G	H	I	J	K	L	M	Output
1	1	1	1	1	1	0	1	0	1	1	0	0	J 1
0	1	0	0	1	0	1	0	1	1	0	1	0	J&G 0
Z	1	0	0	1	1	0	0	0	0	0	0	1	M –

produces a zero, and the NAND gate produces a 1 as shown. This goes to the OR gate again to light up segments a and b; it also goes to Tr3 in order to light up segments c, d, e and f in addition producing the ZERO. The NOR gate detects the absence of either 1 or 0 and the central segment, g, is lit up on the display via Tr2. The device will detect a logic 1 when it sees voltages between 3.5V and 5V. Anything below 0.8V will be regarded as logic 0. Values in between are 'floating'.

Construction of the logic probe

This is a tried and tested circuit which can be constructed using only two ICs, a 7400 quad NAND and a 7432 quad OR. The AND function is implemented using a NAND followed by an inverter – another NAND gate with its inputs connected together. The NOR function is derived in a similar way, by following an OR gate with a NAND wired as an inverter. Further precise details are withheld since construction of this device forms one of the end of chapter assignments.

Power supplies

Before looking at digital and micro techniques in detail, let's consider fault-finding in power supplies. It is true to say that most faults occur here since all the power to the equipment must pass through its power supply; this is where most heat is generated and heat is the biggest enemy of electronics. Figure 18.12 shows a simple basic power supply circuit.

All the components in this circuit could be changed within 15 minutes, but at best this is very wasteful and at worst is *not* logical fault-finding – or any

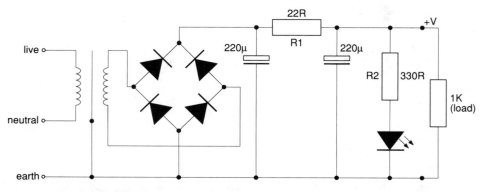

Figure 18.12 *Basic simple power supply circuit.*

Table 18.2   Suggested fault-finding procedure for simple power supply

Test	Result and action
(Visual check)	NOF (No obvious fault)
(Check input resistance)	8kΩ – safe to switch on
1a Does LED light?	YES – see later NO – suspect LED or R2
2 Measure voltage across LED	0V – suspect S/C LED or O/C R2 +V – LED is O/C
3 Measure across load	Normal – check LED and/or 330R on meter 0V – fault further back in circuit
4 Check at 'hot' end of 22R	+V – R1 is O/C 0V – fault further back in circuit
5 Check transformer secondary	OK – suspect primary and/or plug fuse
6 Check primary	OK – suspect fuse or wiring
1b Does LED light?	YES – assume low voltage
2b Measure across load	Low volts – suspect O/C diode or cap
3b Put CRO across load	Measure ripple frequency:
	100Hz – check smoothing capacitors 50Hz – one or more diodes O/C

kind of fault-finding for that matter. There are two things that a good technician would do first:

1   Make a visual check.
2   Measure the input resistance of the circuit.

In 1, we are looking for breaks in the panel or print, missing or damaged
components, or wires frayed or hanging off. In 2 we are looking for short
circuits which may betray the faulty component(s) directly, or which would
reveal a resistance so low that it would be unwise or dangerous to connect
power to the circuit. Items under 1 are fairly obvious, 2 requires some explanation.

The secondary of the transformer will measure only a few ohms at the most
and probably a lot less, so it must first be disconnected (one of the leads will
do). A reading of about 8kΩ or so (in this circuit) across the AC input to the
bridge rectifier would indicate a normal reading and the unit safe to power
up. A reading of less than 6kΩ should make you suspicious, and around 3kΩ
suggest a short circuit diode or smoothing capacitor. A reading of a few
hundred ohms in such a circuit should set the alarm bells ringing – you would
not even consider powering the circuit up.

If an S/C is suspected (because of an abnormally low reading), it's a simple
matter to check all four diodes (as described earlier), and the smoothing capacitors on a resistance range. Short circuit components usually reveal themselves
in circuit, but, because of parallel paths, the only way to be certain is to
remove the component (again by disconnecting one end only) and test it out
of circuit.

If all is well, the circuit can be powered up and checks made roughly
according to Table 18.2.

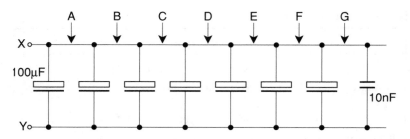

Figure 18.13   *Seven smoothing capacitors and an HF decoupling capacitor in parallel.*

The half-split method

This method involves the isolation of part of a circuit (not necessarily half) in order to determine in which part of it the fault may lie. To take a simple example, suppose a power supply contains several smoothing capacitors in parallel as shown in Figure 18.13.

If a measurement at X–Y indicates an S/C, then it could be the fault of any of the components. Making a break at point D verifies the integrity of half the capacitors immediately (unless two or more are S/C which, fortunately, is very rare). The half-split process is then continued until the faulty component is located. In reality, this technique may be confounded by any of the following:

1   It may not be practicable to isolate parts of the circuit in order to make the measurements, although some electronic equipment is made up from units ('panels') which simply plug in, making the process much easier. Cutting printed circuit tracks is not usually a good idea.
2   All the components may not be equally reliable.
3   All checks may not take a similar amount of time.

Common sense and some experience will help you to decide when the method is appropriate, and when it is not.

Logic circuits

Let's start by considering an AND gate as shown in Figure 18.14 with its truth table.

Suppose that input A is permanently at logic 1, because of an internal fault. This would be described as being 'stuck at one' or 'S-A-1'. The truth table would then be as shown in Table 18.3.

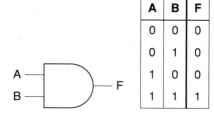

A	B	F
0	0	0
0	1	0
1	0	0
1	1	1

Figure 18.14   *An AND gate with its truth table.*

Table 18.3   Truth table for an AND gate as it would appear if input A were S-A-1

A	B	F
0	0	0
0	1	1
1	0	0
1	1	1

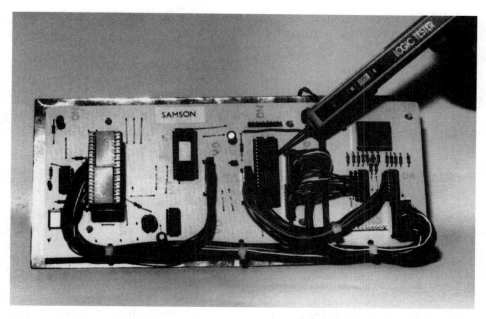

Figure 18.15 *Logic probe in use. A logic clip has been placed over the Z80 CPU on the left.*

Applying a zero to A and a logic 1 to B will reveal this fault. If the output is S-A-1, then the output column, (F), would contain all ones regardless of the inputs applied.

Microprocessor fault-finding

We'll use the three systems which have dominated this book so far – MARC, EMMA and (more recently) SAMSON – to describe some simple fault-finding procedures. First, let's review the equipment available:

1   Logic probe

We've mentioned this most useful and versatile of servicing aids already. The logic probe gives a quick visual (and sometimes audible) representation of the logic level to which it is connected. Many different versions exist, the basics have been outlined; for individual probe designs, the reader will need to consult the instructions for the particular probe available.

The logic probe supplied by Rapid Electronics is designed to measure logic levels in systems employing TTL/CMOS technology. Pulse enlargement capability allows pulse detection down to 25ns. One-shot narrow pulses which are nearly impossible to be seen with a fast 'scope are easily detectable and visible. The intensity of the 'pulse' LED is directly proportional to the duty cycle of the signal observed. LED indicators and a two-tone sounder indicate 'HI' or 'LO' logic states. Figure 18.15 shows the logic probe in use.

2   Logic pulser

This is a device which can be used to force a test point to a particular logic level; it's usually for a very short time, to prevent possible damage to preceding logic elements. Once the desired logic level has been inserted, the output from a suspect gate can be monitored with a logic probe.

3   Logic clip

This is a simple device which clips onto a line or over the pins of an IC in order to reveal their logic states. This is usually achieved by lighting up small LED indicators. It is of limited use where the logic levels are changing rapidly, though it can at least show that activity is occurring in these circumstances.

4   IC test clip

This is a device which clips over the pins of an IC like the logic clip but which has no electronics attached to it. The clip simply provides a series of terminals which can be used for connections to a logic analyser or for easy inspection with a CRO or logic probe.

There are several versions available; Maplins offer a comprehensive range of these double row clips for gaining access to the connector pins of dual-in-line ICs, from 8-pin DIL devices up to 40-pin DIL. The spring-loaded tool is simply clipped over the IC whose signal activity is to be monitored and connections are made to the pins along the top. A logic clip is shown over the CPU in Figure 18.15.

5   Logic analyser

There are many different types of logic analyser but they all do basically the same thing. The test leads are connected to the CPU data or address lines and the system clock. For other tests, further connections may be required which we describe later.

The analyser is triggered by a control word which may be either a memory location (if the address bus is being monitored) or a particular instruction in the program (if the data bus is being monitored). Once the logic analyser is 'armed' and with the CPU running a program, the logic analyser is storing address or data information as required. When the analyser reaches the control word, it stores a sequence of bytes. The unit we shall be describing has a memory 128 words long; it stores 64 words prior to the trigger, the trigger word itself and 63 words after the trigger. These can then be analysed, rather like single-stepping through a program. In the case of a newly built microelectronic system like SAMSON (described in the last chapter) this is an excellent facility.

The logic analyser is made even more versatile by the inclusion of 'qualifiers'. This allows data to be stored only when the $\overline{M1}$ CPU pin is low, or only on memory $\overline{RD}$ or $\overline{WR}$ operations. We describe this in more detail later.

6   Logic comparator

This is a device which can compare a logic circuit under test with a similar circuit known to be in good working order. It is usually based upon an EX-OR gate which gives an output only when its inputs are different. Figure 18.16 shows a simple representation of a logic comparator.

The diagram shows four possible inputs to each device. The counter is designed to step through all possible input sequences in the truth table. Only one EX-OR gate is shown at the output but in a real unit several would be used, one for each output.

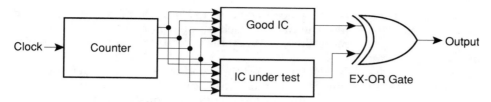

Figure 18.16   *Simple representation of a logic comparator. The output is analysed as the counter steps through its binary sequence.*

A logic comparator may be constructed using the 7485 logic comparator chip. This has four 'A' inputs and four 'B' inputs. The device compares the two quadruple input signals and provides separate outputs depending on whether

A > B, A < B or A = B.

7   Current tracer

This device is particularly useful in tracing low impedance faults and short circuits. The tracer detects current flow in a wire or pcb track by sensing the magnetic field generated by a varying current. This may be a natural function of the system being tested or derive from the use of a logic pulser.

Typical probe sensitivities vary from 1mA to 1A, whilst the device itself usually requires less than 75mA from the supply. One application might be to trace a faulty IC which has a V_{cc} to ground short. The positive supply line should be disconnected and the current tracer used in conjunction with a logic pulser. As an example, study Figure 18.17 which shows part of a microelectronic system.

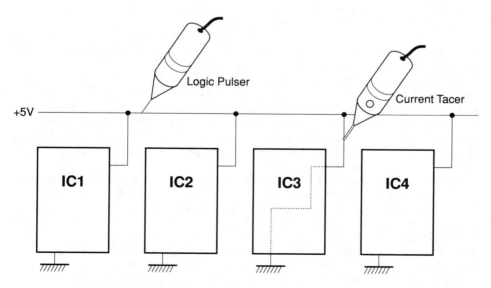

Figure 18.17   *The current tracer; if IC3 has a V_{cc} to ground S/C, the current falls to zero at that point and the tracer indicator goes out.*

The current tracer is first placed adjacent to the logic pulser and adjusted until the indicator lights. The track to each IC V_{cc} connection is then tested in turn, by running the tracer along it. When the indication disappears, the faulty IC has been located.

In practice, when ICs develop a short they tend to exhibit very definite signs of having done so; bubbles and cracks may appear and in severe cases the device can 'explode'. A visual inspection would obviously reveal such a fault; less obvious, and much more common, are S/C decoupling capacitors. These are normally fitted along the supply line close to the IC V_{cc} pins and have a value of between 4.7 and 100nF (generally 10nF), and every so often 10μF 10V tantalum capacitors may also be present. The current tracer is ideal for locating such faults although the correct technique will require some practice to perfect.

The tip of the tracer is electrically insulated as the device relies upon inductive pick-up; it is therefore important to ensure correct alignment. The tracer must be placed perpendicular to the wire or pcb track under test and must remain so even when corners are turned or where junctions are encountered. Short circuits in logic devices revealed to be S-A-0 (stuck at zero volts) and shorts between neighbouring pcb tracks can also be detected using the logic pulser and current tracer.

8 Signature analyser

A 'signature' is often described as the digital equivalent of an analogue waveform. For example, when examining the Schmitt trigger circuit given in Figure 18.1, an engineer will know that for a given sine wave input, the output will be a square wave (usually with a mark to space ratio less than 1:2). The output waveform is the signature. The analogous test points in a microelectronic system – called nodes – will also have characteristic values and the signature analyser displays these, not as a CRO waveform, but as a four-digit hex code.

In order to make a signature analysis, it is common to put the CPU into 'free-run' mode. This frequently takes the form of forcing the CPU to execute the NOP instruction so that it cycles repeatedly through all the available addresses (0 to FFFFH in an 8-bit system).

A simple program can be written to impose the free-run test on a system, but this cannot easily be applied where ROM/EPROM etc. elements exist. It's usually much more convenient to make use of a 'test card'. This is a device which has a 40-pin DIL holder mounted on it, with flying leads to a 'header' plug which fits in place of the CPU. The processor itself is mounted on the test card which has all the data lines disconnected. The NOP instruction is then imposed onto the data line by wiring in the appropriate logic levels. For the Z80, this is very easy because the NOP instruction code is 00H. Hence, connecting all the data lines together and then to 0V will produce the required instruction. This is shown in Figure 18.18.

In theory, such a test card can be made up quite easily in the workshop. The difficulty arises in making so many soldered connections. An easier way is to use a couple of 40-pin DIL wirewrap sockets. These components have very long pins, so that one socket can be slotted on top of the other, with the data lines intercepted in between. These are then connected together and to the 0V line (pin 29), the CPU is slotted into the upper socket and the pins of the lower one go into the original CPU socket. Any other ICs which are connected to the data lines (the EPROM and PIA in SAMSON, for example) are removed and the free-run test can take place. Figure 18.19 shows a simple representation of a signature analyser.

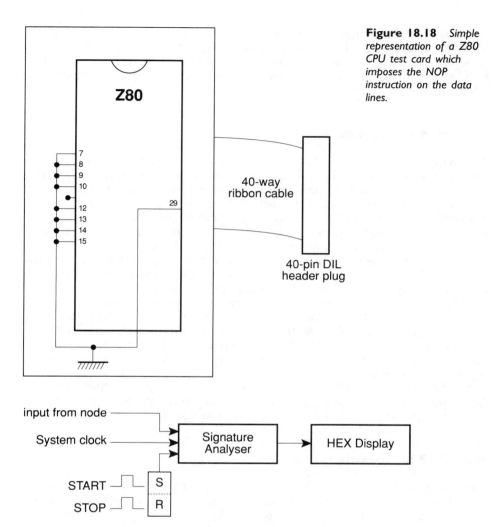

Figure 18.18 *Simple representation of a Z80 CPU test card which imposes the NOP instruction on the data lines.*

Figure 18.19 *Simple representation of a signature analyser.*

The hex display code is often slightly different from the standard hex code to avoid, for example, ambiguity in reading b as 6, so the values used tend to be 0–9 as usual and then A, C, F, H, P and U. The actual values have no meaning anyway; they are simply compared with standard values obtained from a known good system. The signature for a 0V connection should always be 0000 but there is no fixed value for V_{cc}. However, once determined for a particular node in a given system, a signature should remain constant.

The start and stop pulses are usually derived from the msb of the address line, i.e. A_{15}. Only one pulse appears on this line during one cycle through all the addresses. While this is going on, it is useful to check address line activity with a CRO and we discuss this and the use of the oscilloscope generally next.

9   The oscilloscope

The CRO or 'scope is probably one of the most useful and versatile of all servicing instruments. In microelectronic work, it's possible to check the integrity (and frequency) of the clock waveform and to monitor the $\overline{\text{WR}}$, $\overline{\text{RD}}$, $\overline{\text{M1}}$, data and address lines, to name just a few. The technician should use good quality probe leads with miniature crocodile clips fitted. One or two BNC to BNC leads should also be available for connection to other equipment, a logic analyser, for example.

Fault-finding procedures

The main purpose of a text book is to provide the reader with all the necessary facts and procedures required for the course he or she is following. This is not a workshop manual, although we have tried to include many practical aspects of microelectronics. This section on fault-finding, however, though reasonably comprehensive, is necessarily limited; we have to present the material as it would appear in a course of instruction and not as it may be required in a practical situation. In fact, we try to do both.

So, before getting underway, it's worthwhile pointing out that in the real world, faults in microelectronic systems are usually associated with intermittent connections, circuit board cracks and dry joints. The switched faults in the MARC and EMMA systems employ open and short circuits. Although in practice, this kind of fault does sometimes occur, it's only a small part of the story. Nevertheless, looking at these 'switched faults' is extremely useful in gaining familiarity with microelectronic systems, getting used to computer circuit designs and becoming acquainted with the equipment used in servicing and repairing them.

The following notes will give you some confidence and expertise in tackling faulty microelectronic systems in both learning and working environments. As always, we'll assume the use of MARC and EMMA systems – Z80 first.

MARC (Z80) investigations

Start with a correctly working system, with the MARC monitor running and the editor-assembler loaded. You will need pin-out diagrams for all the ICs on the panel; a workshop manual is ideal but not entirely necessary at this stage.

I Power supplies

Check that 5V exist at the input to the panel; locate the IC power supply pins and measure voltages on the following: CPU, CTC, Z80 PIO, ROM, RAM, address decoder, floppy disc interface (1770) and keyboard encoder.

2 Clock signal

Check for the 2MHz clock signal on CPU pin 6 and at the appropriate pin just above top right of the keyboard. Sketch a few cycles of the waveform on graph paper and compare this clock waveform with the CK signal on pin 15 of the Z80A CTC, pin 25 of the Z80A PIO and pin 18 of the 1770 floppy disc interface IC.

3 Reset

Press the reset button whilst observing the voltage on CPU pin 26. Check voltages on $\overline{\text{HALT}}$, $\overline{\text{INT}}$ and $\overline{\text{NMI}}$ pins.

Table 18.4 Program TEST1 for checking the ASCII characters and monitoring the RD line

Address	Machine code	Label	Mnemonics	Comments
				;TEST1
			ORG 5C00H	;origin
			ENT 5C00H	;entry point
		CLS:	EQU 01B9H	
		CRSOFF:	EQU 02C9H	
		VIDEO:	EQU 0E02AH	;video RAM
5C00	CD C9 02		CALL CRSOFF	;no cursor
5C03	CD B9 01	START:	CALL CLS	;clear screen
5C06	21 2A E0		LD HL,VIDEO	;
5C09	06 E0		LD B,0E0H	;no of characters
5C0B	3E 20		LA A,20H	;ASCII blank
5C0D	77	LOOP1:	LD (HL),A	;load in memory
5C0E	3C		INC A	;next character
5C0F	23		INC HL	;next mem location
5C10	CD 21 5C		CALL DELAY	;
5C13	05		DEC B	;
5C14	20 F7		JR NZ,LOOP1	;
5C16	0E FF		LD C,0FFH	;long delay
5C18	CD 23 5C		CALL LONDEL	;
5C1B	CD 23 5C		CALL LONDEL	;
5C1E	C3 03 5C		JP START	;repeat all
5C21	0E 05	DELAY:	LD C,5H	;short delay
5C23	16 00	LONDEL:	LD D,0	;
5C25	15	LOOP2:	DEC D	;
5C26	20 FD		JR NZ,LOOP2	;
5C28	0D		DEC C	;
5C29	20 FA		JR NZ,LOOP2	;
5C2B	C9		RET	;

4   Halt

Type in and run the short program below which should cause the HALT LED to come on:

```
ORG    5C00H
ENT    5C00H
LD     A,33H
HALT
```

5   The RD line

Monitor pin 21 of the CPU with a CRO set to 1μs/cm (assuming a 2MHz clock frequency) and press a character key. Careful adjustment of the trig level on the 'scope should show about two 0V to 5V transitions. Pressing RESET should give a 5V DC level on the RD pin which remains whilst the key is pressed, and for a tiny fraction of time after it's been released. Load program TEST1 in Table 18.4 and note the RD transitions. This program also displays the entire set of ASCII characters.

6 The $\overline{\text{WR}}$ line

Load and run program TEST2 given below and monitor the $\overline{\text{WR}}$ pin. Careful analysis of the $\overline{\text{WR}}$ waveform compared to the CK waveform shows that the $\overline{\text{WR}}$ pin goes low every 22 clock cycles:

```
                              ORG    5C00H
                              ENT    5C00H
5C00   3E 20        TEST2:    LD     A,20H
5C02   32 00 5E     START:    LD     (5E00H),A
5C05   C3 02 5C               JP     START
```

7 $\overline{\text{M1}}$

With the above program still running, monitor the $\overline{\text{M1}}$ pin. This goes low for 2 clock cycles, then high for 11 clock cycles, low for 2 more clock cycles, then high for 8 clock cycles and then repeats.

8 Using a logic analyser

With the simple program above still running, one use of the logic analyser can be demonstrated. We used the L J Electronics Model SA1 (632) analyser which is shown in Figure 18.20.

The analyser should be equipped with a ribbon cable terminating with a 40-way IDC socket at one end and spring-loaded clips at the other. There are 16 data lines (colour coded red) and three (black) 0V connections. The clock connection is coded yellow, clock qualifier green and trigger qualifier blue.

The clock is used to strobe data into the analyser's memory. If the CPU clock signal is used, data is stored in the analyser on every clock pulse. Remembering the nature of timing diagrams discussed in Chapter 8, you will

Figure 18.20 *The L J Electronics Model SA1 (632) logic analyser.*

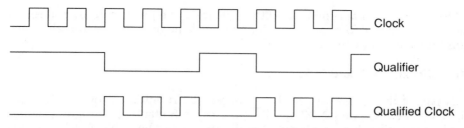

Figure 18.21 *Illustrating the qualified clock signal. The qualifier signal is active LOW (like $\overline{WR}$ and $\overline{RD}$ signals) so it must be inverted before being ANDed with the clock. This can be achieved by setting a switch on the analyser.*

see that this leads to duplication of data. For this reason, a 'clock qualifier' is available. This is a signal which is internally ANDed with the CK to enable data to be recorded selectively. For example, the clock qualifier could be connected to the $\overline{WR}$ or $\overline{RD}$ pins, to obtain data associated with those operations only. See Figure 18.21.

Looking at the TEST2 program

In order to capture the relevant data, the analyser needs to be given a 'trigger' word. In this case, we'll use the instruction 32 and this is loaded into the analyser by pressing the buttons above or below the trigger word windows. The high byte is set to zero and the switches are set to disable; the low byte switches are set to enable; 'Display' to LSB. The trigger qualifier is left unconnected and the associated switch set to 'X'. The CK connection goes to the $\overline{MREQ}$ pin of the CPU and the CK qualifier connection to $\overline{RD}$. Each instruction is then read by the analyser and stored in its memory.

The CK qualifier switch should be set to 0 and the edge to ⌐. The analyser can then be armed and the program RUN.

Inspection of the analyser contents (using DEC or INC) should reveal the following program sequence repeated over and over again: 32   00   5E   C3   02   5C.

9   Checking keyboard operation

The keyboard switches are wired in matrix form and are scanned by a software routine associated with the 8255 PIA. IC41 in the keyboard interface circuit is a 74145, 3- to 8-line decoder. The inputs A, B and C (pins 13, 14 and 15) receive their clock pulses from the 8255 PIA in the system I/O circuitry. If the output is taken high for a given code on the ABC input, a particular key press is detected.

Monitoring the pins below the 74145 chip on a CRO, depress the following keys:

```
KR0    1
KR1    2
KR2    Q
KR3    W
KR4    A
KR5    S
KR6    Z
KR7    X
```

On the CRO, an approximate 100Hz square wave pulse train indicates that the line is intact (KR7 is often much lower in frequency). A low to high transition indicates an open circuit on that line.

10   Checking I/O and interrupts

This part of the system can easily be checked by running a program that uses all bits of the I/O ports, $\overline{NMI}$ and $\overline{INT}$ interrupts. Suitable test programs which can be used directly or with some modifications can be found in Chapter 12.

You should be able to devise a single program which can be used for all the above tests. RAM test programs were hinted at in Chapters 5 and 6. It is a good idea to store a number in a memory location and then read it back immediately in order to verify its integrity. Values of 85 (55H) and 170 (0AAH) are particularly useful because they cover all the combinations in an 8-bit system. Alternatively, engineers favour the use of the 'walking ones' test, where the numbers 1, 10, 100, 1000, etc. are used to check each bit in every memory location.

'Real world' servicing of a MARC

The following notes give guidelines on servicing a faulty microelectronic system in the real world (as opposed to working on switched faults). The basic techniques are appropriate to all systems, whether based upon the Z80, 6502 or any other processor, whether essentially designed for the classroom (like the MARC or EMMA) built in the workshop (like SAMSON) or purchased from a shop (like the Sinclair Spectrum (Z80) or the 6502-based BBC computer).

First steps in servicing

Normally a technician will be told by his superior, or by a customer, what the fault symptom is. 'No sound', 'No display', 'Won't load from disc or cassette' or 'Completely dead' being common examples. We'll start from the premise that the equipment is a completely unknown quantity and outline some logical procedures.

1   Visual inspection

Ensure that all the switched faults are off; check that chips are fitted in the correct sockets, the right way round with no IC pins badly located, bent under the chip or over the holder. Ensure that the disc drive connector is properly fitted and the VDU plug is the right way round. The MARC cold start loader should be in the 'ON' position.

2   Apply power

Check for exactly 5V on the input sockets; reversed polarity tends to damage all the memory chips instantly, so later model MARCS have a diode in series with the positive line to prevent this happening.

3   Clock

If the MARC cold start loader message appears, then both the clock and VDU must be OK. If there's nothing on the screen, it's a useful quick check to 'scope the clock waveform as nothing will work without it.

4   VDU

If there's still nothing on the screen and the VDU is correctly connected, substitute the entire equipment for a known good one. If there's a raster (lines on the screen, but no characters – increase brightness if necessary to see) check the character generator chip power supplies (IC10) and then the device itself (2764) by substitution.

5   'BIOS' EPROM

Check the C000H to DF00H memory chip (IC31). This is a 2716, 2Kb EPROM which acts like a BIOS (basic input/output system) for the MARC and which generates the MARC 'cold start loader' message. All PCs have a BIOS chip which starts everything off when you switch on; it allows the operating system (OS) to be 'booted up' from the hard disc.

6   Disc drive

By now, the 'cold start loader' message should have been obtained. When the switch is returned to OFF, the OS should be down-loaded from disc. If not, check that the drives are set to 40/80 tracks as appropriate to your system. Check that the IDC connector is intact and that none of the pins have been missed, and finally, check the disc drive unit itself by substitution. Repairs to VDUs (apart from simple, power supply faults) are beyond the scope of this chapter. As far as disc drives are concerned, most professional servicing personnel will return faulty units to the manufacturer; specialised knowledge and equipment is required.

7   Other faults

Assuming that the OS has loaded, other faults must be identified before they can be repaired. Generally, chips will betray themselves by running unusually hot and this can be detected by touch (with care!). Other faults may be identified by following the procedures outlined previously.

The Sinclair Spectrum

A huge number of these home computers have been sold in the past and since it is a Z80-based system, a few representative servicing notes are presented here.

Most of the faults reported are associated with loose connections, cracks in the pcb, dry joints and dirty or tarnished contacts. A 28-pin double-sided edge connector is provided for connection to peripherals; connecting or disconnecting any item here without first switching off invariably results in damage to either power supply or memory chips, or both.

The power supply is based upon a 7805 linear regulator IC attached to a substantial heatsink. Breaks in the pcb often occur if the heatsink screws have been fastened too tightly; problems can also occur if they're not tight enough. Tr4 npn transistor (ZTX650) frequently fails if peripherals are connected before switching off the supply. If Tr4 is found to be faulty check the neighbouring coil and Tr5 as well.

Faulty memory devices can be identified because of the excess heat they generate when they go S/C. Apart from the CPU, the major chip in this computer is a 6C001 ULA (sometimes it's a 5C102 or 5C112). A ULA is an 'uncommitted logic array'. This is virtually a complete system on a single chip, and in the Spectrum it controls keyboard interfacing, cassette I/O functions, sound processing and output, memory decoding and memory refresh – the computer uses 4116 Dynamic RAMs. The ULA is frequently responsible for intermittent faults and those that occur when the computer has warmed up.

Typical ULA devices can contain from a few hundred to several thousand gates, so when a fault occurs, it can affect many different parts of the system.

This is a far from comprehensive guide to servicing the Spectrum. Its included mainly to show that many faults on real systems result from user abuse and mechanical (rather than electronic) failures. It also shows how many faults characterise a particular system and how service engineers quickly get used to them. In the system in question, most faults can be cured by tracing and repairing cracks in the print and resoldering dry joints, or replacing a few components in the power supply. Add to this the replacement of the ULA, memory chips, or the CPU itself and more than 90 per cent of all common faults have been covered.

EMMA (6502) investigations

I Power supplies

Check that exactly 5V exists at the input to the panel; locate power supply pins and measure voltages on the following ICs: CPU, 6522 VIA, 2716 Monitor EPROM, 74139 address decoder and 6821 I/O chip (dedicated on the EMMA panel to performing keyboard and display interface functions).

2 Clock signal

Using an oscilloscope, look at the Φ_2 clock waveform on a monitoring pin (it's been buffered by two inverting gates at this point); it can be compared to the unbuffered Φ_2 signal on the crystal itself. It's useful to observe the Φ_1 waveform on pin 3 of the CPU; sketch two or three cycles of each CK signal on graph paper for comparison. From this you should see that the two 1MHz waveforms are out of phase, such that they are never at logic 1 at the same time.

3 Reset

Monitor the reset pin whilst pressing the reset button.

4 R/$\overline{W}$ signal

This can be observed with the monitor running (no user program executing) under which circumstances the waveform is at logic '1' for most of the time. The CRO should be set to 2µs/cm, such that two or three negative going transitions may be seen.

Load and run the following program:

```
0020    A9    00
0022    85    30
0024    85    31
0026    85    32
0028    4C    20   00
```

The program repeatedly stores accumulator contents to memory locations &30, &31 and &32. Examination of the R/$\overline{W}$ line should show the three write transitions; they occupy 14 clock cycles as can be seen by comparison with the Φ_2 clock signal.

5   SYNC signal

This can also be observed with the monitor program running. Examination with a CRO will reveal that the SYNC line is low with transitions high for approximately 30 per cent of the time.

6   I/O, $\overline{\text{NMI}}$ and $\overline{\text{IRQ}}$

These can be checked using programs developed in Chapter 12.

7   Using a logic analyser

Some basic notes about the use of the logic analyser appear in the introduction to this chapter and in the Z80 section, where an example of data capture was given. We now show how the address lines may be monitored on the EMMA. Load the following program into memory:

```
0020        A9      FF
0022        85      30
0024        85      31
0026        85      32
0028        4C      20      00
```

The analyser is connected to the 16 address lines (red plugs). All three 0V probes must be connected to convenient ground pins on the EMMA; the clock (yellow probe) goes to the Φ_2 clock pin and the two qualifiers can be ignored.

The analyser controls are set as follows: trigger word to 0021, with all four switches set to enable. The trigger and clock qualifier switches are set to X and negative edge triggering selected ($\blacktriangledown$).

With the EMMA monitor program running, ARM the logic analyser; the data and cursor displays should go blank. When the program is RUN, the correct address sequence of the program will be stored as it repeatedly cycles through the loop.

8   Seven-segment display check

IC7 (7445 or 74145) is the display decoder/driver. Taking pins 1 to 9 low (by simply shorting to 0V) will light up all segments of each display in turn. *However,* when making this check, do it only momentarily. The devices are caused to light up extremely brightly as they don't normally work independently; when the monitor is running, each display is strobed, so they're only on for one eighth of the time. Taking pin 2 of IC6 (555) temporarily low puts the RH display fully on.

9   RAM

Memory checks can be made using the programs described in Chapters 5 and 6. It is a good idea to store a number in a memory location and then read it back immediately in order to verify its integrity. Values of 85 (&55) and 170 (&AA) are particularly useful because they cover all the combinations in an 8-bit system. Alternatively, engineers favour the use of the 'walking ones' test, where the numbers 1, 10, 100, 1000, etc. are used to check each bit in every memory location.

More hints on microelectronic systems servicing may be gleaned by looking at the end of chapter questions and assignments which now follow.

Questions

1 Suggest ten reasons why the statement 'Never take anything for granted' is a golden rule in electronics fault-finding.
2 What is the possibility of a carbon resistor becoming short circuit?
3 What sort of faults do capacitors tend to exhibit?
4 Describe how an npn transistor can be tested with:
  (a) an 'AVO' meter   (b) a DMM
5 Why should multimeters be turned to the OFF position when not in use?
6 When trying to turn a transistor ON for test purposes, why should V_{cc} be connected to the base of a transistor via a 1K resistor instead of being connected directly?
7 Describe the functioning and use of a logic probe.
8 What is meant by the input (or output) of a logic gate being described as 'floating'?
9 When repairing a faulty power supply, what are the first two checks to make?
10 What is the DC resistance of a typical transformer secondary?
11 Briefly describe the functioning and use of the following test instruments:
  (a) logic pulser
  (b) logic clip
  (c) IC test clip
  (d) logic analyser
  (e) logic comparator
  (f) current tracer
  (g) signature analyser
  (h) oscilloscope
12 When a HALT statement is encountered in a Z80 program, does the CPU HALT pin go low or high?
13 What is a ULA chip?
14 What is a 'walking ones' test?
15 (a) what is meant by putting a CPU in 'free run' mode?
  (b) describe the construction of a piece of hardware which can achieve this.

Assignment 18.1

Construct the logic probe described in the text. When completed, tests should show that a logic 1 produces a '1' on the seven-segment display, a logic 0 produces a '0' and a floating input lights up just the central segment of the display. Constructors might also wish to include facilities for a TTL logic 1 and logic 0 and a switchable slow and fast TTL square wave (perhaps using a 555) to make the unit even more versatile and increase its use as an aid to fault-finding.

Assignment 18.2

Construct a workshop manual for a microelectronic system of your choice (something like 'SAMSON' is ideal). The manual should include such things as:

(a) a brief description of the system
(b) a block diagram
(c) a complete circuit diagram
(d) a layout diagram
(e) a pcb layout
(f) pin-out diagrams for all the ICs used in the system
(g) a test program with a description of its use
(h) clock and other appropriate waveforms obtained at specific test points
(i) typical voltages at various test points
(j) logic analyser connections and test results
(h) signature analyser test results

Assignment 18.3

(a) Suggest ways in which a microelectronic system such as 'SAMSON' (described in Chapter 17) could be constructed in separate modules and hence be used for testing the component parts of a faulty system.
(b) Suggest the design and construction of LED indicating devices which could be incorporated into the modular design in order to give an indication of possible faults in a suspect system.
(c) Given that in business 'time is money', suggest why the installation of a modular microelectronic system would have distinct advantages.
(d) If possible, build a modular microelectronic system and describe suitable test procedures which could be used with it.

Assignment 18.4

A technician has built a microelectronic system based upon 'SAMSON' (described in Chapter 17) and is testing it using the COUNT UP program. This produces a 0 to 9 count continuously on a ULN2803 octal Darlington driver-fed, seven-segment LED display. The numbers 0, 1, 3, 4, 8 and 9 are all displayed correctly, but numbers 2, 5, 6 and 7 appear corrupted as shown in Figure 18.22.

Suggest a starting point for analysing the problem and outline a suitable test procedure for locating and hence correcting the fault.

Hint: the problem has occurred in a system just built and is associated with incorrect connections having been made; it is not a fault which has 'developed' in a system which has at one time worked correctly, although it could occur as a result of replacing the display module.

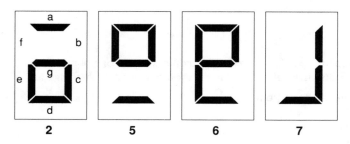

Figure 18.22   *How the numbers 2, 5, 6 and 7 appeared on the seven-segment display of the faulty system.*

Assignment 18.5

(a)   Design and build a dual seven-segment display module (perhaps using two ULN 2803 chips) and describe how it could be tested for correct operation without running a program to light up the segments.

(b)   Write and run a program which will cause an up count from 00 to 99 in a continuous cycle, on a system such as 'SAMSON', using port C of the 8255 for the lsb and port B for the msb.

Appendix I

Answers

Chapter I

Questions

1 A microelectronic system is comprised of a microprocessor (MPU) and suitable support chips (memory and interfacing electronics, for example), which control an operation such as the traffic light sequence, a washing machine, a manufacturing process or a motor car; something which has only one use or purpose. A microcomputer system is also controlled by an MPU, but it also has a VDU, keyboard, floppy disc drives and (usually) a printer. In such a system, the user can run different programs, such as word processing, databases and spreadsheets, etc. and it can be used to develop and run programs for other applications or even simply to play games.

2 A microelectronic system is more flexible than 'hard-wired' electronics. The electronic hardware for a set of traffic lights will be essentially the same for each one; the requirements may vary in terms of the timing, lengths of ON and OFF times, etc. All of them can be programmed independently using a microelectronic system rather than have to design the electronics for each unit separately.

3 Central processing unit. An MPU is simply a very small CPU, fabricated using modern microelectronic technology.

4 (a) BASIC – Beginners All-purpose Symbolic Instruction Code.
   (b) BASIC programs take a relatively long time to execute.
   (c) Slower execution of many programs is of no importance generally. It is only essential in 'real-time' applications, or where the amount of data to be processed is considerable. Generally, home computer users will not need to process so much data that fast speeds are necessary. Where appropriate, home computers will run in machine code directly, anyway.

5 See Figure 1.3.

6 A low-level computer language which uses mnemonics (an aid to the memory) to represent machine code instructions. Assembly language simplifies the process of writing machine code programs.

7 One and zero (1 and 0).

8 Symbol of npn transistor: see Figure 1.6.

9 Transistor as a switch: see Figure 1.7.

10 The binary code uses only the numbers 1 and 0, each of which can be represented electronically by using tiny transistor switches.

11 1010B = 10D.

12 An analogue system can respond to the entire range of signal voltages it is designed to operate on, from the minimum value to the maximum and every tiny value in between. A digital system only responds to certain discrete voltage levels, often only two, as in a digital computer.

13 A bistable circuit (also called a 'flip-flop') contains two transistor elements, each of which can only ever be either ON or OFF, the condition of one always being opposite to the other. The state of the transistor switches can be changed over by the application of a suitable square-wave pulse, the bistable remaining in the new state until another pulse is received, provided the power is not turned off. Bistables can therefore 'remember' a logic 1 or a logic 0, and arranged in sufficient numbers they may form the basis of a computer memory device.

14 (a) The clock is the 'heartbeat' of the system; it enables the CPU to step through and execute program instructions and keep all control signals synchronised.
    (b) See Figure 1.13.
    (c) See Figure 1.10.

15 Typical clock frequency is 2MHz. EMMA runs at 1MHz, MARC runs at 2MHz, most other Z80 systems run at 4MHz, modern CPUs are much faster, e.g. 10MHz, 33MHz, etc.

16 In one form, a CPU register is simply composed of eight bistables. As a unit it can therefore remember or hold a sequence of eight binary digits (bits) ranging from all zeros to all ones (0000 0000 – 1111 1111) which is 0–255 decimal.

17

b_7	b_6	b_5	b_4	b_3	b_2	b_1	b_0

A CPU register.

18 Basically, a computer needs memory in order to hold programming instructions. A ROM (PROM or EPROM) memory chip can be used to hold a program, a series of routines or a complete operating system (OS); it does not lose its contents when power to the system is removed. A RAM memory device holds programs temporarily; the device loses its contents when power is removed. As a user types in a program, it resides in RAM. More is said about this in later chapters.

It should be noted that a computer needs to have some resident programming instructions which tell it what to do – this is the operating system. When programs are created or 'down-loaded' from floppy disc, they are loaded into RAM.

Multiple choice questions

1	d	6	c	11	a	16	c	21	a
2	c	7	d	12	d	17	c	22	d
3	a	8	c	13	c	18	a	23	a
4	a	9	b	14	a	19	d	24	b
5	b	10	d	15	a	20	c	25	c

Chapter 2

Questions

1 (a) 42   (b) 2647   (c) 12   (d) 40   (e) 108   (f) 4019
   (g) 109   (h) 48066   (i) 249

2	(a)	1 1011	(b)	10 1101	(c)	1 0000	(d)	101 0011	(e)	1111 0101
3	(a)	122Q	(b)	21Q	(c)	35Q	(d)	144Q	(e)	41Q
4	(a)	8H	(b)	0BH	(c)	0FH	(d)	10H	(e)	40H
5	(a)	3	(b)	5	(c)	12	(d)	15	(e)	195
6	(a)	6Q	(b)	5Q	(c)	1252Q	(d)	631Q	(e)	56Q
7	(a)	3H	(b)	6H	(c)	0FH	(d)	0D9H	(e)	55H
8	(a)	11	(b)	31	(c)	357	(d)	401	(e)	137
9	(a)	9	(b)	20	(c)	160	(d)	187	(e)	256
10	(a)	bit	(b)	field	(c)	field	(d)	field	(e)	byte

11 In BCD the maximum legal digit is 9; double digits (10, 11, 12, etc.) are illegal. In hexadecimal, the maximum legal number is 0FH (15 in decimal). Hence, 15 = 0FH but in BCD, the 1 and the 5 are treated separately, so that 15 = 1 0101BCD or 0001 0101BCD.

12 A decoder is used, often available on a standard IC.

13 (a) 110110 (54D)   (b) 1001011 (75D)   (c) 111100 (60D)
    (d) 1100111 (103D)   (e) 1111000 (120D)

14 (a) 1010 (b) 111   (c) 100100   (d) 100   (e) 101111

15 1100 1111 (= –79), i.e. a negative number.

Multiple choice questions

1	c	2	b	3	a	4	d	5	d	6	a
7	b	8	b	9	a	10	b	11	c	12	c

Chapter 3

Questions

1 ICs are smaller, cheaper and more reliable than either discrete transistors or valves. When the first electronic calculator appeared in the 1960s, you could hold in your hand more computing power than would fit into a whole room 50 years earlier.

2 (a) 8 bits = 1 byte   (b) 1024 bytes = 1K.

3 Each flip-flop needs two transistors to represent a bit; each cell needs 8 bits, so 8K = 8 × 1024 × 8 × 2 = 131,072.

4 (a) Both RAM and ROM are random access; you can read from or write to a RAM device, but you can only read from ROM.
   (b) A volatile device loses its memory when power is removed.

5 See Figures 3.2 and 3.3.

6 Program instructions from memory are first placed in the instruction register (IR). The CPU decodes the instruction by a series of logic operations, 'decides' what is required and then does it. (See Chapters 7 and 8 for further details.)

7 (a) The ALU performs all arithmetic and logic operations.
   (b) The accumulator.
   (c) The flag or status register.

8 The STACK is a reserved area of RAM used for the temporary storage of data. One simple command 'PUSHes' data onto the stack; you don't have to quote a memory location because the address of the current top of the stack is held in the stack pointer. The special feature of the stack is that it's a LIFO file. The last value to be PUSHed in must be the first one to POP out (Z80) or be PULLed (6502). (See Chapter 9.)

9 It produces a satisfactory compromise between cost and speed.

10 (a) It's the maximum number of bits which can be held in single regis-
        ters or memory locations in a particular system.
   (b) See Figure 3.7.
   (c) See Figure 3.8.

Chapter 4

Questions

1 See Figure A.1.
2 See Figure A.2.
3 See Figure A.3.
4 Number of memory locations = 2^8 = 256 locations (00 to FFH).
5 Zero page in a 6502 system has 256 locations (a quarter of 1K).
6 Hard disc.
7 3.5" DD 100Kb to 800Kb; 3.5" HD 1.44Mb.

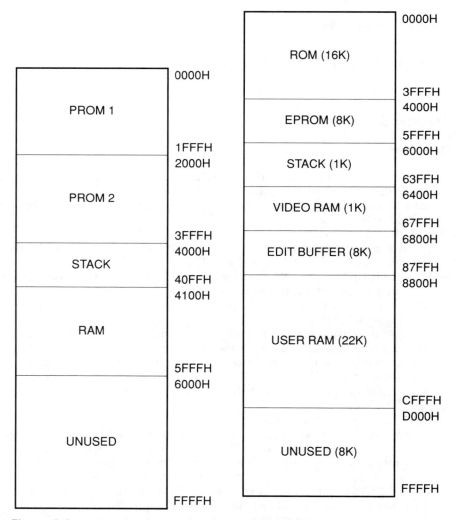

Figure A.1   **Figure A.2**

Figure A.3 (memory map):

EPROM 1 (8K)	0000H – 1FFFH
EPROM 2 (8K)	2000H – 3FFFH
EPROM 3 (8K)	4000H – 5FFFH
EPROM 4 (8K)	6000H – 7FFFH
EPROM 5 (8K)	8000H – 9FFFH
EPROM 6 (8K)	A000H – BFFFH
EPROM 7 (8K)	C000H – DFFFH
RAM	E000H – FFFFH

Figure A.3

Figure A.4 (memory map):

ROM (Operating System)	0000H – 3FFFH
RAM	4000H – 7FFFH
RAM	8000H – BFFFH
SPARE	C000H – FFFFH

Figure A.4

8 (a) Convert between serial and parallel data.
   (b) Read/write data to/from a disc.
   (c) Indicate the status of the disc device.
   (d) Format a disc.
   Basically, the controller is used to provide an interface between the disc drive and the CPU.

9 Slow access time; cassette systems are notoriously unreliable

10 Assume the ROM containing the operating system to be 16K; provide 16K RAM for storing the incoming 6502 code and another 16K in which to place the translated Z80 code – it's got to go somewhere! One solution is given in Figure A.4.

Multiple choice questions

1 c	4 b	7 d	10 d	13 a
2 d	5 a	8 c	11 d	14 d
3 d	6 c	9 a	12 a	15 a

Chapter 5

Questions

1 COBOL; FORTRAN; BASIC; PASCAL; C
   Advantages: Relatively easy to understand for programming. Relatively easy to debug.
   Disadvantages: slower than machine code; require more memory space to store in RAM, ROM or disc, etc.

2 JSP stands for Jackson Structured Programming; it is called a 'top down' approach. The problem is defined as a whole and then broken down into smaller and smaller sections. These include basic components, sequences, selections and iterations.

3 Stages in producing a machine code program:
(a) Write down the algorithm in plain English.
(b) Construct a 'story board', flow chart or JSP design.
(c) Check operation of proposed design against software requirement.
(d) Write the program in assembly mnemonics.
(e) Convert the mnemonics to machine code.
(f) Run and debug as necessary.
(g) Save on disc or firmware.

4 Z80 and 6502 mnemonics

Z80	6502	
LD A,n	LDA #n	Load accumulator with a number
INC A	INX	Increment a register
DEC B	DEY	Decrement a register
LD (5D00H),A	STA &2D00	Store accumulator contents in given memory location
LD A,(5D00H)	LDA &2D00	Load accumulator with contents of memory location
JR NZ,LOOP	BNE LOOP	Branch to part of program labelled LOOP if results of previous ALU operation were non-zero

5 (a) (Z80) LD A,(5E05H) (b) (6502) LDA &2E05

6 Z80 version

(a)	LD B,42H	06 42	;loads B register with 42H
(b)	LD (5E08H),A	32 08 5E	;loads memory location 5E08H with contents of accumulator
(c)	LD A,B	78	;loads accumulator with contents of B register
(d)	INC A	3C	;INCREMENTS: increases the numerical value in the accumulator by one
(e)	DEC C	0D	;DECREMENTS: reduces the numerical value in the C register by one

6502 version

(a)	LDX #&42	A2 42	;loads X register with 42H
(b)	STA &2E08	8D 08 2E	;loads memory location 2E08H with contents of accumulator
(c)	TXA	8A	;loads accumulator with contents of X register
(d)	INX	E8	;INCREMENTS: increases the numerical value in the X register by one
(e)	DEY	88	;DECREMENTS: reduces the numerical value in the Y register by one

7 Example 3 should suggest a satisfactory solution.

8   Some hints are given for a Z80 solution:
    The ASCII code for the letter 'A' is 41H; the blank space is 20H. Load HL
    with the starting address (5E00H) and load the accumulator with 41H.
    Using INC HL and INC A, you should see that it's a simple matter to
    perform the LOAD. The codes for 'MICROPROCESSOR' will need to be
    input separately. There are better ways of writing this program (and many
    of the others) but we wait until Chapter 13 before discussing them.

9   An iterative process is one which repeats itself; where numerical data is
    involved, this is frequently modified by the process. Exit from the routine
    usually occurs when a particular criterion has been satisfied. In computer
    programs this is often when a particular register has decremented to zero,
    though there are many other possibilities.

10 This question is another example of the use of an iterative process. The
    examples given earlier should suggest a solution.

11 (Z80)   There are many possible solutions to this question; the precise
    design of the time delay will depend upon the length of delay required:

Address	Machine code	Mnemonics	Comments
5D00H	06 FF	LD B,0FFH	;put 0FFH into B reg
5D02H	0E FF	LD C,0FFH	;put 0FFH into C reg
5D04H	0D	DEC C	;decrement C register
5D05H	18 FE	JR NZ,5D05H	;inner relative jump
5D07H	05	DEC B	;decrement B register
5D08H	18 F8	JR NZ,5D02H	;outer relative jump

The program would normally be 'CALLed' as a subroutine; we should
end the routine with 'RET' (return) and generally use labels as shown
below. (See Chapter 13 for further details.)

Address	Machine code	Label	Operator and operand
5D00H	06 FF	DELAY:	LD B,0FFH
5D02H	0E FF	LOOP2:	LD C,0FFH
5D04H	0D	LOOP1:	DEC C
5D05H	18 FE		JR NZ,LOOP1
5D07H	05		DEC B
5D08H	18 F8		JR NZ,LOOP2
5D0AH	C9		RET

Now, whenever the delay is required, we simply use the instruction 'CALL
DELAY' and the routine is automatically CALLed and executed; program
execution returning (RET) to where it left off, when the delay has timed out.
(6502)   There are many possible solutions to this question; the precise
design of the time delay will depend upon the length of delay required:

Address	Machine code	Mnemonics	Comments
&2500	A9 FF	LDA #&FF	;put 0FFH into A reg
&2502	85 4D	STA &4D	;store in zero page RAM
&2504	A9 FF	LDA #&FF	;put 0FFH into A reg
&2506	85 4E	STA &4E	;store in zero page RAM
&2508	C6 4E	DEC &4E	;decrement memory
&250A	D0 FC	BNE &2508	;inner relative jump
&250C	C6 4D	DEC &4D	;decrement memory
&250E	D0 F8	BNE &2508	;outer relative jump

The program would normally jump (JSR) to the subroutine; we should end the routine with 'RTS' (return) and generally use labels as shown below. (See Chapter 13 for further details.)

Address	Machine code	Label	Operator and operand
&2500	A9 FF	.DELAY	LDA #&FF
&2502	85 4D		STA &4D
&2504	A9 FF		LDA #&FF
&2506	85 4E	.LOOP	STA &4E
&2508	C6 4E		DEC &4E
&250A	D0 FC		BNE LOOP
&250C	C6 4D		DEC &4D
&250E	D0 F8		BNE LOOP
&2510	60		RTS

Now, whenever the delay is required, we simply use the instruction 'JSR DELAY' and the program automatically jumps to the routine; program execution returning (RTS) to where it left off, when the delay has timed out.

12 (MARC Z80) One solution of many is given here:

Address	Machine code	Mnemonics	Comments
		ORG 5C00H	;ORIGIN
		ENT 5C00H	;ENTRY
5C00H	21 2E E0	LD HL,0E02EH	;first video location
5C03H	3E 31	LD A,31H	;ASCII '1'
5C05H	06 09	LD B,9	;counter
5C07H	77	LD (HL),A	;ASCII '1' to video RAM
5C08H	3C	INC A	;move to '2'
5C09H	23	INC HL	;move to next video
5C0AH	23	INC HL	;location (three spaces)
5C0BH	23	INC HL	
5C0CH	05	DEC B	;decrement counter
5C0DH	20 F8	JR NZ,5C07H	;repeat number up to 9
5C0FH	2B	DEC HL	;go back a space to accommodate '10'
5C10H	0E	LD C,31H	;load registers for 1
5C11H	16 30	LD D,30H	;and zero (10)
5C13H	71	LD (HL),C	;put 1 on screen
5C14H	23	INC HL	;move to next video
5C15H	72	LD (HL),D	;put 0 on screen
5C16H	06 09	LD B,9	;reload counter
5C18H	21 81 E0	LD HL,0E081H	;new video location on next line
5C1AH	71	LD (HL),C	;display a '1'
5C1BH	23	INC HL	;move to next video
5C1CH	77	LD (HL),A	;next digit (1 first)
5C1DH	23	INC HL	;move to next video
5C1EH	23	INC HL	;location (two spaces)
5C1FH	3C	INC A	;next digit
5C20H	05	DEC B	;decrement counter
5C21H	20 F7	JR NZ,5C1AH	;repeat
5C23H	36 32	LD (HL),32H	;first digit of '20'
5C25H	23	INC HL	;next video location
5C26H	72	LD (HL),D	;next digit (0 is already in D register
5C27H	76	HALT	;all done

Running the program produces the following display:

1	2	3	4	5	6	7	8	9	10
11	12	13	14	15	16	17	18	19	20

13 The solution to this will be similar to that for the previous question, you only need to change the ASCII codes and insert a loop to repeat the 'printing'. You might consider using the 'CP' instruction in order to detect when a line of letters has been completed.

14 This is another program where an incrementing (and decrementing) iterative process is required.

15 There are (as always) many different ways of writing this program; you might consider putting all similar letters on the screen first (there are 4 'O's and 3 'E's, for example), but would this be an advantage? More experienced programmers may be aware of the use of 'DEFB' pseudo instructions; this is an example of a 'pseudo-op', an instruction which can be used in an automatic assembler to speed up the assembly process. They are called 'pseudo-ops' because although they have an effect like ordinary opcodes, they are not usually listed in a CPU instruction set. They are only meaningful in the appropriate assembly language program and may vary slightly from one system to another. The more advanced student may wish to experiment, and we say more about pseudo-ops and automatic assembly in Chapter 13.

Chapter 6

Questions

1 The 'operation code' (opcode) is the first byte of a CPU instruction; it is the code which tells the CPU what the instruction is and how long it is. The 'operand' is the data value itself, or the address in memory where the data can be found.

2 Z80 Addressing modes and examples are listed in the text.

3 6502 Addressing modes and examples are listed in the text.

4 (a) LD A,0FH – immediate
  (b) LD A,B – register
  (c) LD A,(HL) – indirect
  (d) ADD A,B – register
  (e) ADD A,0B0H – immediate
  (f) INC A – implied
  (g) LD (BC),A – indirect
  (h) RLA – implied
  (i) INC (IX+5H) – indexed
  (j) LD HL,5DD0H – immediate extended
  (k) JR NZ,LOOP – relative
  (l) NOP – implied

5 (a) LDA #&0F – immediate
  (b) TXA – implied
  (c) LDA &2E00 – absolute
  (d) ADC &33 – zero page
  (e) ADC #&B0 – immediate

(f) INX – implied
(g) STA &2E01 – absolute
(h) ROL A – accumulator
(i) INC (&40,X) – indexed indirect (pre-indexed)
(j) LDA (&44),Y – indexed indirect (post-indexed)
(k) BNE LOOP – relative
(l) NOP – implied

6 The 'PRINT ABCD' program

Label	Mnemonics	Comments
CLS:	EQU 01B9H	
VIDEO:	EQU 0E02AH	
	CALL CLS	;direct
	CALL PRINT	;direct
	HALT	;implied
PRINT:	LD HL,VIDEO	;immediate extended
	LD A,41H	;immediate
	LD B,4	;immediate
LOOP:	LD (HL),A	;register indirect
	INC HL	;implied
	INC A	;implied
	DEC B	;implied
	JP NZ,LOOP	;direct

7 Putting the numbers 1 to 20 on the screen:

Label	Mnemonics	Comments
	ORG 5C00H	;origin
	ENT 5C00H	;entry point
CLS:	EQU 01B9H	;clear screen
CRS:	EQU 02C9H	;cursor off
VID:	EQU 0E02EH	;screen position
VID2:	EQU 0E081H	;new screen position
	CALL CLS	;direct
	CALL CRS	;direct
	LD HL,VID	;immediate extended
	LD A,31H	;immediate
	LD B,9	;immediate
LOOP:	LD (HL),A	;register indirect
	INC A	;implied
	INC HL	;implied
	INC HL	;implied
	INC HL	;implied
	DEC B	;implied
	JR NZ,LOOP	;relative
	DEC HL	;implied
	LD C,31H	;immediate
	LD D,30H	;immediate
	LD (HL),C	;register indirect
	INC HL	;implied
	LD (HL),D	;register indirect
	LD A,31H	;immediate

(continued)

7   *continued*

Label	Mnemonics	Comments
	LD B,9	;immediate
	LD HL,VID2	;immediate extended
LP2:	LD (HL),C	;register indirect
	INC HL	;implied
	LD (HL),A	;register indirect
	INC HL	;implied
	INC HL	;implied
	INC A	;implied
	DEC B	;implied
	JR NZ,LP2	;relative
	LD (HL),32H	;indirect
	INC HL	;implied
	LD (HL),D	;indirect
	HALT	;implied

8   Putting numbers into numerical order. We present a 6502 version for ordering the following numbers:
&15, &27, &00, &80, &45, &18, &A0, &A5, &01, &FF, &33, &72

Address	Machine code	Mnemonics	Comments
0200	A9 15	LDA #&15	;immediate
0202	8D E0 02	STA &02E0	;absolute
0205	A9 27	LDA #&27	;immediate
0207	8D E1 02	STA &02E1	;absolute
.	.	.	.
.	.	.	.
.	.	.	.
0237	A9 72	LDA #&72	;immediate
0239	8D EB 02	STA &02EB	;absolute
			;
023C	A0 00	LDY #&00	;immediate
023E	A2 00	LDX #&00	;immediate
0240	A9 0B	LDA #&0B	;immediate
0242	85 31	STA &31	;zero page
0244	A9 00	LDA #&00	;immediate
0246	85 32	STA &32	;zero page
0248	A5 31	**LDA &31	;zero page
024A	85 30	STA &30	;zero page
024C	B9 E0 02	LDA &02E0,Y	;indexed absolute
024F	DD E1 02	*CMP &02E1,X	;indexed absolute
0252	B0 1C	BCS	;implied
0254	E8	INX	;implied
0255	C6 30	DEC &30	;zero page
0257	D0 F6	BNE *	;relative
0259	C8	INY	;implied
025A	E6 32	INC &32	;zero page
025C	A6 32	LDX &32	;immediate
025E	C6 31	DEC &31	;zero page
0260	D0 E6	BNE **	;relative
0262	4C E0 FE	JMP &FEE0	;absolute

(*continued*)

8 *continued*

Address	Machine code	Mnemonics	Comments
			;
			;SUBROUTINE
0270	85 33	STA &33	;zero page
0272	BD E1 02	LDA &02E1,X	;indexed absolute
0275	99 E0 02	STA &02E0,Y	;indexed absolute
0278	A5 33	LDA &33	;zero page
027A	9D E1 02	STA &02E1,X	;indexed absolute
027D	B9 E0 02	LDA &02E0,Y	;indexed absolute
0280	46 54 02	JMP &0254	;absolute

9   An example was given in the last chapter.
10   Here's a 6502 program that will do the trick:

Address	Machine code	Mnemonics	Comments
		ORG &0200	
		ENT &0200	
0200	A9 14	LDA #&19	;immediate
0202	85 30	STA &30	;zero page
0204	A9 00	LDA #&00	;immediate
0206	85 31	STA &31	;zero page
0208	A2 00	LDX #00	;immediate
020A	9D E0 02	STA &02E0,X	;indexed absolute
020D	BD E0 02	LDA &02E0,X	;indexed absolute
0210	A8	TAY	;implied
0211	18	CLC	;implied
0212	65 31	ADC &31	;zero page
0214	85 31	STA &31	;zero page
0216	C8	INY	;implied
0217	98	TYA	;implied
0218	E8	INX	;implied
0219	C6 30	DEC &30	;zero page
021B	D0 ED	BNE &020A (-19)	;relative
021D	A5 31	LDA &31	;zero page
021F	8D F4 02	STA &02F4	;absolute
			;
			;ADDPROG
0222	A9 00	LDA #00	;immediate
0224	A2 13	LDX #19	;immediate
0226	85 33	STA &33	;zero page
0228	85 34	STA &34	;zero page
022A	A9 01	LDA #01	;immediate
022C	A8	TAY	;implied
022D	18	CLC	;implied
022E	65 34	ADC &34	;zero page
0230	85 34	STA &34	;zero page
0232	C8	INY	;implied
0233	98	TYA	;implied
0234	CA	DEX	;implied
0235	D0 F6	BNE &022E (-10)	;relative
0237	A5 34	LDA &34	;zero page
0239	8D F5 02	STA &02F5	;absolute
023C	00	BRK	;implied

As well as providing a solution to the question we've also tacked on 'ADDPROG'. This calculates the sum for us and places it in the next free memory location (&02F5). This should, of course, duplicate the contents of memory location &02F4. With a bit of thought, this modest program can easily be extended to test a much larger area of RAM.

Guide to tutorial questions

11 Given the programs we have written so far, this one should be quite straightforward. Since the Z80 can't multiply, we would use a section of code to ADD the numbers together four times. One solution is given below. We assume that the number to be multiplied is already in the accumulator.

Label	Mnemonics	Comments
MULT:	LD C,04H	;immediate
	LD E,A	;register
LOOP1:	ADD A,E	;register
	DEC C	;implied
	JR NZ,LOOP1	;relative
	RET	

The final statement 'RET' is necessary if the code is being CALLed as a subroutine. For those not yet familiar with the technique, ignore the 'RET' and simply combine the routine with your own program. If your program has run correctly, memory location 5C5BH should contain 0E5H.

12 Note the comments given above (Question 11) and the following:

Label	Mnemonics	Comments
	LDA #6	;immediate
.MULT4	STA &30	;zero page
	LDX #3	;immediate
.LOOP1	CLC	;implied
	ADC &30	;zero page
.STORE	STA &31	;zero page
	DEX	;implied
	BNE LOOP1	;relative
	RTS	

13 to 17   Work out the expected values by hand and you can check the accuracy of your own programs.

18 Use a program similar to that given in Question 8. Simply convert the letters into ASCII codes, and put the values into the program. You'll have to convert the ASCII codes back to letters if you want to read, print or display them.

19 If you're using a MARC (or similar), you'll need to decide whereabouts on the screen you want to put the numbers and the asterisks. The manual for your system will give you some guiding information; the solution to Question 7 suggests some possibilities.

20 Simply run your program and see if it works.

Multiple choice questions   Z80

1	a	2	d	3	b	4	d	5	c	6	a	7	c		
8	c	9	c	10	a	11	d	12	a	13	d	14	a	15	c

Multiple choice questions   6502

1	a	2	a	3	b	4	a	5	c	6	d	7	c		
8	c	9	b	10	c	11	d	12	d	13	b	14	d	15	a

Chapter 7

No answers are given as solutions can be obtained by inspection or by single-stepping through the programs.

Chapter 8

Questions

1 (a)   The clock is the 'heartbeat' of a CPU. All movements of data and control signals must be precisely synchronised for correct operation and timing. Using a crystal to control the frequency of the clock oscillator ensures precise and constant frequency.

   (b)   The idealised clock waveform is a square wave. Because there will be a finite (small but measurable) rise time, clock (and bus) signals are often shown with less than vertical edges as indicated in Figure A.5.

(i)   (ii)

Figure A.5   *(i) idealised CK waveform – a perfect square wave; (ii) common, exaggerated representation.*

2 Memory read or write; I/O device read or write; interrupt acknowledge.
3 (a)   A clock cycle corresponds to the period of the CLOCK oscillator waveform; it's called a T-cycle. (See Figure 8.1.)

   (b)   A machine cycle defines the number of T-cycles needed for an operation, e.g. a Z80 opcode fetch needs 4 T-cycles. (See Figure 8.2.)
4 See Z80 and 6502 CPU diagrams; Figures 8.5 and 8.12.

5	*Z80*	*6502*
	V_{cc} – 11	V_{cc} – 8
	GND (or 0V) – 29	V_{ss} (or 0V) – 1, 21
	CLOCK (CK) – 6	Φ_0 (in) – 37
		Φ_1 (out) – 3
		Φ_2 (out) – 39
	$\overline{\text{NMI}}$ – 17	$\overline{\text{NMI}}$ – 6
	$\overline{\text{M1}}$ – 27	SYNC – 7

6 (a) and (b)   See text.
7 Control signals used in an opcode fetch:

Z80	*6502*	*Purpose*
$\overline{\text{M1}}$	SYNC	Opcode fetch initialised
$\overline{\text{RD}}$	R/$\overline{\text{W}}$	Memory read
$\overline{\text{MREQ}}$	n/a	Memory request not I/O
CK Φ	$\Phi_0\ \Phi_1\ \Phi_2$	Timing

8 The number of CLOCK cycles required depends upon the instruction. For example, an 'LD A,(5D00H)' requires more machine cycles than an 'LD A,04', because in the first instruction the CPU has to fetch the opcode first and then the two-byte address (in two stages) before the data value (at 5D00H in this case) can be fetched.

9 T = 1/f = 1/1,000,000 = 1µs.

10 See text.

Chapter 9

Multiple choice questions

1	a	2	b	3	a	4	a	5	b
6	a	7	a	8	d	9	c	10	d

Chapter 10

Questions

1 (a) See Figure 10.1.
   (b) See Figure 10.6.
2 (a) See Table 10.1.
   (b) See Table 10.2.
3 (a) The Add/Subtract flag (N), bit 1.
   (b) The BREAK flag (B), bit 4.
4 The ninth bit. (See Figure 10.3.)
5 No. The zero flag is set, so it becomes a logic 1.
6 and 7   Suitable programs are described in the text.
8 A solution is hinted at in the text.
9 (Z80)   JP NC,LABEL   JP C,LABEL
           JR NC,LABEL   JR C,LABEL
   (6502) BCC   BCS

10 *Z80*	*6502*
carry, C	C
zero, Z	Z
overflow, P/V	V
sign, S	negative, N

Multiple choice questions

1	c	2	a	3	b	4	c	5	a
6	d	7	b	8	b	9	a	10	b
11	d	12	c	13	a	14	b	15	c
16	d	17	b	18	a	19	b	20	d

Chapter 11

Questions

1 to 4 These questions may be answered by single-stepping through the programs.
5 (a) A leading zero has been omitted; the instruction by the label 'START:' should be: LD HL,0E224H

(c)   This is a subroutine that should multiply a number by 4. It doesn't because the counter's loaded with 4 and should be loaded with 3. Also, the multiplicand is 6 in the machine code and 3 in the mnemonics. The correct program, with comments, is shown below:

0270	A9	03		LDA #3	;multiplicand
0272	A2	03	.MULT4	LDX #3	;multiplier less 1
0274	85	30		STA &30	;one 3 in there now
0276	18		.LOOP1	CLC	;clear carry flag
0277	65	30		ADC &30	;add another 3 to the accumulator
0279	85	31		STA &31	;store running total
027B	CA			DEX	;reduce counter
027C	D0	F8		BNE LOOP1	;do again
027E	60			RTS	

(d)   At address 5C0AH the instruction should be: LD (HL),A (the brackets have been omitted). The program doesn't work because it doubles up the multiplicand instead of simply adding the value at each pass. This will be revealed by single-stepping.

Chapter 12

Questions

1   Interfacing circuits are necessary at computer I/O ports because the I/O device data registers can neither source nor sink large currents. Without suitable buffers, any bit connected to a low impedance would be destroyed.
2   See text.
3   Control registers (Z80) and DDRs (6502) are used to allow the I/O ports to be configured, generally as either all inputs, all outputs or a combination of the two.
4   (a)   35H = 0011 0101
    (b)   11H = 0001 0001
    (c)   AAH = 1010 1010
    (d)   FEH = 1111 1110
    (e)   88H = 1000 1000
5   to 10   Answers to all these questions are suggested in the text; you'll know when you've obtained the correct solution when your program works. Suitable interfacing circuits are given in Chapter 14.

Chapter 13

Answers are not supplied.

Chapter 14

Questions

1   (a)   See text and Figure 14.3.
    (b)   Consult Figure A.6.
          Figure A.6(a) shows the simplest implementation of a pnp transistor as a switch. Figure A.6(b) shows an alternative design. The 0V line

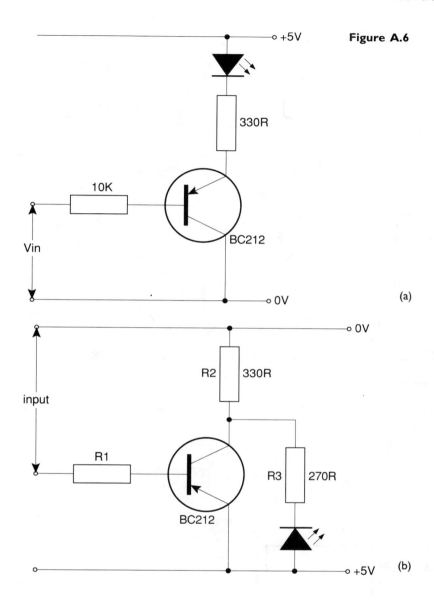

Figure A.6

(a)

(b)

is at the top of the diagram which means that the input from the output port goes between this and the base connection (via R1). The operation is summarised in Table 14.8 below:

Table 14.8   Summary of the action of the pnp transistor switch

Input	Logic	LED
0V	0	OFF
5V	1	ON
none	–	ON

It works like this because of the nature of operation of the pnp transistor. In a way, it's more logical; because if there's no input, it's as if the base connection has 'floated up' to a logic 1, which is what TTL circuits do anyway.

(c)   See Figures A.7 and A.8.

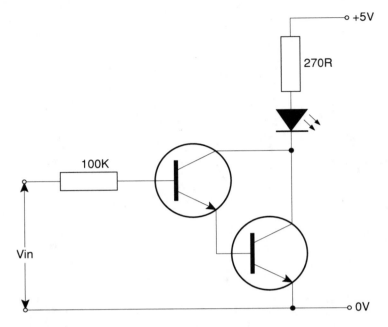

Figure A.7

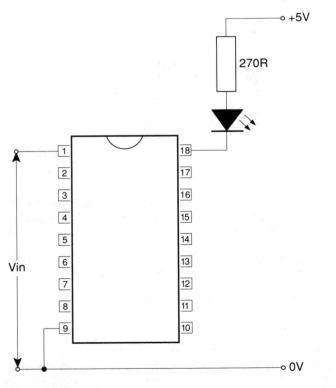

Figure A.8

2 See Figure 14.5.
   (a) See text and/or manufacturers' literature.
   (b) Used in inverse parallel with an inductive load (typically a relay coil)
        to prevent the back EMF (occurring at switch off) from damaging the
        transistor elements.
   (c) See text.
3 Dedicated decoders; latching or non-latching octal buffers. See text.
4 (a) The code for a letter 'P' would be 73H using the convention shown
        in Figure A.9. Some letters of the alphabet can't be reproduced in a
        seven-segment display.

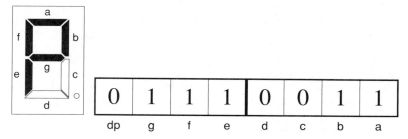

Figure A.9

You'd need to build a matrix of LEDs or use a 5 by 7 red dot matrix array,
available from Maplin Electronics (order code FT61R). The device can
display the complete ASCII character set, and has slots and tongues in the
sides so that rows of displays may be assembled for moving messages,
etc. Exploring the design of microelectronic hardware and software to
produce a practical, working system would make an ideal project or
assignment for an adventurous reader.
5 to 10 All the necessary information is covered in the text, so these
   questions are left open-ended to allow for student investigations.

Chapter 15

Questions

1 The totem pole arrangement refers to the output configuration of transistors and diode in a typical TTL logic gate. It is used to ensure that the
   output of the gate is low impedance regardless of whether it's at logic 1
   or logic 0.
2 WIRED-AND:
   Advantage: saves a gate in certain configurations.
   Disadvantage: can damage the gate by overloading one of the output
   transistors.
3 (a) Tri-state – a device which can have any one of three output condi-
        tions: logic 1, logic 0 or 'high impedance'. The latter indicates that the
        output may take up any state to which it's connected.
   (b) The devices are used to isolate parts of a circuit, typically on data
        buses. An enabling signal may be applied, for example, to select
        either the read or write process associated with a memory device.
   (c) See Figure 15.5.
4 (a) NMOS: N-enhancement metal oxide semiconductor, field effect
        transistor.

(b) NMOS superseded PMOS for three important reasons:
   (i) they have a higher packing density (more gates in a given volume);
   (ii) they work at a higher speed;
   (iii) they are compatible with TTL, having positive gate and drain voltages.

(c) CMOS compared to NMOS:

Advantages	Disadvantages
* faster	* lower packing density
* lower power dissipation	* higher cost

5 Multiplexed data and address lines:
Advantages: needs fewer wires because one bus is shared; fewer pins required on CPU package.
Disadvantage: transmission speed is reduced because of need for switching between the two functions.

Multiple choice questions

1	b	2	b	3	d	4	a
5	c	6	d	7	a	8	d
9	a	10	d	11	d	12	b

Chapter 16

1 (a) $2K = 2^1 \times 2^{10} = 11$ address lines
  (b) $4K = 2^2 \times 2^{10} = 12$ address lines + 1 CS line, i.e. 13 lines altogether

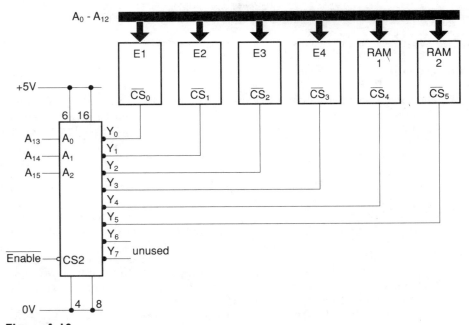

Figure A.10

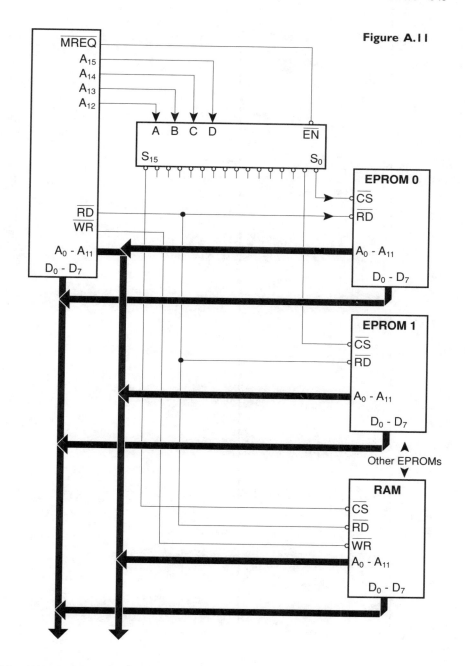

Figure A.11

(c) 2716 EPROMs are 4K devices needing 12 address lines. The 6116 is a 2Kb RAM. A 2- to 4-line decoder may be used for CS signals, so 14 address lines are needed altogether.

(d) 2764 EPROMs are 8Kb devices which need 13 lines ($2^3 \times 2^{10}$). Two more lines are needed for CS making 15 in all. However, it's possible to use only 14 lines if A_{13} selects one chip when it's low and the other when it's high, using an inverter.

(e) 32Kb is $2^5 \times 2^{10}$, i.e. 15 lines, A_0 to A_{14}.

2 (a) Linear address decoding does not involve any form of decoding logic. The address lines are connected directly to the chips.

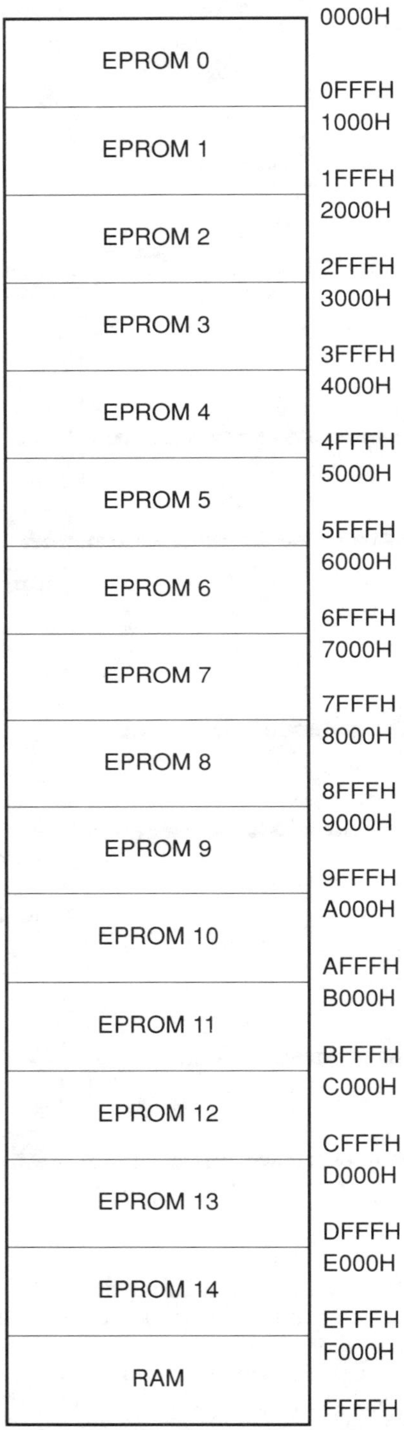

EPROM 0	0000H
	0FFFH
EPROM 1	1000H
	1FFFH
EPROM 2	2000H
	2FFFH
EPROM 3	3000H
	3FFFH
EPROM 4	4000H
	4FFFH
EPROM 5	5000H
	5FFFH
EPROM 6	6000H
	6FFFH
EPROM 7	7000H
	7FFFH
EPROM 8	8000H
	8FFFH
EPROM 9	9000H
	9FFFH
EPROM 10	A000H
	AFFFH
EPROM 11	B000H
	BFFFH
EPROM 12	C000H
	CFFFH
EPROM 13	D000H
	DFFFH
EPROM 14	E000H
	EFFFH
RAM	F000H
	FFFFH

Figure A.12

(b) The great advantages of linear address decoding are simplicity of design and reduction in component count; disadvantages include the production of non-contiguous memory maps and the difficulty in providing later memory expansion.

3   An 8K memory device needs 13 lines to address it. The 16K can be made up from 2 × 6264 8K CMOS static RAM devices, the EPROMs are all 2764 types. The system is completed by using a 74137 3-line to 8-line decoder (active low outputs). Pin connections for the decoder and other details are shown in Figure A.10.
4   8Kb needs 13 address lines. The Z80 and 6502 CPUs can support 16 address lines.
  (a)   Address lines A_{13}, A_{14} and A_{15} can be used to operate the CS on each chip.
  (b) A_{13} and A_{14} only may be used by the addition of a 2- to 4-line decoder.
5   A 4- to 16-line decoder (e.g. a 4515 IC) can be used in the system shown in Figure A.11. The memory map is given in Figure A.12.

Chapter 17

Questions

1   I/O devices are needed to input and output data from a CPU. The actual devices provide a great flexibility in the number of ways in which data may be input and output. Many I/O devices are programmable, so their versatility may be exploited by the use of the appropriate software.
2   (a)   PPI – Programmable peripheral interface.
  (b)   PIO – Programmable input/output device.
  (c)   PIA – Peripheral interface adaptor.
  (d)   VIA – Versatile interface adaptor.
  The PPI is probably the most comprehensive term as it essentially describes all the devices listed. The main differences between them is the number of ports they have and the nature of the control codes necessary to set them up; they may also vary in the nature and mode of interrupts they can handle. Otherwise, they're just different manufacturers' versions of the same thing.
3   Parallel I/O devices handle data 8 bits at a time using 8 parallel wires (in an 8-bit system). Serial devices handle data 1 bit at a time, sequentially, down a single pair of wires; they convert parallel data to serial format for transmission and serial data to parallel format on reception.
4   The Z80 requires a logic 1 to logic 0 transition for resetting; the 8255 needs a logic 0 to 1 transition. A debounced switch with an inverter as shown in Figure 17.4 (the 7404) may be used for the purpose.
5   2MHz (or 4MHz) is a typical clock frequency. The slowest clock frequency you can use with a Z80 is about 100kHz, the maximum is about 4MHz. The mark space ratio should be 50/50, but if you keep the speed down, this becomes less critical. The Z80 does require some sharp edges to the clock, and the simplest way to sharpen up TTL is with a pull-up resistor as shown in Figure 1.13 (680R).
6   The 2764 EPROM contains 64K bits arranged as 8K 8-bit words; it is therefore an 8Kb memory chip. It is a static memory device, being based upon the setting and resetting of tiny flip-flops.
    The device is programmed by the application of +12.5V to pin 21 and an active low pulse to pin 27. UV light of the appropriate intensity and wavelength will completely erase the device in about 20 minutes.
7   CALL instructions use a stack to save the return addresses and a stack cannot be implemented in an EPROM. It is not possible to write to an EPROM once it's been programmed apart from the lengthy process of UV

light erasure of the whole chip contents and subsequent reprogramming. To implement a stack you'd have to include some RAM memory.

8 A look-up table contains data (e.g. ASCII codes) used within a program. The data cannot be loaded into EPROM memory through software because the device is read only; the hex codes must be placed in the appropriate EPROM memory locations.

9 The RAM chip has a R/$\overline{\text{W}}$, $\overline{\text{WR}}$ or separate $\overline{\text{WE}}$ (write enable) pin. The EPROM cannot be written to, so doesn't have a $\overline{\text{WE}}$ pin.

10 (a) 8K EPROM: 0000H to 1FFFH
        8K RAM: 8000H to 9FFFH

   (b) There are 64K possible addresses and only two 8K memory devices are being used in the system.

   (c) In 'SAMSON', EPROM is selected when address line A_{15} is low and RAM when it is high. A_{13} and A_{14} are unused, so EPROM is selected between 0000 0000 0000 0000 and 0001 1111 1111 1111, i.e. 0000H and 1FFFH. RAM is selected over the same range and when A_{15} is a '1', hence between:

   1000 0000 0000 0000   and

   1001 1111 1111 1111   i.e.

   8000H to 9FFFH. These two sets of address ranges are obviously not contiguous.

   (d) This can be achieved by using a 32K EPROM and 32K RAM and making use of address lines A_{13} and A_{14}.

11 Parallel data transmission

Advantages	Disadvantages
1 Increased speed	1 Data corruption over long distances
2 Simple to implement	2 Large number of wires needed

12 Asynchronous data transmissions are not related in time, they can be sent and received as and when required. Synchronous data transmissions are accurately controlled by simultaneous clocking or other synchronising pulses. Television signals are transmitted synchronously – they are displayed on your TV screen in strict time relation to the studio generated pictures. When computer signals are sent to a printer, the second page of information can wait until the first one has been completed, no matter how long it takes.

13 Logic 1 level.

14 See Figure 17.9.

15 To detect when the data is being sent and when a word has been completed.

16 See Figure 17.11.

17 (a) (i)   UART – universal asynchronous receiver transmitter
         (ii) ACIA – asynchronous communications interface adapter

   (b) There is little or no difference between these two devices. They are simply different manufacturers' versions of the same thing.

18 Common baud rates: 600, 1200, 2400 baud. Some serial links employ speeds ranging from 110 baud to 9600 baud and even lower (and higher) speeds are sometimes used.

19 In RS232 links, logic 1 = –12V and logic 0 = +12V.

20 Centronics: A parallel communications system frequently used in computer systems to connect to a printer.

Chapter 18

Questions

1 In fault-finding, never take anything for granted for the following reasons:
  (a) Just because a power supply is turned on, and an internal meter or display indicates 5V, it doesn't necessarily mean that 5V are coming out of it. Check it with a meter.
  (b) Just because a voltage is placed across two wires at one end, it doesn't mean the voltage will automatically be present at the other end. Wires do go open circuit, and screws inside terminal plugs loosen; moulded plugs are not immune from this. The wire inside can still break or – worse still – become intermittent.
  (c) Just because a signal generator is set to 1kHz, it doesn't mean that's what's coming out of it.
  (d) Just because a component drawer says 22K on it, doesn't mean it's full of 22K resistors!
  (e) Just because you've replaced a transistor with a 'new' one, it doesn't mean it can't be faulty.
  You can think of more yourself – always remember, in electronics servicing, never take anything for granted.
2 Millions to one against; we've never known it to happen.
3 S/C, O/C and leaky (low capacitance); electrolytics can leak their electrolyte (dangerous on pcbs because it's a conductive liquid). Under extreme circumstances, capacitors can explode.
4 See text.
5 (a) Analogue – protects meter movement.
  (b) Digital – saves battery.
6 Base/emitter current must be limited otherwise junction will be damaged.
7 See text.
8 Not connected to anything, so it can take up any logic level to which it's connected.
9 (a) Visual inspection.
  (b) Resistance across power supply input leads.
10 An ohm or two at the most and probably a lot less (stepdown).
11 See text.
12 It goes low.
13 See text.
14 A RAM test.
15 (a) Imposing a unique instruction onto the data bus.
  (b) See text (test card or wire wrap IC holder construction).

Appendix II
Selected IC pin-outs

All the information in this section is taken from the Maplin Electronics catalogue with the kind permission of the publishers. All the goods described here (and elsewhere in this book unless otherwise stated) are available from Maplin Electronics (PO Box 3, Rayleigh, Essex SS6 2BR).

This appendix is not intended to provide comprehensive coverage of all ICs available; it is meant to be a quick source of reference for the IC pin-outs mentioned in this book. Where appropriate, supplementary useful information is also included. For further details, the reader is referred to the Maplins Electronics catalogue which is an excellent and thoroughly up-to-date source of technical information on all the commonly used ICs available in Britain today. Maplins also supply data sheets for any of the ICs they stock, the vast majority of which usually contain a good deal of information on the use of the IC, including example application circuits. They are ideal for completing the project and assignment work described in this book.

The 74HC and 74HCT series of ICs

The 74 HC series is recommended for all new designs. It is pin-for-pin compatible with respective types in other ranges. For example, 74HC00 is a direct replacement for 7400 or 74LS00, and 74HC4016 is a direct replacement for 4016BE.

The families may be mixed, but note that the input transition levels on 74HC are different from TTL. Therefore if you are driving 74HC devices from TTL connect a Min Res 4K7 pull-up resistor between the TTL output and V_{cc}. Alternatively you can directly connect a 74HCT device. These devices have input characteristics identical to TTL. However, in order to obtain these characteristics the ultra high input noise immunity of 74HC devices is much reduced and maximum operating frequencies are also lower.

NAND GATES
Quad 2-Input

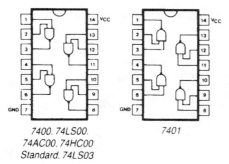

7400. 74LS00.
74AC00. 74HC00
Standard. 74LS03

7401

NOR GATES
Quad 2-Input

A range of ICs with four 2-input NOR gates in a single package. The '02' has standard totem-pole outputs. A CMOS version is available and type '02' is available in the 74HC series.

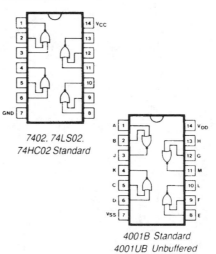

7402. 74LS02.
74HC02 Standard

4001B Standard
4001UB Unbuffered

AND GATES
Quad 2-Input

A range of ICs with four 2-input AND gates in a single package. The '08' has standard totem-pole outputs. A CMOS version is available and type '08' is available in 74HC series.

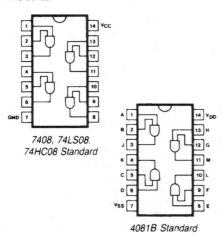

7408, 74LS08.
74HC08 Standard

4081B Standard

OR GATES
Quad 2-Input

Four 2-input OR gates in a single package available in standard TTL, LS, CMOS and HC types.

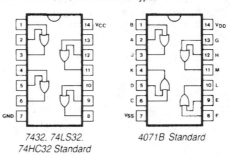

7432. 74LS32.
74HC32 Standard

4071B Standard

Hex Inverters

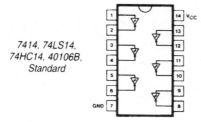

7414, 74LS14,
74HC14, 40106B,
Standard

Six Schmitt trigger inverters in a single package,
available in standard TTL, LS, CMOS and HC types.

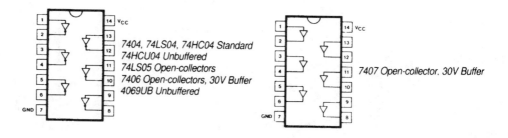

7404, 74LS04, 74HC04 Standard
74HCU04 Unbuffered
74LS05 Open-collectors
7406 Open-collectors, 30V Buffer
4069UB Unbuffered

7407 Open-collector, 30V Buffer

FLIP FLOPS
Dual D-Type

Two D-type positive-edge-triggered flip-flops with set
(preset) and reset (clear). The data at a D-input is
transferred to the Q output and the Q complement
output on the next positive-going edge of the clock
input.

4013B

	5V	10V	15V
Max clock frequency	4MHz	10MHz	14MHz
Propagation delay	175ns	75ns	50ns

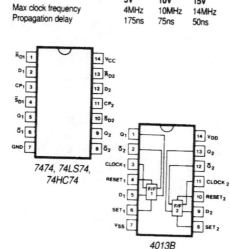

7474, 74LS74,
74HC74

4013B

Octal D-Type Latches

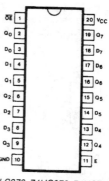

74LS373, 74HC373, 74HCT373
74HCT533

Eight D-type latches in a single package. When the
enable input is high, the outputs follow the inputs and
when low, the outputs are latched. The outputs may be
set to high impedance by applying a high to the output
control pin 1. Type '533' has inverted outputs while the
'373' is non-inverted.

Dual J-K Type

A range of IC's each containing two separate J-K flip-flops. Types '76' and 4027 offer preset (set) and clear (reset) inputs whilst type '73' has only clear (reset). All are negative-edge-triggered except 4027.

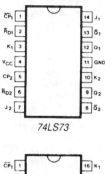

74LS73

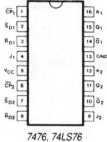

7476, 74LS76

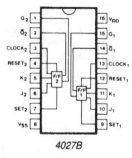

4027B

8-Bit Addressable Latches

A range of IC's each comprising eight latches any one of which may be selected by applying the appropriate address. The data is entered serially and the output is available as 8-bit parallel. A write enable and reset are available.

	LS259	4099
Propagation delay low to high/ high to low 5V	19ns/13ns	200ns
10V		75ns
15V		50ns
Supply current	22mA	

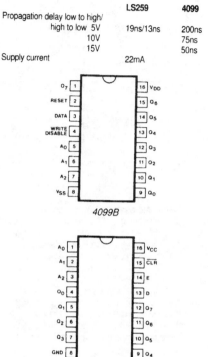

4099B

74LS259

DECODERS AND DEMULTIPLEXERS
Dual 2-Line to 4-Line

The '139' and '4555' each have two fully independent 2-line to 4-line decoders where a specific code on the 'select' inputs will drive one of the four outputs on (low on '139', high on '4555') providing enable is low. The enable input can be used as a data input for demultiplexing. The 4555B is a dual 1-of-4 decoder/demultiplexer. Each has two address inputs (A_0 and A_1), an active LOW enable input (E) and four mutually exclusive outputs which are active HIGH (O_0 to O_3). When used as a decoder, E when HIGH, forces O_0 to O_3 LOW. When used as a demultiplexer, the appropriate output is selected by the information on A_0 and A_1 with E as data input. All unselected outputs are LOW.

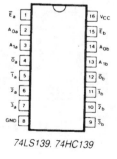

74LS139, 74HC139

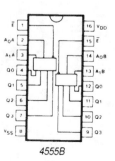

4555B

3-Line to 8-Line

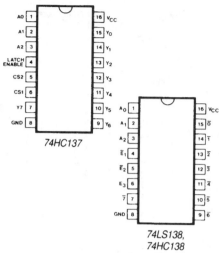

74HC137

74LS138,
74HC138

In the '137' there are two enable inputs and an address latch, whilst in the '138' there are three enable inputs. On the '137' when pin 4 goes from low to high, the address present on the 'select' inputs is stored in the latches and further changes ignored while pin 4 remains high. All outputs are high unless pin 6 is high and pin 5 is low. On the '138' all outputs are high unless pin 6 is high and pins 4 and 5 are low. This enables easy expansion. For demultiplexing an enable input can be used as a data input. On all devices with the chip enabled, a specific code on the three select inputs will drive one of the four outputs on (low on '137' and '138').

4-Line to 16-Line

On all types a specific code on the four input pins will switch one of the 16 output lines on. On type '154' there are two strobe inputs which must both be low. If one or both are high then all outputs are high. For demultiplexing operation, hold one strobe line low and connect data to the other strobe input. The outputs of the 4514 are active high, whilst on the '4515' they are active low. On the '4514' and '4515', an inhibit is provided which when high switches all outputs off. Only one strobe line is provided which when taken from high to low latches the input code. Changes on the inputs will have no effect while strobe is low. Note that two '138' IC's offer higher speed operation than one '154'.

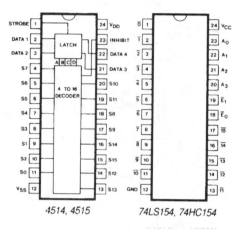

4514, 4515   74LS154, 74HC154

4-Line to 10-Line

A specific code between binary 0 to 9 on the 4 input lines will switch on one of the ten outputs. Binary codes 10 to 15 switch all outputs off.

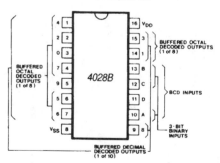

BCD to 7-Segment

On all types a specific code on the four input lines generates an output for driving a 7-segment display. Illegal inputs are suppressed on the '4511' and '74HC4543', and on the other IC's the display is as shown below.

Type '47' has active-low open-collector outputs for driving common anode LED displays or incandescent indicators whilst the '48' has active-high (2kΩ pull-up) outputs for driving common cathode LED displays or lamp buffers. Types '4056' and '74HC4543' are designed for driving liquid crystal displays, and types '4511' and '74HC4511' will directly drive common cathode LED's via a series resistor. Types '47' and '48' have ripple blanking inputs and outputs for leading zero suppression in lamp arrays and these and the '4511' have a lamp test input which lights all segments simultaneously. Types '4056', '4511' and '74HC4543' have 'strobe', 'latch enable' and 'latch disable' respectively which freezes the display regardless of changes on the input. In addition, types '47', '48', '4511' and '74HC4543' have a display blanking input for power saving. The '74HC4543' requires the LCD backplane frequency connected to pin 6. Alternatively the device can drive LED displays: for common cathode connect pin 6 to 0V and for common anode connect pin 6 to pin 16.

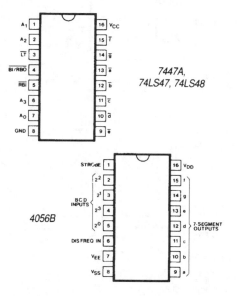

LOGIC COMPARATORS
4-Bit

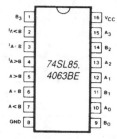

Four-bit magnitude comparators that determine whether the binary code on the four 'A' inputs is greater than, equal to, or smaller than the binary code

8-Bit

The 74LS688 can determine whether the binary code on the eight 'P' inputs is greater than, equal to, or less than the binary code on the eight 'Q' inputs, going low only when P=Q and with chip enable (pin 1) low.

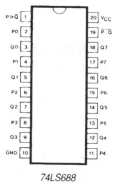

74LS688

Arithmetic Logic Unit & Look Ahead Carry Block

An arithmetic logic unit (ALU) that can perform 16 different binary arithmetic operations on two 4-bit words depending on the code on the select inputs. Operations include addition, subtraction, decrement, 2's complement, straight transfer etc. A fast carry look-ahead permits fast operation even where several devices are cascaded. For lower speeds a carry output on the ALU may simply be connected to the carry input on the next most significant device.

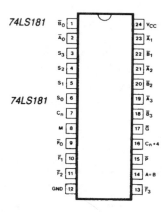

74LS181

MULTIVIBRATORS

The table below shows the basic differences between the different types available.

	121	123	221	4047	4098	4538
Single	★		★			
Dual		★	★	★		★
Schmitt inputs	★		★			
Retriggerable		★		★	★	★
Precision pulse width						★
Basic type		121			4098	

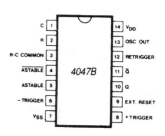

NE555N Timer
SGS-Thomson

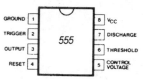

The 555 is a highly stable device for generating accurate time delays or oscillation. Additional terminals are provided for triggering or resetting if desired. In the time delay (monostable) mode of operation the time is precisely controlled by one external resistor and one capacitor. For stable operation as an oscillator, the free running frequency and the duty cycle are both accurately controlled with two external resistors and one capacitor. The circuit may be triggered and reset on falling waveforms and the output structure can source or sink up to 200mA or drive TTL directly. This IC may also be *correctly* supplied marked as MC1455PI. Supply decoupling must be provided close to the IC to counter the 'crowbar' effect of the device's internal discharge switch, a suitable value is 10 to 100µF as shown in the accompanying diagrams.

Monostable Mode

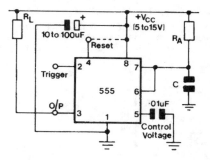

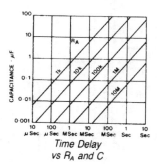

*Time Delay
vs R_A and C*

On time after triggering (i.e. applying a voltage to pin 2 less than $\frac{1}{3}$ supply voltage) is equal to $1.1R_AC$. The load may be connected to V_{CC} for normally-on operation or between pin 3 and ground for normally-off. Connecting reset to ground during on time, drives the output low until a new trigger pulse occurs. Additional trigger pulses during on time have no effect. If reset is not being used, connect it to V_{CC}.

LM741CN
SGS-Thomson

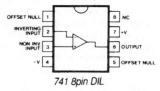

741 8pin DIL

The industry standard general purpose op-amp featuring internal frequency compensation. The amp is overload protected on input and output with no latch-up if common mode range is exceeded.

LM380N
National Semiconductor

An audio amp in a 14-pin DIL package that requires very few external components to make a complete 2.5W power amplifier. In most cases, however, it is advisable to add a Min Res 2.7Ω and Polyester 0.1μF in series from pin 8 to ground and an Axial 4.7μF from pin 1 to ground.

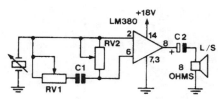

High-Output-Crystal-Cartridge Power Amp
A 2.5W rms power amp the 380 is shown in the circuit driven by a high output crystal pickup. The IC requires only 4 other components (without tone control only two other components! — simply omit C1 and RV1).

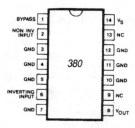

ULN2801A/3A Octal Darlington Driver Arrays
SGS-Thomson

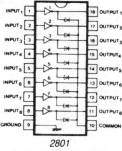

2801

Eight separate darlington amplifiers in one 18-pin package, each capable of supplying 500mA at up to 50V. Outputs may be paralleled to give up to 4A at 50V (at 23% duty cycle and 25°C). Internal diodes are provided for inductive loads. Type 2801 may be used with standard bipolar digital logic or CMOS, while type 2803 has a 2k7 base resistor to enable direct connection to TTL and 5V CMOS.

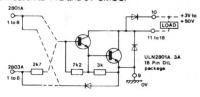

UCN5801A Latching Octal Driver
Sprague

A high current, high voltage driver IC comprising eight CMOS data latches, a bipolar darlington transistor driver for each latch, and CMOS control circuitry. Inputs are CMOS, PMOS and NMOS compatible, and a pull-up resistor is required for TTL. Input speeds up to 5MHz are possible with 5V supply, and much higher rates with 12V supply. Outputs are open collector with integral diodes for inductive loads, and are capable of sinking 500mA at 50V at 25°C. If more than two maximum loads are connected at once, then the duty cycle must be reduced (to 23% for all eight loads at 25°C). Outputs can be paralleled for higher currents.

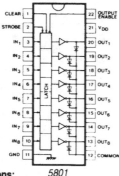

Specifications: 5801

Supply voltage (V_{DD}):	5V to 12V
Input voltage high (min):	$V_{DD} -1{\cdot}5V$ (V_{DD} max.)
Input voltage low (max):	1V (−0·3V min.)
Supply current:	5·6mA @ V_{DD} = 5V
	8mA @ V_{DD} = 12V

Data present at an input is transferred to its latch when pin 2 is high. A high on pin 1 sets all latches to output off regardless. A high on pin 22 sets all outputs off regardless. When pin 22 is low, the output depends on the state of its latch.

Z80-PIO Parallel Interface Controller

This IC provides a universal means of interfacing parallel data to a microprocessor. It can interface the 8-bit data bus of the MPU to two 8-bit peripheral buses e.g. keyboard, VDU, printer etc. Data are able to flow in either direction to and from the peripheral buses under the control of the microprocessor. Features include interrupt driven "handshake" for fast response; byte output, byte input, byte bidirectional bus (port 'A' only), and bit modes of operation; programmable interrupts on peripheral status conditions; daisy chain priority interrupt logic included to provide automatic interrupt vectoring without external logic; eight outputs capable of driving Darlington transistors (−1.5mA at 1.5V); and all inputs and outputs fully TTL compatible.

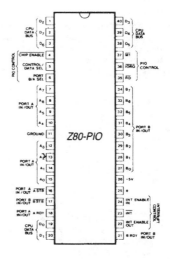

6402 Universal Asynchronous Receiver/Transmitter

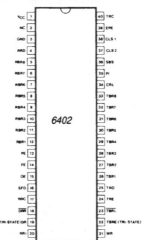

This industry standard UART will interface computers or microprocessors to asynchronous serial data channels. The receiver converts serial start, data parity and stop bits to parallel data, verifying proper code transmission, parity and stop bits. The transmitter converts parallel data into serial form and automatically adds start, parity and stop bits. The data word length can be 5, 6, 7 or 8-bits. Parity may be odd or even. Parity checking and generation can be inhibited. The stop bits may be one or two, or one and a half if transmitting five bit code. This IC is sometimes supplied coded CDP1854ACE. These two parts are identical.

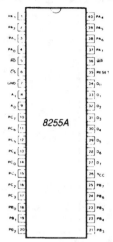

8255A Peripheral Interface Adaptor

RC6522P Versatile Interface Adaptor (VIA)
Rockwell

A very flexible I/O device that contains a pair of very powerful 16-bit interval timers, a serial-to-parallel/parallel-to-serial shift register and input data latching on the peripheral ports. Expanded handshaking capability allows control of bi-directional data transfers between VIA's in multiple processor systems. Control of peripheral devices is handled primarily through two 8-bit bi-directional ports. Each line can be programmed as either an input or an output. Several peripheral I/O lines can be controlled directly from the interval timers for generating programmable frequency square waves or for counting externally generated pulses. To facilitate control of the many powerful features of this chip, an interrupt flag register, an interrrupt enable register and a pair of function control registers are provided.

A general purpose I/O device having 24 I/O pins which may be individually programmed in two groups of 12 and used in 3 major modes of operation. In mode 0 each group of 12 I/O pins may be programmed in sets of four to be input or output. In mode 1 each group may be programmed to have 8 lines of I/O, and of the remaining 4, three are used for handshaking and interrupt control signals. Mode 2 is a bidirectional bus mode which uses 8 lines for the bus and 5 lines (one borrowed from the other group) for handshaking.

MEMORY IC's
2114 4K Static Random Access Memory

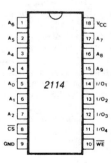

A 4096-bit static random access read/write memory (RAM) organised in 1024 x 4-bit words. The IC operates from a single 5V supply at typically 80mA. Access time is <450ns and thus the chip is suitable for use with all our microprocessors. The input/outputs are 3-state and TTL compatible and there is chip enable input for memory expansion.

6116 16K CMOS Static RAM

A 2048 x 8-bit static RAM built in CMOS. Pin compatible with 16K EPROM'S the device offers access times of 150ns and data retention at voltages down to 2V with standby currents as small as 10nA at 3V. The chip operates from a single +5V supply.

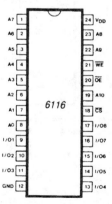

Characteristics (typical)
Supply voltage	5V
Supply current	5mA static
	25mA @ 150ns cycle
Data retention voltage	2V min
current	10nA (10µA max)
Access time	100ns max

6264 64K CMOS Static RAM

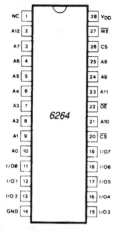

An 8192 x 8-bit static RAM built in CMOS. Pin compatible with 64K EPROM's, the device offers access times of 100ns or 150ns and data retention at voltages down to 2V with standby currents as small as 20μA at 3V. The chip operates from a single +5V supply.

M2716-1F1 16K Erasable, Programmable Read Only Memory
SGS-Thomson

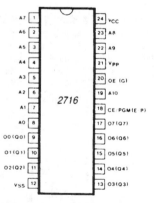

A 16,384-bit electrically programmable and ultra- violet erasable read only memory (EPROM) organised as 2048 x 8-bit words. The IC operates on a single +5V supply in read mode. Access time is 350ns and the IC is fully static. The outputs are 3-state and inputs and outputs are TTL compatible. Programming is achieved by applying +25V to pin 21 and with the address and data lines stable apply a +5V pulse to pin 18. Note that only one pulse is required for each location. A transparent window on top of the IC allows the user to erase the bit pattern by exposing the chip to ultraviolet light at 253.7nm with an incident energy of 15W-seconds/cm². Thus with a 12mW/cm² UV tube and the device positioned one inch from it and with no intervening filter or glass, the IC will be completely erased in about 20 minutes.

62256 256K CMOS Static RAM

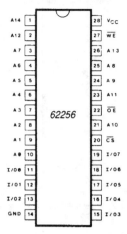

A 32,768 x 8-bit static RAM built in CMOS. The device offers access times of 100ns and data retention with standby currents as low as 40μA. The chip operates from a single +5V supply.

Characteristics (typical)
Supply voltage: 5V±10%
Supply current: 8mA
                    33mA @ 150ns cycle
Standby current: 40μA
Access time: 100ns

M2732A-2F1 32K Erasable, Programmable Read Only Memory
SGS-Thomson

A 32,768-bit electronically programmable and ultra-violet erasable read only memory (EPROM) organised as 4096 x 8-bit words. The IC operates on a single +5V supply in read mode. Access time is 200ns and the IC is fully static. The outputs are 3-state and inputs and outputs are TTL compatible. Programming is achieved by applying +21V to pin 20 and with the address and data lines stable apply a +5V pulse to pin 18. Note that only one pulse is required for each location. A transparent window on top of the IC allows the user to erase the bit pattern by exposing the chip to ultraviolet light at 253.7nm with an incident energy of 15W-seconds/cm². Thus with a 12mW/cm² UV tube and the device positioned one inch from it and with no intervening filter or glass, the IC will be completely erased in about 20 minutes.

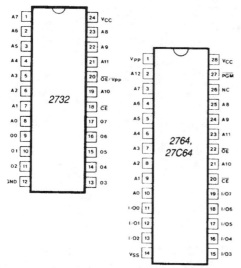

M2764AF1 64K Erasable, Programmable Read Only Memory

SGS-Thomson

A 65,536-bit electronically programmable and ultra-violet erasable read only memory (EPROM) organised as 8192 x 8-bit words. The IC operates on a single +5V supply in read mode. Access time is 250ns and the IC is fully static. The outputs are 3-state and inputs and outputs are TTL compatible. Programming is achieved by applying +12.5V to pin 21 and with the address and data lines stable apply an active low pulse to pin 27. Note that only one pulse is required for each location. A transparent window on top of the IC allows the user to erase the bit pattern by exposing the chip to ultra violet light at 253.7nm with an incident energy of 15W-seconds/cm². Thus with a 12mW/cm² UV tube and the device positioned one inch from it and with no intervening filter of glass, the IC will be completely erased in about 20 minutes.

TMS27C128-25JL 128K CMOS EPROM

Texas Instruments

Identical to the 27128 above, but offering a much lower power supply current and very low standby current. Access time is 250ns.

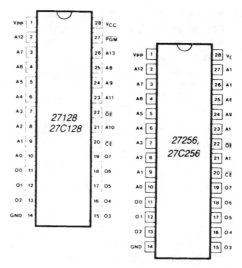

M27256F1 (250ns) 256K Erasable, Programmable Read Only Memory

SGS-Thomson

A 32,768 x 8 bit ultra-violet erasable PROM, featuring 250ns access time and high-performance programming at only 12.5V. Inputs and outputs are TTL compatible in READ and program modes. For READ operation, V_{CC} and V_{PP} must be +5V ±5%. Supply current is 105mA max. (45mA typical), standby 40mA max. For programming mode, V_{CC} must be taken to 6V ±0.25V and V_{PP} to 12.5V ±0.3V (*NOT 21V*). With address and data stable (2µs), a 1ms ±5% active low pulse is applied to pin 20. An average program time is $1\frac{1}{2}$ minutes per chip. The erasure procedure and timings are the same as for the 128K EPROM.

M27C256B-12XF1 256K CMOS EPROM
SGS-Thomson

Similar to the 27256, but offering a much lower power supply current (30mA max.) and very low standby current. Access time is 120ns.

M27512 F1 512K Erasable, Programmable Read Only Memory
SGS-Thomson

A 65,536 x 8-bit ultra-violet erasable PROM, featuring 250ns access time and high performance programming at only 12.5V. Inputs and outputs are TTL compatible in read and program modes. For read operation V_{CC} and V_{PP} must be +5V ±5%. Supply current is 90mA, standby 20mA typical. For programming mode, V_{CC} must be taken to 6V ±0.25V and V_{PP} to 12.5V ±0.3V. With address and data stable (2μs) a 1ms ±5% active low pulse is applied to pin 20. An average program time is 6 minutes per chip. The erasure procedure and timings are the same as for the 128K EPROM.

M27128AF1(250ns) 128K Erasable, Programmable Read Only Memory
SGS-Thomson

A 131,072-bit electrically programmable and ultra-violet erasable read only memory (EPROM) organised as 16,384 x 8-bit words. The IC operates on a single +5V supply in read mode. Access time is 250ns and the IC is fully static. The outputs are 3-state and inputs and outputs are TTL compatible. Programming is achieved by applying +12.5V to pin 1 and with the address and data lines stable apply an active low pulse to pin 27. Note that pin 22 must also be high. A transparent window on top of the IC allows the user to erase the bit pattern by exposing the chip to ultra-violet light at 253.7nm with an incident energy of 15W-seconds/cm^2.
Thus with a 12mW/cm^2 UV tube and the device positioned one inch from it and with no intervening filter or glass, the IC will be completely erased in about 20 minutes.

Pin configuration for 27512, 27C512:

Pin	Signal		Pin	Signal
1	A15		28	Vcc
2	A12		27	A14
3	A7		26	A13
4	A6		25	A8
5	A5		24	A9
6	A4		23	A11
7	A3		22	$\overline{OE}$ / V_{PP}
8	A2		21	A10
9	A1		20	$\overline{CE}$
10	A0		19	O7
11	O0		18	O6
12	O1		17	O5
13	O2		16	O4
14	GND		15	O3

M27C512-15XF1 512K CMOS EPROM
SGS-Thomson

Similar to the 27512, but offering a much lower power supply current (30mA max) and very low standby current. Unlike the 27512 the programming voltages are V_{PP} = 12.75V ±0.25V and V_{CC} = 6.25V ±0.25V. Access time is 150ns.

M27C1001-15XF1 1M CMOS EPROM
SGS-Thomson

A 131,072 x 8 bit ultra-violet erasable PROM< featuring 150ns access time. For READ operation, V_{CC} and V_{PP} must be +5V ±5%. Supply current is 35mA max, and standby current 200μA max. For programming mode, V_{CC} must be taken to 6.25V ±0.25V and V_{PP} to 12.75V ±0.25V. The erasure procedure and timings are the same as for the 128K EPROM. Programming can be completed in 12 seconds. 150ns access time.

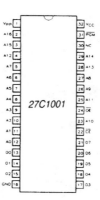

Pin configuration for 27C1001:

Pin	Signal		Pin	Signal
1	Vpp		32	Vcc
2	A16		31	PGM
3	A15		30	NC
4	A12		29	A14
5	A7		28	A13
6	A6		27	A8
7	A5		26	A9
8	A4		25	A11
9	A3		24	$\overline{OE}$
10	A2		23	A10
11	A1		22	$\overline{CE}$
12	A0		21	O7
13	O0		20	O6
14	O1		19	O5
15	O2		18	O4
16	GND		17	O3

DAC0801LCN 8-Bit D/A Converter
National Semiconductor

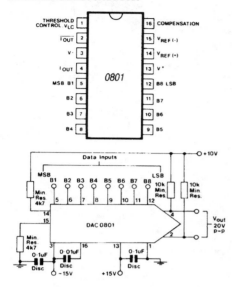

An 8-bit digital to analogue converter with a full scale error of less than ±0.39%. The DAC has high compliance complementary current outputs to allow differential output voltages of 20V peak-to-peak with simple resistor loads.

Specification

Supply voltage:	±4.5V to ±18V
Settling time:	100ns (typical)
Output voltage compliance:	−10V to +18V
Full scale current (V_{ref} = 10V, R14,15 = 5kΩ):	1.99mA
Output current range (V^- = −5V):	0 to 2.1mA
(V^- = −8V to −18V):	0 to 4.2mA
Reference bias current (I15):	−1µA
Reference input slew rate:	8mA/µsec

Power supply current (V_S = ±5V, I_{ref} = 1mA)

	I⁺:	2.3mA
	I⁻:	4.3mA

Power supply current (V_S = ±15V, I_{ref} = 2mA)

	I⁺:	2.5mA
	I⁻:	−6.5mA

ZN425E8 8-Bit D/A and A/D Converter
GEC-Plessey

An 8-bit D/A converter also containing a counter and a 2.5V precision voltage reference. By including an 8-bit counter, analogue to digital conversion can be obtained simply by adding an external comparator and clock inhibit gating (7400). By simply clocking the counter, the IC can be used as a self-contained precision ramp generator.

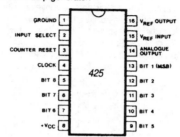

Characteristics (typical)

Supply voltage	4.5 to 5.5V
Settling time	1µs
Voltage reference	2.55V
Non-linearity	±0.5 LSB
Analogue output resistance	10kΩ
Counter clock frequency	5MHz max
Supply current	25mA

ZN426E-8 8-Bit D/A Converter
GEC-Plessey

An 8-bit D/A converter also containing a 2.5V precision voltage reference. Binary weighted voltages are produced at the output, the value depending on the digital number applied to the input bits.

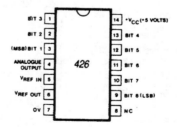

Characteristics (typical)

Supply voltage	4.5 to 5.5V
Settling time	1µs
Voltage reference	2.55V
Non-linearity	±0.5 LSB
Analogue output resistance	10kΩ
Supply current	5mA

ADC0804LCN 8-Bit A/D Converter
National Semiconductor

A CMOS 8-bit analogue to digital converter with output latches that can directly drive a microprocessor data bus. The IC looks like a memory location or I/O port to the microprocessor so no interfacing logic is required. The analogue input voltage range is 0V to 5V with a single 5V supply, and 2.5V applied to pin 9. However, the voltage reference on 9 can be any voltage under 2.5V so that any voltage span cam be converted with a full 8-bits of resolution. In addition, by connecting pin 7 to a voltage other than ground the span need not start at 0V. For example if the span was 0.5V to 3.5V (a span of 3V) 0.5V would be applied to pin 7 and 1.5V to pin 9 (i.e. $\frac{1}{2}$ of 3V). No zero adjustment is needed with this IC.

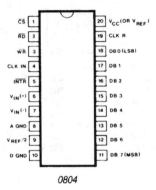

0804

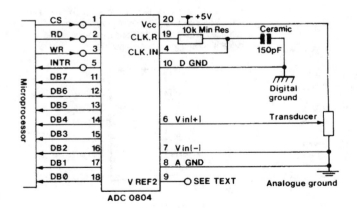

ADC 0804

Specification

Supply voltage:	+5V (V_{CC})
Max error:	±1 bit
Input resistance at pin 9:	1.3kΩ
Analogue input voltage range:	Ground to V_{CC}
Conversion rate:	8770/second max
Supply current:	1.3mA

Appendix III
Glossary

Accumulator   The main register in a CPU where traditionally the results of ALU operations 'accumulate'. The name is still used to refer to the A register; it's normally the way in and out of the CPU. See Chapter 1 et seq.

Address   A unique numerical value which labels each memory location in the system and permits access to it by placing the value on the address bus. See Chapters 4 and 16.

Addressing mode   One of several different methods of specifying an address in an instruction. See Chapter 6.

Algorithm   A statement in plain English which describes the function of a computer program. See Chapters 5, 6 and 13.

ALU   Arithmetic and logic unit – a circuit built up from many logic gates which can execute mathematical and logical instructions; the device can typically add, subtract, negate, and complement, and logically AND, OR and EXOR. See Chapter 3.

Architecture   The internal structure of a CPU or other device of similar complexity; the structure of a microelectronic system. See Chapter 3.

ASCII code   The American Standard Code for Information Interchange; most commonly used code for representing alphanumeric characters in a microelectronic system. See Chapter 2.

Assembler   The combined electronics and computer program which converts source code (mnemonics constructed from operator and operand) into object code (machine code). See Chapter 1 et seq.

Astable   An electronic circuit consisting of two transistors which alternately turn on and off to produce square wave oscillations. Also called a multi-vibrator because it produces multiple vibrations (many different frequencies); in theory, a square wave is made up of an infinite number of sine waves, harmonically related. See Chapter 1.

Asynchronous   An electronic system which conveys information such that each data bit is independent of another in time. See Chapter 17.

Baud rate   The number of bits transmitted per second in a particular system. Named after French engineer, Baudot. See Chapter 17.

Binary   A mathematical system of counting which uses only two numbers, usually one and zero; a mathematical system using the base 2 so that 100 equals 4, for example. See Chapter 2.

Binary coded decimal (BCD)   A system in which the decimal numbers 0 to 9 are coded in binary. See Chapter 2.

Bistable   A simple electronic circuit, consisting of two transistors, each of which may be either on or off. The states are stable and only change over on the application of a suitable pulse; also called a 'flip-flop'. The basis of many memory chips as the state is retained as long as power is supplied. See Chapter 1.

Bit   Short for a binary digit, one or zero. See Chapters 1 and 2.
Buffer   An electronic circuit or device which separates one circuit from another. The isolation achieved prevents the action of one circuit affecting another in an adverse way. Some buffers also increase voltage, current or power without affecting the information contained in the signal. See Chapters 12, 14 and 16.
Bug   A hardware or software fault. See Chapter 11.
Byte   A sequence of 8 bits, often used to represent a character.

Carry flag   The bit in the flag or processor status register which indicates whether or not the previous ALU operation produced a carry or not. See Chapter 10.
Clock   A square wave oscillator which provides the timing waveform for the CPU, allowing it to step through program instructions and synchronise control functions and bus activity.
Control bus   The wired connections which allow control signals to be transmitted around a microelectronic system.
CPU   A central processing unit; the electronic device which is at the heart of any computer. It usually contains all the registers and timing and control logic necessary for executing programs. The MPU (micro processor) is the modern equivalent, much reduced in size and which is frequently referred to as the 'silicon chip'. Some of the later designs also have some on-board memory; some have their own clock. See Chapters 1 and 3.

Decoder   A logic device which selects any one of a number of outputs depending on the binary code applied to its input. See Chapters 15 and 16.
Decrement   To reduce a number by one.
DRAM   Dynamic random access memory – a computer memory system based upon the charges held by tiny capacitors within the memory matrix. DRAMs need to be constantly refreshed by reading the content of each memory cell and writing it back again immediately afterwards. Compare static RAM devices which consist of an array of transistor elements built into bistables. See Chapter 3.

EAROM   Electrically alterable read only memory – a memory device which can be user-programmed and which will retain the program even after power has been removed from the device. It can be erased by the application of a suitable voltage. Individual bits may be modified in this way whereas most similar devices may only be modified a byte at a time. See Chapter 3.
EPROM   Erasable programmable read only memory – a non-volatile memory chip which can be programmed a byte at a time. The whole chip contents may be erased by the application of ultra violet light and then reprogrammed. See Chapter 3.
Execution   The carrying out of program instructions. See Chapter 7.

FET   Field effect transistor – an amplifying device which operates according to the voltage (as opposed to the current) applied to its input. Characterised by having a high input impedance, relatively low voltage gain (around 25) and employing only one type of charge carrier.
Fetch–execute cycle   Process by which a CPU fetches and then executes instructions in a computer program. See Chapter 7.
Flag register   Also called a processor status register, it is a special-purpose register whose individual bits may be automatically set or reset according to the results of a previous ALU operation. See Chapter 10.
Flip-flop   See bistable.

Flow chart   A chart using recognised symbols (which determine their meaning) and which summarise the action of a computer program; compare with JSP. See Chapters 1, 5, 6, 7 and 13.

Hand assembly   The conversion of instruction mnemonics to machine code without the use of an assembler. See Chapters 1, 2 and 5 et seq.

Hexadecimal system   The base 16 counting system containing the numbers 0 to 15. Useful because the binary digit 1111 is equal to 15 so that a byte may be represented by only two hex digits. In order to represent each number in the hexadecimal system by one character, 0–9 are retained as in the decimal system, but the double numbers 10 to 15 are replaced with the first five letters of the alphabet. 1011 is therefore B in hexadecimal. See Chapter 2.

High level language   A computer language containing instructions in English or near English (e.g. PRINT, INPUT, GOTO, etc. in BASIC) thus making it easier to use and understand. See Chapter 1.

Increment   The process of increasing a number by one.

Input/output port   An external microelectronic system connection which allows two-way communication with the outside world. See Chapters 12, 14 and 17.

Instruction   A single command, often only one byte long, which tells the CPU what it is required to do. See Chapters 5, 6 and 7.

Instruction register   A special purpose register within the CPU which decodes an instruction and initiates the sequence necessary to execute it. See Chapters 5, 6 and 7.

Instruction set   All the instructions available to a programmer using a specific processor.

Interface   An electronic circuit which gives power to a signal. See Chapters 12, 14 and 17.

Interrupt   A method of temporarily halting normal CPU program execution to allow external devices to be serviced and then allowing a return to normal execution. See Chapter 12.

Iteration   The process of repeating a sequence of operations for a preset number of times.

JSP   A top-down approach to software planning in which a problem is defined and then broken down into smaller elements. It is seen as a viable, more structured approach to writing software than using flow charts. See Chapter 13.

JUMP   A program instruction which diverts execution to a different part of that program.

Light emitting diode   An electronic component which emits light when current flows through it in a particular direction. It's based upon a phenomenon known as 'electroluminescence' – the direct conversion of electricity into light. An LED contains no filament and draws relatively little current compared to light bulbs of similar light intensity. Light of limited wavelength is emitted because an applied voltage allows electron transitions in discrete energy packets only, associated with the material from which the device is made. Red ones are most common because least energy is needed to produce visible photons of that wavelength. See Chapters 12 and 14.

Look-up table   A table of codes or other data stored in contiguous memory locations. The codes for lighting up appropriate sections of a seven-segment LED display may be stored in this way.

Machine code   Instructions and data generally converted into hex code for convenience which represent the binary digits a computer can actually 'understand'. A machine code program consists of streams of hex data and instructions interleaved; if correctly sequenced, the CPU instruction register can recognise the instructions, decode and then execute them. See Chapters 1, 2, 5, 6 and 7.

Memory   That section of a computer or other microelectronic system which consists of elements such as RAM, ROM, PROM, etc. which stores programs, instructions and data for use within the system. Frequently the data are held for future use as in ROM or PROM devices or on magnetic tape or disc. See Chapters 3 and 4.

Memory map   A pseudo-pictorial map which indicates what memory is available within a system, what it consists of (RAM, ROM, EPROM, etc.) and what it is reserved for (OS, USER RAM, VIDEO, STACK, etc.). See Chapters 4 and 16.

Memory-mapped I/O   The system used in a microelectronic system whereby the I/O ports are accessed just as if they were parts of the main memory. See Chapters 12, 14 and 17.

MPU   Micro processor unit – a CPU fabricated using microelectronics, often referred to as a silicon chip. See Chapter 1.

Open collector   An electronic device (often a gate or buffer) whose output transistor has no internal collector load resistor enabling the user to connect one and hence use a different supply voltage. It's possible, for example, to use a 24V relay in a TTL system, by using an open collector buffer. See Chapter 15.

Operand   That part of an instruction which supplies the data to be acted upon or an address in memory where it may be found (directly or indirectly). See Chapters 5 and 6.

Operator   Also called **opcode** – that part of an instruction which defines the operation to be performed as opposed to a data value or address. See Chapters 5 and 6.

Parallel data transfer   A method of transmitting data along several parallel wires, one byte at a time. Compare with serial data transfer. See Chapter 17.

Port   The route into and out of a computer or microelectronic system; data may typically be sampled by the system via a port and control signals output in a similar manner. Some kind of interfacing is nearly always required. See Chapters 12, 14 and 17.

Program counter   A special-purpose register in a CPU which holds the address of the instruction to be executed. See Chapters 1, 3, 7 and 8.

Programmable I/O device (PIO)   An input/output device which can be user-programmed to determine its characteristics. See Chapters 12, 14 and 17.

Random access memory   A matrix consisting of many bistable elements which can store machine code in binary representation. Random because any memory location may be accessed with equal speed compared to the lengthy serial process associated with magnetic tape. See Chapter 3.

Read   The process of accessing data from a memory device.

Refresh   The process of reading each memory location in a dynamic RAM system, every 20ms or so, and then immediately writing it back to the same location in a continuous cycle. Without memory refresh, the tiny capacitive elements which store the digital codes would discharge and the information would be lost. See also 'DRAM'.

Register   A set of bistables arranged in a group of 8 (in an 8-bit system) which can store any number from zero to 255 in binary, representing the data or instructions used in a computer program. A CPU may contain several such registers. See Chapter 1.

Relative addressing A mode of addressing used only for relative JUMP instructions. A single byte of data following the opcode is used to specify a displacement forwards or backwards from the point at which the jump occurs. A backwards displacement must be given in two's complement form; all displacements are given in hexadecimal. See Chapters 2 and 6.

Relay An electromagnetic device which can operate various kinds of switches. The coil is often rated at a much lower current or voltage than that associated with the contacts performing the switching function. See Chapters 14 and 15.

Scale of integration Method by which the number of gates or circuits of similar complexity may be expressed for a particular integrated circuit. See Chapter 15.

Serial data transfer A method of transmitting data down a wire, one bit at a time, one after the other. Compare with parallel data transfer. See Chapter 17.

Shift register A register which is designed to allow data to be moved in and out, either serially or in parallel, under clock control. See Chapter 17.

Stack A reserved block of RAM to allow the automatic storage of data or addresses on a temporary basis using simple instructions. See Chapters 3 and 9.

Stack pointer A special-purpose register which keeps track of stack contents and holds the address of the next available location. SP contents vary from one system to another; some point to the last used location, others point to the next one available. See Chapter 9.

Switch bounce What happens when an electrical switch is opened or closed. A clean make or break rarely occurs and the pulses which result may cause problems in a fast-acting electronic system. Switch debouncing is employed to correct the problem. See Chapter 12.

Synchronous A system where events occur in time relation, often under clock control. In a synchronous binary counter, for example, all transitions occur simultaneously. Compare with asynchronous systems, where signals 'ripple through'; succeeding events must wait until those preceding them have been completed. See Chapter 17.

Trace table A table which shows the contents of registers and memory locations for each step in a particular program. Used as an aid to debugging a faulty program. See Chapter 11.

Transistor A three electrode, semiconducting device which can switch at high speeds efficiently or amplify a signal. It is a current device characterised by having a low input impedance and utilising two types of charge carriers (holes and electrons).

Tri-state device One in which a logic 1, logic 0 or floating input or output may occur. Generally, the device is sent out of the tri-state condition by the application of an enable (EN) signal.

TTL (transistor-transistor logic) A logic system in which one logic level is conveyed to another via transistor elements. Bipolar transistors are used in the 7400 series of TTL ICs. See Chapters 1, 12, 14 and 15.

Two's complement The method by which a negative number may be represented in a microelectronic system. See Chapter 2.

Vector The memory location(s) where a particular routine, or the address of a particular routine, may be located. An interrupt service routine, for example, may be located by an interrupt vector. See Chapter 12.

VLSI Very large scale integration – relating to the number of circuits which may be constructed in a single device. See Chapter 15.

Word   The number of bits which can be simultaneously processed by a computer or microelectronic system. In Z80 and 6502 systems, the word length is a byte (8 bits). In the Intel 8086 series and the Motorola 68000 series it is 2 bytes (16 bits). See Chapters 2 and 3.

Write   The process of storing a word into a memory location or to a peripheral control device (a PIO data register, for example).

Appendix IV
Logic symbols with their truth tables

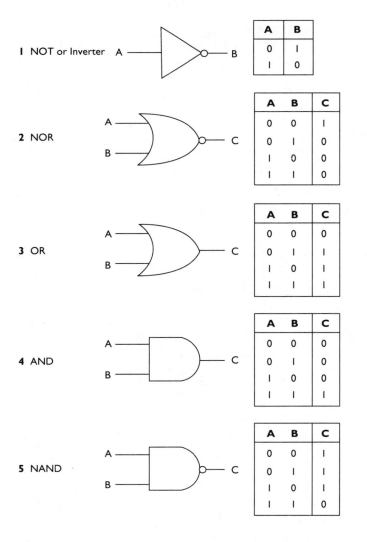

A	B
0	I
I	0

I NOT or Inverter

A	B	C
0	0	I
0	I	0
I	0	0
I	I	0

2 NOR

A	B	C
0	0	0
0	I	I
I	0	I
I	I	I

3 OR

A	B	C
0	0	0
0	I	0
I	0	0
I	I	I

4 AND

A	B	C
0	0	I
0	I	I
I	0	I
I	I	0

5 NAND

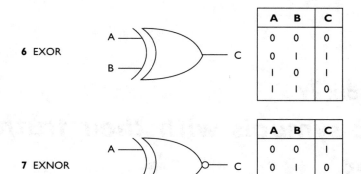

6 EXOR

A	B	C
0	0	0
0	1	1
1	0	1
1	1	0

7 EXNOR

A	B	C
0	0	1
0	1	0
1	0	0
1	1	1

Z80 instruction set

Using the instruction set

In the first table the column down the left is the destination, all the other columns contain sources. For example, to load the accumulator (the destination) with the contents of the B register (the source), use LD A,B. A is in the left-hand column so read along to the B column finding the code 78 there. Hence, LD A,B is 78H.

Similarly, LD A,(HL) is 7EH whilst LD (HL),A is 77H.

A blank space indicates that the instruction does not exist, for example it is not possible to load the B register with the contents of the memory location pointed to by BC, i.e. LD B,(BC) is an illegal instruction for which there is no code. However, LD B,(HL) would be 46H, whilst LD (BC),A is 02H.

(Z80 8-bit load group)

Destination	Register							Register indirect			Indexed		Ext addr	Imm	Implied	
	A	B	C	D	E	H	L	(HL)	(BC)	(DE)	(IX+d)	(IY+d)	(mm)	b	I	R
A	7F	78	79	7A	7B	7C	7D	7E	0A	1A	DD 7E d	FD 7E d	3A mm	3E b	ED 57	ED 5F
B	47	40	41	42	43	44	45	46			DD 46 d	FD 46 d		06 b		
C	4F	48	49	4A	4B	4C	4D	4E			DD 4E d	FD 4E d		0E b		
D	57	50	51	52	53	54	55	56			DD 56 d	FD 56 d		16 b		
E	5F	58	59	5A	5B	5C	5D	5E			DD 5E d	FD 5E d		1E b		
H	67	60	61	62	63	64	65	66			DD 66 d	FD 66 d		26 b		
L	6F	68	69	6A	6B	6C	6D	6E			DD 6E d	FD 6E d		2E b		
(HL)	77	70	71	72	73	74	75							36 b		
(BC)	02															
(DE)	12															
(IX)+d	DD 77 d	DD 70 d	DD 71 d	DD 72 d	DD 73 d	DD 74 d	DD 75 d							DD 36 d b		
(IY)+d	FD 77 d	FD 70 d	FD 71 d	FD 72 d	FD 73 d	FD 74 d	FD 75 d							FD 36 d b		
(mm)	32 mm															
I	ED 47															
R	ED 4F															

mm = 2 byte (4 hex digit) address; d = 1 byte (2 hex digit) displacement; b = 1 byte (2 hex digits) of immediate data

8-bit arithmetic and logic group

	Operand address										
	Register							Reg ind	Indexed		Immed
	A	B	C	D	E	H	L	(HL)	(IX+d)	(IY+d)	b
ADD	87	80	81	82	83	84	85	86	DD 86 d	FD 86 d	C6 b
ADC	8F	88	89	8A	8B	8C	8D	8E	DD 8E d	FD 8E d	CE b
SUB	97	90	91	92	93	94	95	96	DD 96 d	FD 96 d	D6 b
SBC	9F	98	99	9A	9B	9C	9D	9E	DD 9E d	FD 9E d	DE b
AND	A7	A0	A1	A2	A3	A4	A5	A6	DD A6 d	FD A6 d	E6 b
XOR	AF	A8	A9	AA	AB	AC	AD	AE	DD AE d	FD AE d	EE b
OR	B7	B0	B1	B2	B3	B4	B5	B6	DD B6 d	FD B6 d	F6 b
CP	BF	B8	B9	BA	BB	BC	BD	BE	DD BE d	FD BE d	FE b
INC	3C	04	0C	14	1C	24	2C	34	DD 34 d	FD 34 D	
DEC	3D	05	0D	15	1D	25	2D	35	DD 35 d	FD 35 d	

16-bit load group

Desti-nation	Source							Imm ext	Ext addr	Reg indir
	Register									
	AF	BC	DE	HL	SP	IX	IY	nn	(nn)	(SP)
AF										F1
BC								01 nn	ED 4B nn	C1
DE								11 nn	ED 5B nn	D1
HL								21 nn	2A nn	E1
SP				F9		DD F9	FD F9	31 nn	ED 7B nn	
IX								DD 21 nn	DD 2A nn	DD E1
IY								FD 21 nn	FD 2A nn	FD E1
(nn)		ED 43 nn	ED 53 nn	22 nn	ED 73 nn	DD 22 nn	FD 22 nn			(POP)
(SP)	F5	C5	D5	E5		DD E5	FD E5	(PUSH)		

nn = 2 byte (4 hex digit) address

16-bit arithmetic group

Destination		Source					
		BC	DE	HL	SP	IX	IY
ADD {	HL	09	19	29	39		
	IX	DD 09	DD 19		DD 39	DD 29	
	IY	FD 09	FD 19		FD 39		FD 29
ADC	HL	ED 4A	ED 5A	ED 6A	ED 7A		
SBC	HL	ED 42	ED 52	ED 62	ED 72		
INC		03	13	23	33	DD 23	FD 23
DEC		0B	1B	2B	3B	DD 2B	FD 2B

Transfer of control group

	Address mode		NO CND	NON ZER	ZER	NO CAR	CAR	PAR ODD	PAR EVN	POS	NEG	REG B=0
							Condition					
JP	Immediate extended	nn	C3 nn	C2 nn	CA nn	D2 nn	DA nn	E2 nn	EA nn	F2 nn	FA nn	
JR	Relative	PC+e	18 e-2	20 e-2	28 e-2	30 e-2	38 e-2					
JP		(HL)	E9									
JP	Register indirect	(IX)	DD E9									
JP		(IY)	FD E9									
DJNZ	Relative	PC+e										10 e-2
CALL	Immediate extended	nn	CD nn	C4 nn	CC nn	D4 nn	DC nn	E4 nn	EC nn	F4 nn	FC nn	
RET			C9	C0	C8	D0	D8	E0	E8	F0	F8	
RETI (INT)	Register indirect		ED 4D									
RETN (NMI)			ED 45									

			Restart number									
RST	Modified base page		0 C7	8 CF	16 D7	24 DF	32 E7	40 EF	48 F7	56 FF		

nn = 2 byte (4 hex digit) address
e = 1 byte (2 hex digit) signed offset from the beginning of the
instruction to the destination address
None of the instructions in the transfer of control group affect any of the flags.

Input and output group

	Destination							Register indirect
	Register							
Source port	A	B	C	D	E	H	L	(HL)
IN A,(n) n	DB n							
IN r,(C) C)	ED 78	ED 40	ED 48	ED 50	ED 58	ED 60	ED 68	
INI (C)								ED A2
INIR (C)								ED B2
IND (C)								ED AA
INDR (C)								ED BA

	Source							Register indirect
	Register							
Destination port	A	B	C	D	E	H	L	(HL)
OUT (n),A n	D3 n							
OUT (C),r (C)	ED 79	ED 41	ED 49	ED 51	ED 59	ED 61	ED 69	
OUTI (C)								ED A3
OUTIR (C)								ED B3
OUTD (C)								ED AB
OUTDR (C)								ED BB

n = 1 byte (2 hex digit) address

During input or output instructions, using the accumulator as destination or source, the contents of the accumulator appear on address lines A_8–A_{15}.

During all other input or output instructions, the contents of register B appear on A_8–A_{15}.

Rotate and shift group

	A	B	C	D	E	H	L	Register indirect (HL)	Indexed (IX+d)	Indexed (IY+d)
RLC	CB 07	CB 00	CB 01	CB 02	CB 03	CB 04	CB 05	CB 06	DD CB d 06	FD CB d 06
RRC	CB 0F	CB 08	CB 09	CB 0A	CB 0B	CB 0C	CB 0D	CB 0E	DD CB d 0E	FD CB d 0E
RL	CB 17	CB 10	CB 11	CB 12	CB 13	CB 14	CB 15	CB 16	DD CB d 16	FD CD d 16
RR	CB 1F	CB 18	CB 19	CB 1A	CB 1B	CB 1C	CB 1D	CB 1E	DD CB d 1E	FD CB d 1E
SLA	CB 27	CB 20	CB 21	CB 22	CB 23	CB 24	CB 25	CB 26	DD CB d 26	FD CB d 26
SRA	CB 2F	CB 28	CB 29	CB 2A	CB 2B	CB 2C	CB 2D	CB 2E	DD CB d 2E	FD CB d 2E
SRL	CB 3F	CB 38	CB 39	CB 3A	CB 3B	CB 3C	CB 3D	CB 3E	DD CB d 3E	FD CB d 3E
RLD								ED 6F		
RRD								ED 67		
RLCA	07									
RRCA	0F									
RLA	17									
RRA	1F									

d = 1 byte (2 hex digit) offset

Bit test set and reset group

					Register				Register indirect	Indexed	
		A	B	C	D	E	H	L	(HL)	(IX+d)	(IY+d)
All codes preceded by		CB	CB	CB	CB	CB	CB	CB	CB	DD CB d	FD CB d
BIT	0	47	40	41	42	43	44	45	46	46	46
	1	4F	48	49	4A	4B	4C	4D	4E	4E	4E
	2	57	50	51	52	53	54	55	56	56	56
	3	5F	58	59	5A	5B	5C	5D	5E	5E	5E
	4	67	60˙	61	62	63	64	65	66	66	66
	5	6F	68	69	6A	6B	6C	6D	6E	6E	6E
	6	77	70	71	72	73	74	75	76	76	76
	7	7F	78	79	7A	7B	7C	7D	7E	7E	7E
RES	0	87	80	81	82	83	84	85	86	86	86
	1	8F	88	89	8A	8B	8C	8D	8E	8E	8E
	2	97	90	91	92	93	94	95	96	96	96
	3	9F	98	99	9A	9B	9C	9D	9E	9E	9E
	4	A7	A0	A1	A2	A3	A4	A5	A6	A6	A6
	5	AF	A8	A9	AA	AB	AC	AD	AE	AE	AE
	6	B7	B0	B1	B2	B3	B4	B5	B6	B6	B6
	7	BF	B8	B9	BA	BB	BC	BD	BE	BE	BE
SET	0	C7	C0	C1	C2	C3	C4	C5	C6	C6	C6
	1	CF	C8	C9	CA	CB	CC	CD	CE	CE	CE
	2	D7	D0	D1	D2	D3	D4	D5	D6	D6	D6
	3	DF	D8	D9	DA	DB	DC	DD	DE	DE	DE
	4	E7	E0	E1	E2	E3	E4	E5	E6	E6	E6
	5	EF	E8	E9	EA	EB	EC	ED	EE	EE	EE
	6	F7	F0	F1	F2	F3	F4	F5	F6	F6	F6
	7	FF	F8	F9	FA	FB	FC	FD	FE	FE	FE

d = 1 byte (2 hex digits) offset

6502 instruction set

Load and store group

Mnemonic	Immed	ABS	Z PAGE	(IND,X)	(IND),Y	Z PAGE,X	ABS,X	ABS,Y	Z PAGE,Y
LDA	A9	AD	A5	A1	B1	B5	BD	B9	
LDX	A2	AE	A6					BE	B6
LDY	A0	AC	A4			B4	BC		
STA		8D	85	81	91	95	9D	99	
STX		8E	86						96
STY		8C	84			94			

Arithmetic and logic group

Mnemonic	Immed	ABS	Z PAGE	ACCUM	(IND,X)	(IND),Y	Z PAGE,X	ABS,X	ABS,Y
ADC	69	6D	65		61	71	75	7D	79
AND	29	2D	25		21	31	35	3D	39
ASL		0E	06	0A			16	1E	
BIT		2C	24						
CMP	C9	CD	C5		C1	D1	D5	DD	D9
CPX	E0	EC	E4						
CPY	C0	CC	C4						
DEC		CE	C6				D6	DE	
EOR	49	4D	45		41	51	55	5D	59
INC		EE	E6				F6	FE	
LSR		4E	46	4A			56	5E	
ORA	09	0D	05		01	11	15	1D	19
ROL		2E	26	2A			36	3E	
ROR		6E	66	6A			76	7E	
SBC	E9	ED	E5		E1	F1	F5	FD	F9

Implied instructions

Clear flags

Mnemonic	Implied
CLC	18
CLD	D8
CLI	58
CLV	B8

NOP and set flags

NOP	EA
SEC	38
SED	F8
SEI	78

Increment and decrement

Mnemonic	Implied
DEX	CA
DEY	88
INX	E8
INY	C8

Break

BRK	00

RTI and RTS
– see below

Push, pull and transfer

Mnemonic	Implied
PHA	48
PHP	08
PLA	68
PLP	28
TAX	AA
TAY	A8
TSX	BA
TXA	8A
TXS	9A
TYA	98

JUMP and branch instructions

Mnemonic	Absolute	Relative	Indirect	Implied
BCC		90		
BCS		B0		
BEQ		F0		
BMI		30		
BNE		D0		
BPL		10		
BVC		50		
BVS		70		
JMP	4C		6C	
JSR	20			
RTI				40
RTS				60

DFB is a pseudo-opcode which instructs an assembler to put the following byte into the next memory location.

Appendix VII
Some Z80 and 6502 instruction equivalents

Z80 instructions		6502 instructions	
LD A,n	Load accumulator (A) with a number (n)	LDA #n	Load accumulator (A) with a number (n)
LD B,n	Load B register with a number (n)	LDX #n	Load X register with a number (n)
LD C,n	Load C register with a number (n)	LDY #n	Load Y register with a number (n)
LD HL,nn	Load HL register pair with a 16-bit number	No equivalent instruction	
LD A,(nn)	Load accumulator with contents of memory location nn	LDA nn	Load accumulator with contents of memory location nn
LD (nn),A	Load contents of A to memory location nn	STA nn	Store contents of A in memory location nn
LD A,B	Register B to accumulator	TXA	Transfer register X to accumulator
LD B,A	Accumulator to Register B	TAX	Transfer accumulator X register
LD A,(HL)	Memory (pointed to by HL) to accumulator	No equivalent instruction	
LD (HL),A	Accumulator to memory (pointed to by HL)	No equivalent instruction	
LD (HL),n	Load n in memory location pointed to by HL register pair	No equivalent instruction	
LD A,(IX+d)	Accumulator loaded with IR contents plus displacement integer d	LDA &0D0A,X	Load A with the contents of memory location 0D0A plus contents of X reg
ADD A,n	Add number n to A	ADC #n	Add to A with carry

ADD A,B	Add B contents to A	No equivalent instruction (including X or Y to A)	
SUB B	Subtract B from A	No equivalent instruction	
INC A	Increment accumulator	No equivalent instruction	
INC B	Increment B register	INX	Increment X register
DEC B	Decrement B register	DEX	Decrement X register
INC (HL)	Increment contents of memory location pointed to by HL register pair	INC &0D08	Increment contents of memory location 0D08
CP 10	Compare contents of A with the number 10	CPX #10	Compare X register with the number 10
OR 10	OR accumulator with 10	ORA #10	OR acc. with 10
AND 10	AND acc. with 10	AND #10	AND acc. with 10
XOR 10	Exclusive OR acc. with 10	EOR #10	Exclusive OR acc. with 10
BIT 6,A	Tests bit 3 of accumulator. If bit is zero, Z flag is set	BIT &0D08	Tests bits in address 0D08 using mask in A
SET 4,A	Sets bit 4 in A to 1	No equivalent instruction	
RES 4,A	Resets bit 4 in acc.	No equivalent instruction	
DAA	Decimal adjust acc.	SED	Set decimal mode
RRCA	Rotate right with carry contents of accumulator	ROR	Rotate one BIT right memory or accumulator
JP nn	Jump unconditionally to specified address	JMP nn	Jump unconditionally to specified address
JP Z,nn	Jump to specified address on zero	No equivalent instruction – use a branch instruction	
JR nn	Jump relative always to address nn	No equivalent instruction	
JR Z,nn	Jump relative to address nn on zero	BEQ nn	Branch to address nn on zero
JR NZ,nn	Jump relative to address nn, non zero	BNE nn	Branch to address nn if result non zero
CALL	Jump to subroutine	JSR	Jump to subroutine
RET	Return from subroutine	RTS	Return from subroutine
OUT (00H),A	Output to port A contents of accumulator (00H is port A address)	STA &FE61	Output to port A contents of accumulator (FE61 is port A address – memory-mapped I/O)

IN (01H),A	Input to accumulator contents of port B (01H is port B address)	LDA &FE60	Input to acc. the contents of port B (FE60 is port B address – memory-mapped I/O)
LD SP,4300H	Load stack pointer with 4300H; position in RAM where stack begins	TXS	Transfer X to stack pointer (single byte; 01FF to 0100 assumed)
PUSH AF	Push A and F onto stack simultaneously	PHA	Push A only onto stack
		PHP	Push flag register onto stack
POP AF	Pop top two values from stack and put in AF	PLA	Pull from stack to A
		PLP	Pull from stack to P (P = processor status register, same as flag register in Z80)
NOP	No operation	NOP	No operation
HALT	Halts program execution	BRK	Software interrupt

Index